Vertical Flow Constructed Wetlands

Eco-engineering Systems for Wastewater and Sludge Treatment

Vertical Flow Constructed Wetlands

Eco-engineering Systems for Wastewater and Sludge Treatment

Alexandros Stefanakis
Helmholtz Center for Environmental Research - UFZ, Leipzig, Germany

Christos S. Akratos
Department of Environmental and Natural Resources Management, University of Patras, Patras, Greece

Vassilios A. Tsihrintzis
Centre for the Assessment of Natural Hazards and Proactive Planning, Laboratory of Reclamation Works and Water Resources Management, Department of Infrastructure and Rural Development
School of Rural and Surveying Engineering
National Technical University of Athens, Athens, Greece

ELSEVIER
Amsterdam • Boston • Heidelberg • London • New York • Oxford
Paris • San Diego • San Francisco • Singapore • Sydney • Tokyo

Elsevier
Radarweg 29, PO Box 211, 1000 AE Amsterdam, Netherlands
The Boulevard, Langford Lane, Kidlington, Oxford OX5 1GB, UK
225 Wyman Street, Waltham, MA 02451, USA

First edition **2014**

Library of Congress Cataloging-in-Publication Data
A catalog record for this book is available from the Library of Congress

British Library Cataloguing in Publication Data
A catalogue record for this book is available from the British Library

Printed and bound in China
14 15 16 17 18 10 9 8 7 6 5 4 3 2 1

ISBN: 978-0-12-404612-2

Dedication

To my lovely parents
(Alexandros Stefanakis)

To Anastasia and Zoi
(Christos S. Akratos)

To my wife Alexandra and two sons Andreas and
Konstantinos for their support and patience
(Vassilios A. Tsihrintzis)

Contents

Author Biography

Dr. Stefanakis is an Environmental Engineer, with Diploma and Doctoral degrees from the Department of Environmental Engineering, Democritus University of Thrace, Greece. He also holds an M.Sc. degree in the field of Hydraulic Engineering from the Department of Civil Engineering of the same university. His M.Sc. and Ph.D. theses focused on the ecological treatment of wastewater and sludge using natural systems, mainly constructed wetlands. He has published 13 papers in international peer-reviewed journals, several papers in conference proceedings, technical reports, lecture notes, and book chapters. His work includes numerous experiments with CW systems of different types. He has participated in many national and EU research projects in Greece, Portugal, and Germany. He has taught at the undergraduate level on the topic of environmental protection, delivered seminars in postgraduate studies programs, and has been involved in international conference organizing. He has also supervised and trained various undergraduate students. He has been awarded twice during his career for his research work. Finally, Dr. Stefanakis is also a practicing environmental engineer in Greece, dealing with the design of constructed wetlands facilities.

Dr. Akratos received his Diploma and Doctoral degrees from the Department of Environmental Engineering, Democritus University of Thrace, Greece, and now is an Assistant Professor at the Department of Environmental and Natural Resources Management, University of Patras. He has an extensive experience in constructed wetlands starting from his doctoral dissertation. The majority of his publications deal with wastewater treatment in constructed wetlands. He has published 25 refereed journal papers and has participated as a research team member in 15 research programs, most of them in the area of wastewater treatment in constructed wetlands.

Dr. Tsihrintzis is a Professor of Ecological Engineering and Technology at the School of Rural and Surveying Engineering, National Technical University of Athens, Greece. He is a Civil Engineer and has M.Sc. and Ph.D. degrees in Hydrosystems Engineering from the University of Illinois at Urbana-Champaign, Illinois, USA. Dr. Tsihrintzis' research interests concentrate, among others, in use of natural systems for wastewater treatment with emphasis on constructed wetlands, and in water resources engineering and management with emphasis in water quality and pollution control, ecohydrology and ecohydraulics. His published research work includes more than 100 papers in peer-reviewed scientific journals and over 250 papers in conference proceedings. He has also authored or coauthored books/book chapters on operations research, urban hydrology and runoff quality management, and natural systems for wastewater and runoff treatment, among others. He has participated as a PI or team member in various research projects in the USA, the EU, and Greece. Dr. Tsihrintzis has supervised more than 60 undergraduate and postgraduate student theses and 12 doctoral dissertations. He regularly teaches, among others, the course Natural Wastewater Treatment Systems, which is directly related to the proposed book. He has also served as a Professor and the Head of the Department of Environmental Engineering, Democritus University of Thrace, Greece, for several years. Finally, Dr. Tsihrintzis has an extensive professional experience as a practicing civil environmental engineer both in the USA (he was a registered Professional Engineer in California) and Greece, having designed several water management, wetland restoration, and constructed wetlands systems.

Foreword

Within the last decades, it became obvious that freshwater resources in many countries are increasingly overused and even limited. Thus, the treatment of polluted water and its reuse possibilities are in focus as a main action to solve this problem.

Water technology nowadays is generally able to "transform" highly polluted wastewater into water of varying quality up to drinking water quality. The main limitation is the needed energy input and the related costs for realization of this goal. Because of rising energy prices, especially during the last two decades, technologies with low energy demands are gaining in importance. Concerning energy demand for wastewater treatment, it can be assumed that the—still nowadays—widely applied technology of activated sludge system will be partially replaced by other less energy-consuming and less complex technologies.

Within the last decades, it was made gradually clear that using near-nature technically modified ecosystems, like ponds and wetlands, is an appropriate and attractive option for an economic and environmentally friendly treatment of wastewater. So, especially the technology of constructed wetlands (CWs) made and still makes a fast progress in research development and also in full-scale applications.

The main argument for the overall limited application of CW technology is the high area demand compared to conventional treatment systems. However, this is not an actual limitation in many regions around the world; it can be overcome by the various combinations in hybrid systems and can be argued as an activity toward the compensation of general natural ecosystem devastation in many industrialized countries.

In the context of land area demands, vertical flow CWs appear as a good compromise between highly intensive treatment systems, usually connected with a high energy input but extreme low area demand like activated sludge technology, and horizontal flow CWs, which can have almost no external energy input demand—similar to ponds—but also a higher-area demand because of the low oxygen input into the system. In general, it is expected the application of vertical-flow CWs will extend within the next years and, depending on effluent quality demands, it can be combined with other treatment systems.

Because of the fast development of vertical-flow CWs and their huge potential for implementation into practice, a book dedicated only to this technology will help to foster the distribution of knowledge about this technology, not only to scientists but especially to engineers involved in practical aspects of wastewater treatment.

In contrast to some other textbooks, the topic of wastewater treatment is here combined with the sludge treatment because of the very similar technological approach. Like vertical flow CWs for wastewater treatment, and maybe even more, sludge treatment wetlands (STWs) are still a not widely considered option to handle sludge materials, like sludge obtained from the activated sludge process. The availability of summarizing literature about the technology of STWs is even more limited. The main advantage, in comparison to the conventional applied sludge treatment technologies, is also, like in case of CWs for wastewater treatment, the low energy input and the nonuse of chemicals.

The goal of this book is to provide an extensive review of the existing scientific, technical, and economic practices of using vertical flow CWs. The authors have indeed long-term experience by working for many years in this field with many publications in refereed scientific-engineering journals. Their book summarizes the latest knowledge of two technological systems, vertical flow CWs for wastewater and sludge treatment, which have a broad future application potential. The book comes at the right time, while we get increasingly aware of the need to protect one of our main basis of life, the water, and offers us a potential to treat water and sludge in a more sustainable way, even when financial sources are limited. What is mainly needed is the knowledge about alternatives to the conventional applied technologies. The reader can get this knowledge from this book. In this book, the reader can find a synthesis of current available literature on both technological systems, which provides a deep understanding of the current state-of-the-art knowledge and use of Vertical Flow Constructed Wetlands.

Peter Kuschk
Helmholtz Center for
Environmental Research - UFZ, Leipzig, Germany
March 2014

Preface

The idea of this book was born about 2 years ago. Working for several years on the experimentation and modeling of constructed wetlands for wastewater and sludge treatment, and having designed and implemented several systems in Greece, we have realized the dynamics of the vertical flow constructed wetland systems and the need for a book on this subject. Based on the promising results of several research and design projects, and the extended literature review conducted over the years, it became clear to us that these systems offer a series of significant benefits concerning their technical efficiency, environmentally friendly character, and economic viability. However, at the same time, we realized that a single and comprehensive reference for these systems simply did not exist. Although there are several books for other types of constructed wetlands, presenting their efficiency, general operation and construction parameters, and case studies and experiences from several countries, a respective reference for vertical flow systems was not available in the literature. Existing books on the subject are mainly edited books, containing book chapters by experts in the field; even there, again, limited information is given about vertical flow constructed wetlands. Until today, the only way for someone to find relative information is to search and collect several scientific papers and perhaps some chapters from various books.

We were happy to see that Elsevier Publishing was positive and willing to publish such a book. Thus, the idea of this book became a reality.

The result of our efforts over the last two years is the book at hand. This book represents the first single reference for vertical flow constructed wetlands, which is a relatively new and still developing technology in the field of ecological and environmental engineering. This book gathers and presents the current status of knowledge and experience on vertical flow constructed wetlands for wastewater treatment and sludge dewatering. It provides information on the design, construction, operation, and maintenance of these systems, as it also thoroughly describes their treatment performance. We have made a great effort to gather and present the state-of-the-art in knowledge on this subject. The book contains a brief introduction to constructed wetlands technology, theory, fundamental knowledge of the processes taking place within these systems, applications, and design considerations.

The wetland technology today is continuously evolving with new and innovative applications. Therefore, the book addresses not only municipal wastewater treatment but also various types of industrial and agro-industrial wastewaters and, of course, one of the most promising applications, sludge dewatering and drying. Additionally, the current level of modeling of vertical flow constructed wetlands is also briefly presented. Finally, in order to present a global overview of these systems, the economics of various applications are included to describe the level of investment and operational costs of such facilities, and the relative economic benefits. Finally, the environmental footprint of implementing such systems is also addressed, in terms of greenhouse gas emissions and global environmental impact, and is compared to that of conventional treatment methods.

It is our hope that this book will be a helpful reference material for undergraduate and graduate students in civil engineering, environmental engineering, environmental science, chemical engineering, rural engineering, agricultural engineering departments, professionals dealing with wastewater treatment facility designs, and researchers in the field of wastewater treatment using constructed wetlands. For this reason, the book is written in a way to address both the scientific (providing the necessary information and justifications) and the professional point of view. It contains information from existing treatment systems and existing guidelines for designing constructed wetlands, which will be useful for both academic and professional use. It also highlights present treatment limitations, gaps in fundamental knowledge, and areas which need further investigation.

We hope that this book will be a useful and essential reference for the wetland community, and not only, and that it will assist in the better understanding and growing worldwide interest for these excellent treatment systems.

Alexandros Stefanakis
Christos S. Akratos
Vassilios A. Tsihrintzis

Chapter 1

Introduction

1.1 NATURAL V. CONSTRUCTED WETLANDS

1.1.1 Definitions

Natural wetlands are transitional areas between terrestrial and aquatic systems, integrating characteristics of both dry and wet environments. They can be fully or partially covered by water for extended periods of time or during the whole year. They are dynamic systems, continuously evolving and changing their characteristics with time. The level of water saturation is a main factor that determines the nature of the soil and the types of plant and animal species that live in wetlands. The characteristics of natural wetlands are affected by a variety of local/regional parameters, including climate, hydrology, topography, water chemistry, vegetation, and human disturbance nowadays, among others. Due to these exact characteristics and parameters that regulate their status and appearance, natural wetlands can be found on every continent except Antarctica.

There is a variety of wetland types, which makes it difficult to formulate a precise, internationally accepted definition. One of the best and recognized definitions for natural wetlands was provided by the Ramsar Convention on wetlands in 1971 (Ramsar, 2012). This Convention adopted an international, intergovernmental definition for wetlands, based on a broad approach to describe to the best the main wetland characteristics. Wetlands are defined as "*areas of marsh, fen, peatland or water, whether natural or artificial, permanent or temporary, with water that is static or flowing, fresh, brackish or salt, including areas of marine water, the depth of which at low tide does not exceed six meters*." Natural wetlands include areas like estuaries, mangroves, tidal flats, floodplains, deltas, freshwater marshes, lakes, lagoons, swamps, and springs of underground aquifers. As their name indicates, they are created without any human intervention. According to the same Convention, the same terms also include riparian and coastal zones adjacent to natural wetlands or islands or sea ponds that are deeper than 6 m, but located within the boundaries of the wetland. Moreover, Section 404 of the US Clean Water Act defines wetlands as

☆"To view the full reference list for the book, click here"

follows: "*Those areas that are inundated or saturated by surface or ground water (hydrology) at a frequency and duration sufficient to support, and that under normal circumstances do support, a prevalence of vegetation (hydrophytes) typically adapted for life in saturated soil conditions (hydric soils). Wetlands generally include swamps, marshes, bogs, and similar areas*" (USEPA, 1972).

The IUCN (International Union for the Conservation of Nature) Natural Heritage Program, which was established in 1996, prepared a list of 77 World Heritage wetland sites with major and secondary values in 50 different countries (Thorsell et al., 1997). The world area of wetlands is difficult to estimate. Some estimates report a present total wetland area of 5.7 million km^2 (6% of Earth's surface), of which 30% are bogs, 26% fens, 20% swamps, 15% floodplains, and 2% lakes, with the addition of 0.24 and 0.6 million km^2 of remaining mangroves and coral reefs, respectively (Thorsell et al., 1997). Another estimate increases the total wetland area up to 6.9 million km^2, including 1.5 million km^2 of rice paddies (Matthews and Fung, 1987). The Global Review of Wetland Resources and Priorities for Wetland Inventory in 1999 increased the estimated global wetlands area from national inventories up to 12.80 million km^2 (Finlayson and Spiers, 1999). The data for this estimate were obtained from several sources and include inland and coastal wetlands (including marshes, lakes, and rivers), near-shore marine areas (to a depth of 6 m below low tide), and human-made wetlands such as reservoirs and rice paddies (Table 1.1).

1.1.2 Function and Values of Natural Wetlands

It is only during the last 50 years that humanity began to realize the multiple benefits of wetlands to human society. Wetlands are of special ecological importance, due to the diversity of species and population densities they support,

TABLE 1.1 Ramsar Sites Number and Area per Region (MEA, 2005)

	Total Area of Ramsar Sites (million ha, 2011)	Number of Ramsar Sites (2011)	Wetlands of International Importance (Ramsar Sites)
Africa	85	310	160
Asia	14	254	174
Europe	25	939	805
Neotropics	33	161	126
North America	23	191	117
Oceania	8	77	74
World	189	1932	1456

their high productivity rate, and the particular habitats they include. They have even been called as the "biological supermarket," since they are among the most productive natural environments on Earth (Barbier et al., 1997). They have often been described as the "Earth's kidneys" because they operate as a filter, retaining the pollutants from water that flows through on its way to lakes, streams, and oceans (Kadlec and Knight, 1996; USEPA, 2004). Finally, they provide substantial social and economic benefits to humans.

Natural wetlands fulfill a series of multiple functions, based on their hydrological regime, i.e., the water recycling rate, the water budget, etc. From these functions arise various values for humanity. These two terms, functions and values, are often confused and are considered identical (Barbier et al., 1997). Probably this is due to the fact that some functions are beneficial for humankind without any human interference, while others add benefits only after human effort. For example, carbon dioxide absorption by wetland plants is beneficial for the global climate and takes place without any human effort, while supporting of food chains possesses a value (e.g., fishing) only after a respective human activity. Good knowledge of wetland functions is important in order to clearly determine their values for humanity and set the framework for their proper management. Wetlands generally offer the following functions (MEA, 2005; De Grot et al., 2006; Ghermandi et al., 2010):

- Enrichment of groundwater aquifers
- Control/amendment of flood incidents (protective buffers)
- Trapping of sediments and other substances
- Absorption of carbon dioxide
- Heat storage and release
- Absorption of solar radiation and respective support to food chains

The term "value" defines the services and goods that wetlands offer to humanity. These values derive from the above-mentioned natural functions. A general distinction is often made among consuming and social values. Wetland values are not independent but a change or deterioration of one causes a respective change-upgrade or deterioration of another. The importance and the range of each value is not the same for every wetland system. The values of wetlands can be distinguished into three main types: ecological, socio-cultural, and economic. These together define the Total Value of wetlands (Figure 1.1). Each type of value has its own set of criteria and value units (De Grot et al., 2006).

On the whole, the values of wetlands can be summarized to the following (Schuyt and Brander, 2004; MEA, 2005): (1) biodiversity (biological value), (2) water supply, (3) irrigation, (4) fishery, (5) livestock, (6) water quality improvement, (7) flood protection, (8) recreation, (9) culture, (10) protection against the anthropogenic enrichment of the atmosphere with carbon dioxide, (11) climate improvement, (12) prey value, (13) scientific, (14) educational, (15) timber, (16) hydroelectrical power, (17) salt production, (18) sand provision, (19) anticorrosive, (20) healing – thermal, (21) transportation.

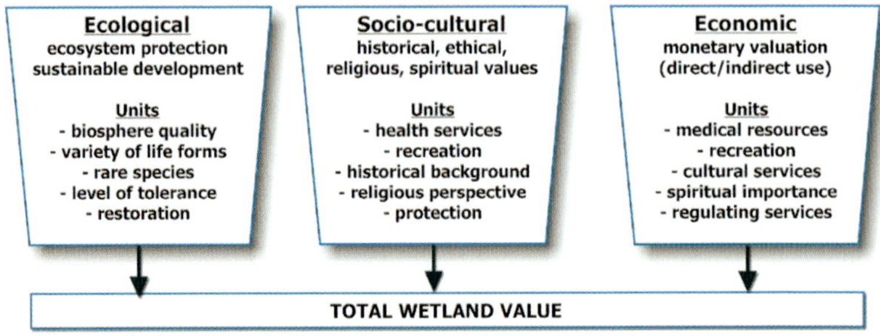

FIGURE 1.1 Individual type of values and respective criteria and value units for the determination of the Total Wetland Value (De Grot et al., 2006).

One of the most important values of wetlands, historically, is their significant contribution to the viability and development of several cultures in the human history. Even the first human civilizations used to live close to wetland environments, which provided a series of multiple economical and vital sources. It is known that early civilizations, like the Mesopotamian and Egyptian, developed near rivers and marshes. Various discoveries could be attributed to this adjacency: paper and ships were made of papyrus, reeds were used to build houses, etc. An interesting division between hydraulic and aquatic civilizations has been proposed (Dugan, 1993): the first ones controlled water flow (dams, pumps, dikes) in order to bring water to regions where water was not available during the entire year; and the second ones exploited the surrounding floodplains and deltas. The exploitation of wetlands by humans is, thus, very old (aquatic societies), without neglecting the negative effects of this use, especially in hydraulic societies.

In other words, wetlands represent a critical life source for humans and wildlife and their contribution to our life quality is of great importance. However, wetland values are often undervalued, since the range of wetland products (food, medicinal plants, building materials, etc.) is not practically realized by most people. Simultaneously, wetland areas are subjected to several threats and pressure from human activities. The need to increase agricultural productivity resulted in the draining and drying of vast areas globally, and their conversion into agricultural fields and farmlands (Davis and Ogden, 1994). Thus, a large number of wetland sites have been destroyed, while others have been degraded due to the construction of irrigation works, the spread of pollution, and the inflow of solid and liquid wastes. Additionally, the need for more electricity and the construction of hydroelectric power plants brought irreversible damage to the natural wetland environment. Since 1900, more than 50% of the world's wetlands have disappeared (Schuyt and Brander, 2004).

1.1.3 Economic Value of Natural Wetlands

The better understanding of the Total Value of Wetlands can be represented in terms of economic assessment. Wetlands represent a considerable section of national and local economies, contributing with the provision of resources, recreational activities, and other benefits, such as pollution control and flood protection. Although the calculation of the actual economic value of a wetland is a hard task to assess, it is possible to evaluate the range of services derived by wetlands and assign a monetary value. There are several studies and reports providing overviews of economic values of wetlands, but a global comprehensive overview does not exist. Costanza et al. (1997) attempted to measure the global economic value of wetlands and estimated it to be up to US$14.9 trillion. A relative WWF/SAEFL report (Schuyt and Brander, 2004) calculated the total economic value of the world's wetlands equal to US$70 billion per year, based on the global wetland area (12.8 million km^2) estimated by the Ramsar Convention. These estimates indicate the magnitude of wetland economic value comprising their biodiversity, scientific, ecological, sociocultural, and other important wetland values. The same study reports that coastal wetlands in the United States provided storm protection services with an estimated value of US$23.2 billion per year. Moreover, New York City estimated a reduction of the cost for new wastewater treatment plants from US$3-8 billion to US $1.5 billion if the upstate land around the reservoirs was purchased and used to purify for free the water supply.

1.1.4 From Natural to Constructed Wetlands

Although natural wetlands offer all these benefits, their value has only been recognized in the recent years. It is understood today that natural wetlands have the ability to receive and control flood incidents and alleviate the possible negative impacts to the society. They have been used for the discharge of wastewater for centuries, although the main reason for this was the convenience they offer as disposal sites rather than for treatment purposes. It is remarkable that during the Minoan times in the Greek island of Crete, advanced sewerage collection systems were constructed in Knossos and Zakros Palaces (Angelakis et al., 2005); nevertheless, nearby torrents were used as effluent disposal sites. The continuous use of wetlands, until recently, as disposal sites resulted in their degradation in many areas of the world.

Over the last decades, the water purification capacity of wetland systems was gradually more and more recognized. It is today identified that wetlands are able to eliminate and transform various pollutants (organics, nutrients, trace elements, etc.) through a series of natural, biological, and chemical processes, thus improving water quality. This overall realization of the wide range of ecological and economic benefits of wetlands stimulated the interest regarding the possibility to exploit these wetland capacities for a series of specific technological applications. This observation led to the investigation of human-made wetland ecosystems, aiming at exploiting the purifying functions of natural wetlands. Natural wetlands have been used for the disposal and respective treatment of secondary and tertiary wastewater effluents for many years, while many wastewater treatment facilities discharge their effluents to natural wetlands worldwide (Mander and Jenssen, 2002). Several studies revealed the potential of wetland ecosystems for pollutant assimilation. However, the principle of natural environment protection does now allow for the use of natural wetlands for treatment purposes, since this could cause irreversible damage.

The basic concept of building constructed wetland (CWs) systems is to replicate the various wetland processes in a more beneficial for humans way and under controlled environmental conditions. The functions of special interest are flood protection, water storage, and water quality improvement. Generally, constructed wetlands are made by humans and they are built in a way to operate similar to a natural wetland. A common definition of these systems reports that constructed wetlands are "man-made complexes of saturated substrate, emergent and submerged vegetation, animal life and water that simulate natural wetlands for human use and benefits" (Hammer and Bastian, 1989).

1.1.5 Evaluating the Benefits of Constructed Wetlands

Constructed wetlands do provide most of the previously mentioned benefits related with natural wetlands. Similar to natural wetlands, there are efforts undertaken towards the economic evaluation of the benefits of constructed wetlands, as already done for natural wetlands. Generally, the problems for this are more or less the same; to identify the social benefits and costs, and express them in monetary terms, since the majority of the goods and services provided are usually not market priced. Additionally, the existing studies database is larger for natural than for constructed wetlands. Benefits arising from constructed wetlands include, besides treatment capacity, provision of wildlife and habitat diversity, ability for recreational activities (e.g., bird-watching), water storage, and aesthetic upgrade of the surrounding (urban or rural) environment, among others (Knight et al., 2001). Thus, it appears that both natural and constructed wetlands provide similar ecological functions (Campbell et al., 2002), although constructed wetlands could be characterized as a more ecologically encumbered environment. Ghermandi et al. (2010) investigated the comparative values of 186 constructed and natural wetland sites worldwide and reported that constructed wetlands possess higher value compared to other wetland types, especially concerning flood and stormwater control and water quality improvement, while they provide the possibility to restore and enhance the local biodiversity, e.g., in modern urbanized areas.

1.2 DEVELOPMENT OF CONSTRUCTED WETLAND TECHNOLOGY

Taking advantage of wetland systems for water quality improvement is not something new, since many wetland areas have been used as discharge sites in the past. Even if there was not a wetland, water discharge usually resulted in the formation of a wetland (Brix, 1994a). However, the term "Constructed Wetlands" today refers to engineered systems designed in a way to exploit the processes that appear in the natural environment, but under controlled conditions. This treatment technology represents a very interesting development in the field of ecological engineering during the twentieth century. Two different but simultaneous facts led to the development of this technology. First of all, for more than 50 years, wastewater treatment in the western world – the so-called developed countries – has been performed by conventional biological methods, with respective construction of centralized facilities. These heavy installations are energy consuming with a mean lifetime of 30 years; thus, the following issue has emerged: the initial investment problem of building new facilities to replace the old ones. On the other hand, developing (low-income) countries lack both the funds to construct centralized facilities and the technical expertise to manage and operate them, which means that even today disposal of untreated wastewater to the various receivers continues. This contradiction between developed and developing countries has been converted into a common need for alternative treatment techniques, which would combine the acceptable effectiveness with the minimum capital or investment cost. Taking also into account the increasing environmental concerns and awareness worldwide, another parameter was added to the prerequisite conditions for new treatment techniques: it has to be sustainable, i.e., with lower environmental impact as well.

At this point, natural treatment systems can play a crucial role. These alternative systems do not depend on intensive biological processes that demand high amounts of energy and expensive, synthetic raw materials, as conventional treatment plants do. On the contrary, they are based on the exploitation of natural processes and natural components (e.g., plants, aggregate materials), while the major portion of their energy demands is covered by renewable sources. In other words, they provide an ecological way of treatment. Moreover, the idea of using plants to purify wastewater was always innovative and attractive not only for scientists but also for regular people; this means that treatment with natural systems can also increase social acceptability. Thus, CW technology attracts increasing worldwide interest. It is worth noting that treatment of wastewater with nature components was known already from the antiquity, since the Minoan hydraulic technicians were aware of the basic principles of hydraulic and sanitation engineering, while land application of wastewater was already utilized in the Minoan civilization (Angelakis et al., 2005).

The first use of plants in an artificial treatment system took place in the 1950s in the Max Planck Institute in Germany by Dr. Seidel (1966). This innovative experiment aimed at the counterbalance of overfertilization, pollution from sewage, and silting up of inland waters using appropriate plants (Seidel, 1953). Brix (1994a) reports even an oldest use of CWs which goes back to 1904; even so, it is generally accepted that the initiation of CW scientific study and technological evolution begins in the 1950s. Experiments by Seidel (1953) revealed that some reed species (e.g., bulrush) were able to remove various pollutants from wastewater and acted differently from those grown in other unpolluted aquatic environments. She also noticed that plant density of these species was 10-20 times higher compared to that of similar plants grown in uncontaminated water. The cooperation of Drs. Seidel and Kickuth from the University of Göttingen resulted in the development of the so-called "Root Zone Method", which was a horizontal flow (HF) CW system planted with common reeds (Brix, 1994a).

The first CW system in Europe was built in the Netherlands in the 1960s and was a free water surface (FWS) CW (Vymazal, 2011), while the first research activities in the United States occurred in the 1970s and 1980s. The first systems were installed in the 1970s and their number increased during the 1980s. Applications of CWs in Europe and North America were gradual and their number increased rather slowly at the beginning. The creation of the Environmental Protection Agency in the United States enhanced creative thinking on alternative methods for wastewater treatment. While use of subsurface flow (SF) CWs (wastewater flow below surface) was more common in Europe, in the United States FWS systems were mostly used (Brix, 1994a). The first installations in Germany and Denmark demonstrated the treatment efficiency of these systems, but soon clogging problems occurred due to low soil hydraulic conductivity. Respective problems due to false design were observed in many facilities during the 1970s, thus the need for further research toward improvement of the system performance and operation became obvious.

CWs in North America are used mainly as large-scale facilities for municipal wastewater treatment. On the other hand, subsurface flow CWs in Europe, particularly of vertical flow (VF), were promoted due to their relatively reduced surface area needs mostly for small-scale applications, e.g., domestic wastewater in small settlements. With time and as technology advanced, more CWs were designed and used for wastewater treatment, and at the end of the twentieth century the number of CW applications increased rapidly; the range of applications was also expanded and included treatment of stormwater runoff and industrial, mining, and agricultural wastes (Vymazal, 2011). Some of the most interesting applications are the treatment of wastewater from refineries and fuel storage tanks (petroleum industry; Eke and Scholz, 2008; Stefanakis et al., 2014), the systems for sludge dewatering and drying (Stefanakis and Tsihrintzis, 2012c,d), paper and pulp industry effluent treatment (Ranieri et al., 2011), etc. Today, more than 10,000 systems are in operation in Europe. A detailed review of the historical timeline of CW technology is already available (Kadlec and Knight, 1996; Kadlec and Wallace, 2009; Vymazal, 2011).

On the whole, this treatment technology is a relatively recent development (i.e., about 30-40 years) when compared to conventional treatment methods which have been used for more than 80 years now. This relatively short period implies a respective limited level of experience in both design and operation of CWs. Although the technology exists for about four decades, it is only during the last 15-20 years that a tremendous increase of interest concerning CWs has occurred. This is not unrelated with the fact that during the same period the actual value of wetlands has been realized, as a result of the gradual increase of environmental concerns. Thus, research on CWs and observation of the operation of the first full-scale systems have been intensified in order to improve the fundamental knowledge and basic understanding of the processes taking place within the system and to optimize their efficiency.

A simple search on the Internet reveals the rapid increase of research studies and projects produced during the last 15-20 years mainly. Table 1.2 presents the number of publications that appear in Google Scholar for the various keywords used. It is clear that, during the last two decades, research on CWs met a significant increase. Specifically, the number of published scientific papers and reports during the period 2000-2010 was more than double compared to that of the period 1990-2000, for both keywords "constructed wetlands" and "Treatment Wetlands," the second being a term also widely used. This difference reflects the respectively enhanced interest that CWs attract with time, as a relatively new rising technology, and also the need for further research on CW performance. It is also interesting that during the last decade (2000-2010) the published papers on Vertical Flow CWs (VFCWs) and Horizontal Subsurface Flow CWs (HSF CWs) increased about three and four times, respectively, compared to the 1990s. This is also an indication of the sharp increase that occurred on the interest on subsurface systems, a result of realizing the advantages of this CW type. On the other hand, the respective increase on free water surface CWs (FWS CWs) was less than double, which again implies the movement of research interest towards subsurface flow systems.

Vertical Flow CWs are considered as the latest development in this technology. Although systems with vertical flow exist for many years, the use of this type was not very widespread. The gradual realization of the specific advantages of this CW type (e.g., less surface area requirements, better aeration of the bed, etc.) attracted the interest of researchers and engineers. Thus, today the so-called second generation of VFCWs represents probably one of the most promising alternatives for decentralized wastewater treatment.

1.3 CONVENTIONAL V. CONSTRUCTED WETLANDS SYSTEMS

It is important to clearly state the benefits of CW usage compared to conventional treatment systems, in order to promote their dissemination and better understand what they offer. Conventional systems have been proved quite effective in wastewater treatment for several decades, but they come along with various undesired effects and prerequisites. Their philosophy relies on the building of extensive collection and transport systems in order to gather wastewater and treat it in a centralized facility. This fact possesses negative impacts, both environmental and economic. Conventional treatment plants are usually industrial looking, unattractive facilities, placed away from residential areas. Their equipment includes large mechanical parts (ventilators, pumps, etc.) and extensive use of concrete and steel. As a result, they require large energy

TABLE 1.2 Number of Publications per Keyword (scholar.google.com) and Period

Period	Keyword				
	Constructed Wetlands	*Treatment Wetlands*	*Vertical Flow Constructed Wetlands*	*Horizontal Subsurface Flow Constructed Wetlands*	*Free Water Surface Constructed Wetlands*
Total	155,000	168,000	27,200	10,800	37,500
2010-2013	**28,700**	**26,300**	**15,500**	**6770**	**16,400**
2000-2010	**45,700**	**50,100**	**16,000**	**6970**	**17,100**
2005-2010	25,600	23,500	13,300	4870	16,100
2000-2005	22,500	20,300	8250	2750	13,500
1990-2000	**17,400**	**20,400**	**4880**	**1690**	**9950**
1995-2000	15,600	16,900	3640	1280	7930
1990-1995	8840	11,500	1650	542	3740
1980-1990	**4720**	**6590**	**965**	**291**	**1670**
1985-1990	3450	4740	700	213	1170
1980-1985	1620	2390	323	92	634
1970-1980	**1380**	**2080**	**245**	**85**	**543**
1975-1980	1020	1540	183	66	397
1970-1975	485	719	83	29	208

amounts for their operation, with respective high carbon dioxide emissions, while they produce odors and noise. In addition, the initial investment cost is usually high, as also the necessary costs for the continuous and proper operation, including salaries for the necessary specialized staff. Because of the mechanical parts, damages and malfunction are common phenomena, which increase the cost for maintenance (labor and spare parts). And of course, the operation of conventional biological treatment plants comes together with the daily production of by-products, such as sludge, whose handling and management increase significantly the total operational costs. Finally, they also have a relatively limited useful lifetime (usually up to 20-30 years), while they cannot easily manage sudden flow increases. Generally, conventional treatment facilities possess a negative environmental footprint and image, although their exclusive function deals with improving the environment.

1.3.1 Sustainable Wastewater Treatment

Adoption and application of constructed wetlands possess quite different characteristics compared to conventional treatment methods in various aspects, i.e., technical, design, philosophy of dealing with issues. The main difference is that they serve the same function (wastewater treatment) but with a different approach, i.e., decentralized treatment although centralized facilities also exist. In our days, with increased environmental concerns, the multiple benefits of decentralized systems are gaining more attention as an alternative approach to treat wastewater at or near the source. This approach is also in agreement with the philosophy of sustainable development, meaning that the same function (here, effective wastewater treatment) is achieved in a more economic, environmentally friendly, and energy-efficient way.

Constructed wetlands provide an effective and reliable treatment of various types of wastewater (domestic, municipal, industrial, agricultural), resulting in a high-level effluent quality. The treatment relies on the combined action of natural means, such as wetland plants (hydrophytes), microorganisms, and aggregate materials. Among the ecological characteristics are also the potential they offer for integration into parks and other recreational areas, the fact that they can operate as a host for wildlife, and the possibility to recycle the high quality effluent for landscape irrigation or pond creation for educational and environmental purposes. Their appearance is dominated by the dense cluster of reeds, which makes them aesthetically accepted, pleasing to nearby neighborhoods, and easily adapted to the natural landscape. Based on these, they can be characterized as a green, i.e., environmentally friendly, technology.

1.3.2 Economic and Technical Benefits—Feasibility

In financial terms, constructed wetlands are an inexpensive treatment technology for wastewater treatment. Usually, the initial investment for the facility is smaller compared to conventional plants, while the costs for operation and maintenance are practically eliminated. This technology is suitable for the sustainable use of locally available resources. The facility is easy and simple to build, since no complex infrastructure is needed, and there are low needs for expensive materials such as concrete and steel, while local labor can be used for the construction. Energy requirements are very low and most of them are covered by renewable sources such as solar energy and wind power, which are used by the plants. The low amount of the external energy input needed is usually consumed for the lighting of the facility, and, possibly, for the operation of few pumps for wastewater feeding and lifting; however, pump use can be minimized or even totally eliminated with proper cascade design and exploitation of the natural ground slope to have gravity flow.

Moreover, there is no need for specialized personnel to run the facility. Periodic onsite visits for observation, problem fixing, and maintenance are adequate, compared to continuous attention to ensure the proper operation for long periods of conventional treatment facilities. As passive treatment systems, CWs do not require regular routine maintenance; this makes them appropriate for locations with no infrastructure support or reduced funding for permanent personnel. Additionally, the operation of CWs is not accompanied by the production of hazardous by-products, contrariwise to the production of sludge in the conventional plants. Odors are easily controlled, particularly in subsurface flow CW systems, and their operation does not produce annoying noises. These systems are also capable of receiving occasionally increased hydraulic and pollutant loads, without significant adverse effects on the effluent quality. CWs are tolerant systems which ensure the necessary reliability for the treatment efficiency. Furthermore, they have a prolonged useful lifetime, showing also the tendency to increase their treatment capacity with time. Since the treatment is based on the plants, their presence at high density contributes to air quality improvement through oxygen production and absorption of carbon dioxide.

Constructed wetlands provide unique appropriateness for a series of installations of various scales. Based on their flexible design they can be easily built on most sites. They can be installed to treat the wastewater from single household or residential complexes, replacing this way other methods, like septic tanks, which possess potential risks of groundwater contamination and need regular emptying. For rural, remote, and mountainous areas, where usually no sewer system exists nor the population is served by a centralized plant, CWs are an extremely appropriate alternative from both economic and environmental point of view. Moreover, they can be effectively applied for villages and small or medium cities/settlements up to several thousands of population. Table 1.3 summarizes the main differences between conventional treatment plants and CWs.

1.3.3 Limitations of Constructed Wetlands

The major limitation of these systems is the relatively large area requirements for their installation compared to conventional plants, in order to achieve the same level of wastewater purification. This fact alone makes them, thus,

TABLE 1.3 Comparison Between Conventional Treatment Systems and Constructed Wetlands

	Conventional Treatment Systems	Constructed Wetlands
Performance	Continuous effluent of high quality	Satisfying, small fluctuations with temperature variations
Facility	Many and large mechanical parts	No mechanical parts (maybe pumps)
Energy	High energy consumption	Low energy demand
Raw materials	Use of nonrenewable sources during construction (concrete, steel, etc.) and operation (electrical power, chemicals)	Almost exclusive use of renewable sources (solar, wind, etc.)—"ecological" systems
Cost	High costs for the construction and operation	Lower construction cost (especially if there is available land), minimum to zero operational costs
Operation	Demand for continuous monitoring Useful lifetime up to 30 years	Only periodical check, prolonged lifetime (>30 years)
Staff	Demand for specialized personnel	No specialized personnel needed
Maintenance	Mechanical equipment → high maintenance cost, regular damages	Small mechanical parts → low maintenance cost
Land area	Low demands	High demand (e.g., 3-10 m^2/p.e.)
Odors	Open air tanks, odor production	Possible only in free water surface systems
Insects	Usually no significant problems	Possible only in free water surface systems
Flow variations	High/shock inflow rate usually results in reduced performance	Robust to high flow variations
Robustness	Toxic pollutants may lead to system breakdown	Robust to some toxic constituents
By-products (sludge)	Large volumes of by-products which demand handling and management on a daily basis	Zero production
Appearance	Unattractive	Aesthetically accepted, green view

unsuitable for areas with high population density, e.g., large cities and capitals. In such regions, the high existing flows to be treated, the limited available free space, and the high cost of land may be obstacles for the application of CWs. Therefore, it is clear that CWs cannot fully replace existing conventional plants. The diversity of advantages they offer is valid for small- and medium-scale installations. However, based on the different approach (i.e., decentralized facilities instead of centralized) of the same issue (i.e., effective wastewater treatment) the application of CW technology could significantly contribute to the gradual decrease of centralized plant size and number, and contribute to the significant limitation of the need for centralized treatment facilities. Redesigning the wastewater treatment strategy to promote decentralized facilities could practically reduce to the minimum the necessity for centralized plants, thus significantly reducing their negative environmental impact.

Other difficulties that need to be taken into account are that the performance of CWs is more or less affected by climatic conditions. As for most treatment methods, the system efficiency varies between winter and summer months. Warm climates are more suitable for CW applications. However, lower efficiency in cold climates can be compensated by increasing the surface area. Moreover, CW systems need a start-up period of several months or even 1-2 years in order to reach full treatment capacity and optimum performance. This period is necessary for the plants to reach an adequate level of development and density. Problems arising due to the possible presence of mosquitoes or other insects usually do not exist, but false design or overload may lead to their appearance. Also, when designing a CW system, the location to install it should be carefully selected to avoid steep topography and places where the underground water table is high, in order to prevent potential leaching of wastewater.

To summarize, although the design of CWs still remains empirical at some level and the research toward the better understanding of the processes and the efficiency optimization is ongoing, the current status of knowledge is adequate to provide an acceptable and quite satisfying performance. Usually, the first and probably most important issue for the implementation of a CW facility is to find the necessary available area. If this requirement is met, then the limitations mentioned above are more or less manageable and do not represent in any case an insurmountable obstacle.

1.4 SCOPE OF THIS BOOK

Constructed wetlands are today considered as a relatively new and still growing alternative technology in the field of ecological and environmental engineering. The various advantages they provide have placed them among the most promising green technologies. Among the various types of CWs, vertical flow constructed wetlands (VFCWs) represent probably the state-of-the-art design which attracts increasing interest worldwide. The main benefits of this type are the lower area demands compared to horizontal flow systems and the fact that they provide sufficient oxygen within the bed for nitrification. Although the literature on CWs includes a series of interesting books, there is not yet a published book focusing and presenting exclusively the performance, design, and operational guidelines for this CW type. The information about VFCWs is still based mostly on empirical knowledge, while the research carried out remains fragmentary and not concentrated. It is essential to produce a book that will gather all the current knowledge on these systems that appear as one of the most attractive alternatives in the field of ecological engineering. Therefore, this book aims to cover this gap and provides all the so far gained knowledge and experience on the design, construction, operation, and maintenance of these systems, as also information on their performance. Information contained in this book addresses both scientific (providing the necessary information and justifications) and professional point of view, useful for academicians, practicing engineers, and students. A brief overview of the contents is provided below.

1.4.1 Part A—VFCWs for Wastewater Treatment

This part contains information about the VFCW systems for the treatment of various types of wastewater. First, a brief description of all CW types is given in Chapter 2, in order to provide a general picture of CW treatment technology. In Chapter 3, emphasis is given to the various variants of vertical flow systems. Chapter 4 presents information about the basic components of VFCWs; vegetation and filter media. Chapter 5 focuses on the various processes that take place within VFCWs which affect and regulate the transformation and removal of various pollutants. Chapter 6 provides the reader with an overall overview of the main characteristics of VFCWs for municipal wastewater treatment and with all the necessary background information for their design and operation. Chapter 7 summarizes the existing information and experience about the performance of VFCW systems concerning the treatment of industrial (tannery, textile, landfill leachate, mine drainage) and agricultural (animal farm, dairy industry, olive mill) wastewater. Chapter 8 presents the current status of modeling the VFCW performance.

1.4.2 Part B—Sludge Treatment Wetlands

This part of the book presents one of the most promising and state-of-the-art applications of VFCWs: wastewater sludge treatment. Chapter 9 includes a brief review of sludge characteristics and the crucial issue of sludge management. Chapter 10 describes the basic design and operational parameters and the key elements of the system setup. Chapter 11 gives an overview of the processes and mechanisms involved in the dewatering performance in sludge treatment wetlands. Chapter 12 presents the efficiency of VFCWs in sludge dewatering and drying, combined with the description of the dewatering processes and their contribution to the water budget. Moreover, the system performance on the improvement of the sludge quality in terms of various pollutants concentration (e.g., heavy metals, nutrients) is provided. It also presents existing data on the quality characteristics of the final sludge product (biosolids) and examines its suitability for reuse and recycle, based on the requirements and limit concentrations of national and international legislation.

1.4.3 Part C—Technoeconomical Aspects

Chapter 13 includes the economic aspects of VFCWs (investment and operation costs). This chapter of the book also investigates the environmental impact of VFCWs as expressed by economical studies and life cycle assessment.

Chapter 2

Constructed Wetlands Classification

In the previous chapter, we presented the functions and benefits of natural and constructed wetlands. Here, we classify CWs according to their function and purpose in the following three major areas of application:

1. *Constructed wetlands for habitat creation*: these systems aim to provide a wildlife habitat. The main goal is to exploit the main ecological benefits of CWs and not only their function as a treatment facility (Knight, 1997). The main characteristics of CWs (presence of water and vegetation) make them quite suitable for the creation of an ecological habitat, by attracting wildlife species, especially birds, and establishing a green area. Generally, there are four types of CWs that can be created: (a) ponds, which can also have an appropriate depth for fish; (b) marshes, usually shallow aquatic area with herbaceous plants; (c) swamps, with woody vegetation; and (d) ephemeral wetlands, which collect water on a seasonal basis. These systems can also be utilized as a source of food and fiber, and as public recreation sites (Knight, 1997; Sundaravadivel and Vigneswaran, 2001).
2. *Constructed wetlands for flood control*: these wetland systems are built to receive the runoff during flood events (Bueno et al., 1995; Tsihrintzis et al., 1995a,b; 1996; Tsihrintzis and Hamid, 1997; Tsihrintzis et al., 1998). Their implementation may increase the stormwater storage capacity and the infiltration volumes, while reducing the volume of water reaching the sewer system and the treatment plants. Within the urban hydrologic cycle, these systems may contribute to Integrated Urban Water Management and also provide the ability to recycle the stored water volume (Sundaravadivel and Vigneswaran, 2001; Shutes et al., 2010).
3. *Constructed wetlands for wastewater treatment*: the purpose of these engineered wetlands is to receive and purify wastewater of various types, based on the naturally occurring treatment processes. The emphasis of this book is on this category of constructed wetlands.

☆"To view the full reference list for the book, click here"

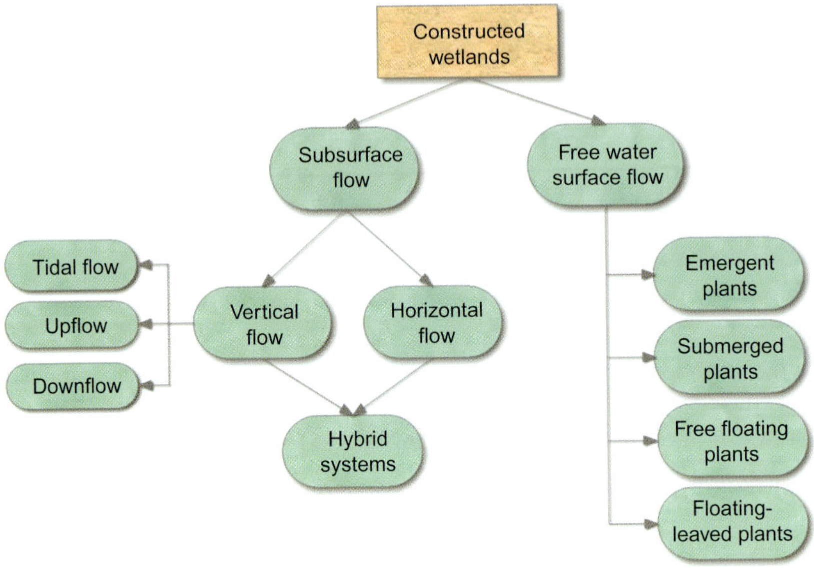

FIGURE 2.1 Classification of constructed wetlands for wastewater treatment.

CWs for wastewater treatment can be further divided into other categories, depending on the specific characteristics of the system, e.g., the type of vegetation or the direction of water flow through the system (Figure 2.1). Depending on the flow path in the system there are two broad types (Sundaravadivel and Vigneswaran, 2001; Vymazal, 2007; Kadlec and Wallace, 2009):

A. Free water surface constructed wetlands (FWS CWs), and
B. Subsurface flow constructed wetlands (SF CWs)

In FWS CWs, the water slowly flows above a substrate medium, thus creating a free water surface and a water column depth usually of a few centimeters. On the contrary, in SF CWs the water flows inside a porous substrate. Depending on the direction of the flow path, SF CWs can be subdivided into horizontal (HSF) or vertical flow (VF). A detailed description of these systems follows.

Another classification can be made based on the growth characteristics of the vegetation. Thus, one can distinguish (Vymazal et al., 1998):

A. Floating treatment wetlands (FTWs) (Floating Islands),
B. Emergent macrophyte wetlands, and
C. Submerged macrophyte wetlands.

Usually, CW systems are planted with rooted emergent macrophyte species. A general description of the basic characteristics of the various CW types follows.

2.1 FREE WATER SURFACE CONSTRUCTED WETLANDS

The use of FWS CW systems is more common in North America (Kadlec and Wallace, 2009). They are used almost exclusively for municipal wastewater treatment (Kadlec and Wallace, 2009). FWS CWs comprise shallow channels or basins, with a sealed bottom (e.g., geotextile or clay) to prevent wastewater seepage to the underlying aquifer. They contain a soil layer of up to 40 cm thick, where the macrophytes (usually emergent, but also submerged or floating can be used) are planted. Water flows nearly horizontally at low velocity above the soil layer along the system, creating a water column depth of 20–40 cm (Vymazal et al., 2006) or even up to 80 cm (Akratos et al., 2006; Crites et al., 2006); therefore, it is exposed to the atmosphere and to partial sunlight (Figure 2.2). The water level is maintained with an appropriate outlet level control arrangement. The water flows through the wetland bed and comes into contact with the soil grains and plant parts, thus enabling a series of physical, biological, and chemical processes to take place, which contribute to the degradation and removal of various pollutants. However, nearly standing water increases the possibility of mosquito breeding.

FWS CWs have been proved to be effective in the removal of suspended solids (SS) and biochemical oxygen demand (BOD_5). Removal of nitrogen (N), pathogens, and other pollutants (e.g., heavy metals; HM) is high, while phosphorus (P) removal is limited (Kadlec and Knight, 1996; Vymazal, 2007; Kadlec and Wallace, 2009; Kotti et al., 2010; Tsihrintzis and Gikas, 2010). Dissolved oxygen (DO) concentration in the water column varies from high (near the surface) to practically zero (near the bottom). Typical applications of FWS CWs include advanced treatment of secondary effluents and stormwater runoff treatment (Shutes et al., 1997; Carleton et al., 2001; Kadlec and Wallace, 2009). Before treating raw wastewater, these systems are usually equipped with a pre-treatment stage. Usually, FWS CWs demand higher surface area compared to other CW types for the same wastewater flow and characteristics, since the porous media and the plant roots in SF CW systems provide a greater contact specific area

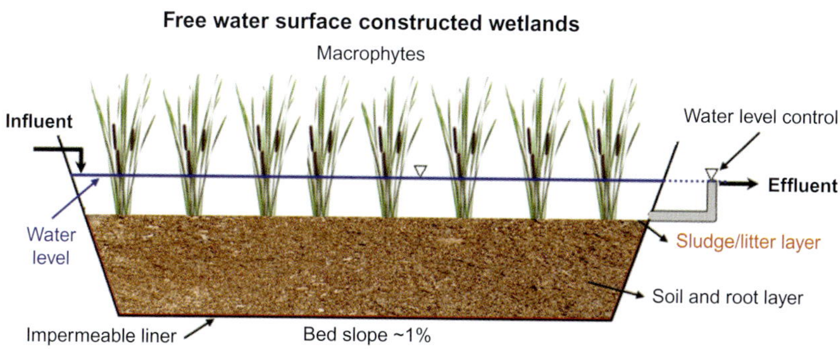

FIGURE 2.2 Free water surface constructed wetlands (FWS CWs; schematic representation).

(USEPA, 2000a,b; Tsihrintzis et al., 2007; Kotti et al., 2010). On the other hand, SF CWs tend to demand higher initial investment (Vymazal et al., 2006; Tsihrintzis et al., 2007). Due to the presence of a water surface, FWS systems tend to better resemble natural wetlands, and thus, provide more wildlife habitat benefits (Vymazal et al., 1998).

2.2 HORIZONTAL SUBSURFACE FLOW CONSTRUCTED WETLANDS (HSF CWs)

HSF CW systems are gravel or soil beds usually planted with common reeds (Vymazal et al., 2006). The main difference with FWS CWs is that there is no water surface exposed to the atmosphere, since the water flows horizontally along the bed below the substrate surface through the pores of the porous media and the plant roots (Vymazal et al., 1998). Thus, health risk for wildlife habitat and humans is minimized, and mosquito breeding is not favored (Kadlec and Wallace, 2009). The substrate, which usually is gravel or a mixture of sand and gravel, supports the growth of the vegetation. The depth of the substrate varies between 30 and 80 cm (USEPA, 1988; Crites et al., 2006; Vymazal et al., 2006; Akratos and Tsihrintzis, 2007), depending on plant root depth, while the bottom of the bed is covered by an impermeable geo-membrane (Figure 2.3). A small slope (1–3%) of the bottom is usually applied to promote gravitational water flow (USEPA, 1988; Kadlec and Wallace, 2009). If the system is properly designed, no surface flow is visible and the water level is maintained about 5–15 cm below the top of the substrate with an appropriate outlet structure (Vymazal et al., 2006). The uniform distribution of the influent wastewater across the bed width is crucial. For this, the wastewater feeding in the inlet is made through a perforated pipe placed across the entire bed width to cover the entire bed surface (Akratos and Tsihrintzis, 2007).

This type of CWs is most commonly used in the United States and quite commonly in Europe (Vymazal et al., 1998; 2006). Compared to FWS systems, HSF CWs possess higher investment costs, although the area demand is smaller

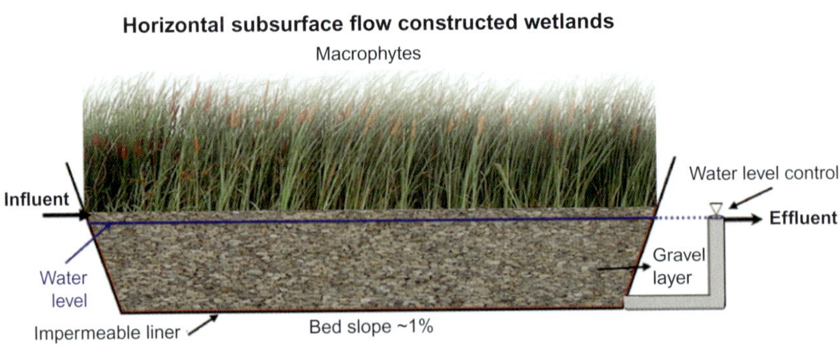

FIGURE 2.3 Horizontal subsurface flow constructed wetlands (HSF CWs; schematic representation).

(Tsihrintzis et al., 2007; Kadlec and Wallace, 2009). HSF systems are effective in municipal wastewater treatment. The presence of plant roots and porous media favors the development of the biofilm, which enhances the removal of organic matter (OM) and SS, while nutrient removal (N and P) usually reaches lower levels (Kadlec and Knight, 1996; Vymazal et al., 2006; Akratos and Tsihrintzis, 2007; Kadlec and Wallace, 2009; Gikas and Tsihrintzis, 2010). The good performance of this CW type has enabled research efforts, in order to determine possible modifications toward further efficiency enhancement, e.g., wastewater step-feeding (Sundaravadivel and Vigneswaran, 2001; Stefanakis et al., 2011a), effluent recirculation (Reese, 2005; Stefanakis and Tsihrintzis, 2009a), and outlet water level raising (USEPA, 1988; Stefanakis and Tsihrintzis, 2009a).

2.3 VERTICAL FLOW CONSTRUCTED WETLANDS

Vertical subsurface flow constructed wetlands (VFCWs) were initially developed by Seidel (1965) as a middle stage after an anaerobic septic tank and before HSF CWs (Vymazal et al., 2006). At the initial applications of the CW technology, focus was given on the other CW types, since VFCWs generally possess higher operation cost. The gradual increase in the use of VFCWs was a result of the realization that HSF systems possess relatively low oxygen transfer capacity (OTC) for the demands of a secondary treatment, which respectively limits the ammonia nitrogen (NH_4^+-N) oxidizing capacity (Cooper, 1999). Therefore, the need to enhance the oxygen amount that reaches the wetland body led to more intensive research on VFCWs during the last two decades. The main advantage of VFCW systems is their higher OTC due to the feeding regime of nearly instantaneously flooding the bed surface of the VFCW. Moreover, VFCWs possess smaller area demands (up to 2 m^2/p.e.) compared to HSF CWs (usually 5-10 m^2/p.e.), which also implies lower construction costs (Vymazal et al., 1998; Cooper, 1999). VFCWs are mainly used in Europe, and especially in Denmark, Austria, Germany, France, and the United Kingdom, but also in the United States (USEPA, 1995; Kadlec and Wallace, 2009).

The most common system setup includes a bed filled with gravel/sand layers with increasing gradation with depth (Vymazal et al., 2006) (Figure 2.4). The depth of the bed varies from 0.45 to 1.20 m and the bottom of the bed has a small slope (1-2%), which allows collection of treated water and drainage out of the unit. Similar to the other CW types, the bottom is covered by a geo-membrane or may be made of reinforced concrete. Common reeds (*Phragmites australis*) are more often used and are planted at the top of the bed. The French version of VFCWs (the so-called French system) includes no pretreatment (e.g., settling tank), while the porous media of the first-stage beds are coarser compared to the common design in other European countries and the United States (Boutin et al., 1997; Molle et al., 2005).

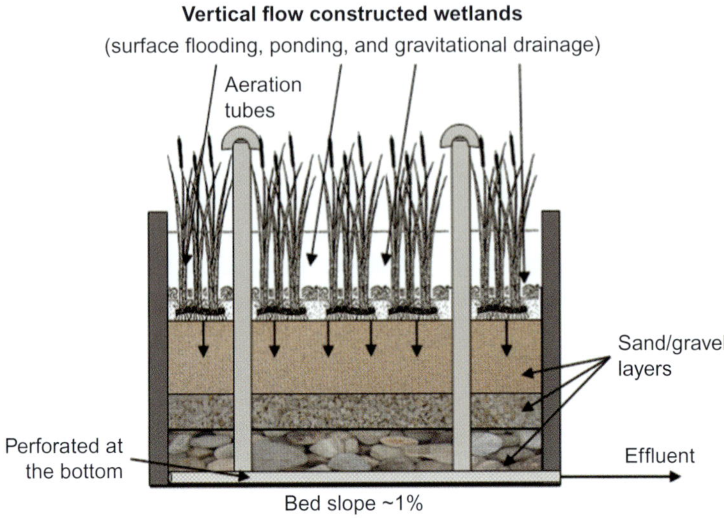

FIGURE 2.4 Vertical flow constructed wetlands (VFCWs; schematic representation).

The wastewater is applied in large batches onto the bed surface, flooding the entire surface, creating for a short time water ponding of 3-5 cm, and then, percolating and draining vertically by gravity through the porous media. With this mode of operation, the wastewater spreads on the entire CW surface and moves downward by gravity, pushing out the trapped air and sucking fresh air into the bed, thus enhancing aeration. Passive aeration is further enhanced when providing aeration pipes in the bed, and also, employing a wet-dry cycle of operation. This way of feeding is important in the treatment process, since increased oxygen transfer within the bed provides enhanced aerobic conditions for the oxidation of ammonia nitrogen (nitrification) and decomposition of the OM, compared to HSF CWs (Cooper et al., 1996; Vymazal et al., 1998; Cooper, 1999; Vymazal, et al., 2006; Vymazal, 2007; Kadlec and Wallace, 2009; Stefanakis and Tsihrintzis, 2009b, 2012a). However, these conditions do not favor denitrification, while the removal of P is limited, mainly due to inadequate contact time between the porous media and the wastewater as the later flows down by gravity (Brix and Arias, 2005; Stefanakis and Tsihrintzis, 2012a).

VFCWs have been mainly used for municipal and domestic wastewater treatment. Due to their increased nitrification capacity they have also been used for the treatment of other wastewater types, mostly those with high ammonia nitrogen concentration, e.g., landfill leachate, dairy wastewater, and food processing wastewater, among others (Kadlec and Wallace, 2009).

2.4 HYBRID CONSTRUCTED WETLANDS

Hybrid systems are combinations of various CW types, mainly VFCWs and HSF CWs, aiming at improving the overall efficiency (Vymazal et al., 1998; Cooper, 1999; Cooper et al., 1999; Vymazal et al., 2006; Vymazal, 2011).

The concept is to exploit the advantages of the one type to counterbalance the disadvantages of the other. Thus, the fact that HSF systems have lower nitrification capacity due to limited OTC can be offset with VFCWs which are more effective in nitrification (higher OTC) (Vymazal et al., 1998). On the other hand, HSF CWs provide good conditions for denitrification, contrariwise to VFCWs. The first attempt to combine various CW types was made by Seidel, who designed a two stage system: parallel VFCWs followed by HSF CWs in series (Seidel, 1965). Generally, there are two common types of hybrid systems: (a) a stage with VF units followed by HSF units in series and (b) a HSF stage followed by VF units (Cooper et al., 1996; Cooper, 1999; 2001). Today, the first type is the most widely used hybrid system (Vymazal, 2005a; Kadlec and Wallace, 2009).

The first combination (VF → HSF; Figure 2.5a) includes VF systems placed first in order to remove OM and SS, as also to provide good oxidative conditions

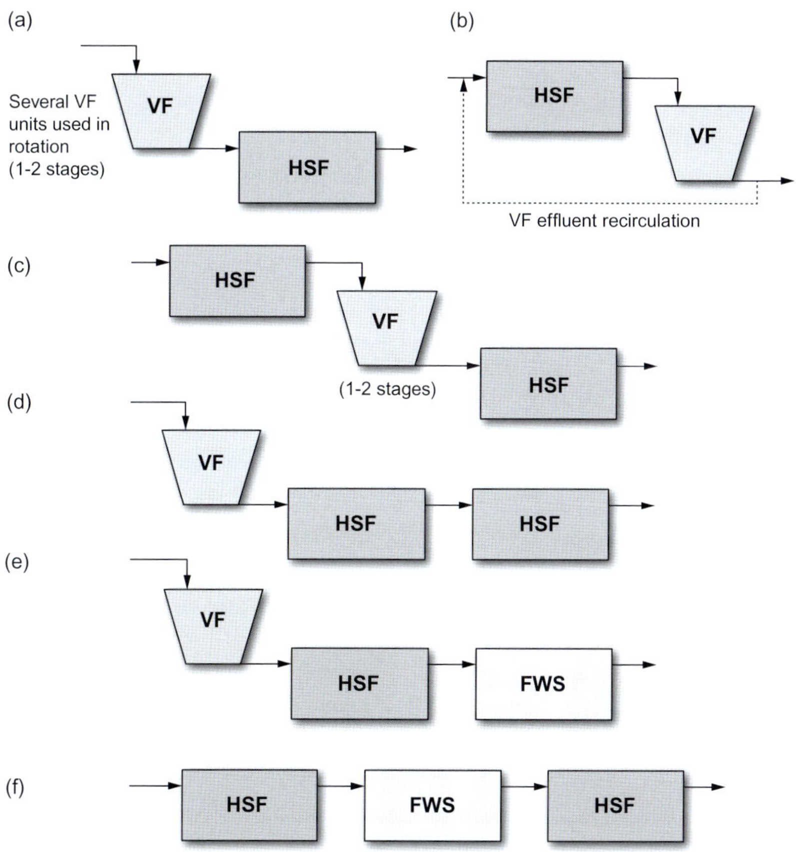

FIGURE 2.5 Various combinations of different CW types in hybrid Constructed Wetlands: (a) VF-HSF, (b) HSF-VF, (c) HSF-VF-HSF, (d) VF-HSF-HSF, (e) VF-HSF-FWS, (f) HSF-FWS-HSF.

(nitrification) for ammonia nitrogen. In the HSF system that follows, the oxidized nitrogen forms are further transformed via denitrification, while additional removal of OM and SS takes place (Cooper, 2001). According to the Seidel model, the first two stages include several VF units operating in parallel, followed by a third stage of two or three HSF units (Seidel, 1965; Vymazal, 2005a). Various modifications of this type have been tested with different number of stages, different beds per stage, different filter media, and different wastewater types (O' Hogain, 2003; Öövel et al., 2008; Molle et al., 2008; Melián et al., 2010; Serrano et al., 2011; Wen et al., 2012).

The second combination is the HSF→VF system (Figure 2.5b). The HSF bed is placed first to remove OM and SS, and to provide denitrification. It is followed by a VF system of smaller surface to enhance OM and SS removal, as also to provide good conditions for nitrification (Vymazal, 2005a; Xinshan et al., 2010). Due to increased nitrate concentration in the VF effluent, as a result of the ammonia nitrogen oxidation, the effluent has to be recirculated back to the HSF inflow or to the pretreatment stage (sedimentation tank), in order to achieve a higher Total Nitrogen (TN) removal (Cooper, 2001; Vymazal, 2005a). These systems have also been proved to be very effective (Gómez Cerezo et al., 2001; Vymazal, 2005a; Cui et al., 2006; Masi et al., 2007).

Various combinations are also under research (Vymazal, 2005a) (Figure 2.6c and d). There are studies which investigate 1–4 treatment stages in series with alternating CW types, for domestic (Tuszynska and Obarska-Pempkowiak, 2008; Gaboutloeloe et al., 2009) or greenhouse wastewater (Seo et al., 2008). The system with HSF-VF beds has also been examined with different media depths after a hydroponic ditch (Ren et al., 2011) or anaerobic baffled reactor for domestic wastewater treatment (Singh et al., 2009) and after steel slag filters for dairy wastewater treatment (Lee et al., 2010). Another type of hybrid CWs also includes FWS CWs as a final (third) stage after VF and HSF beds (Ávila et al., 2012) (Figure 2.5e) or as a middle stage between HSF beds (Ye and Li, 2009) (Figure 2.5f).

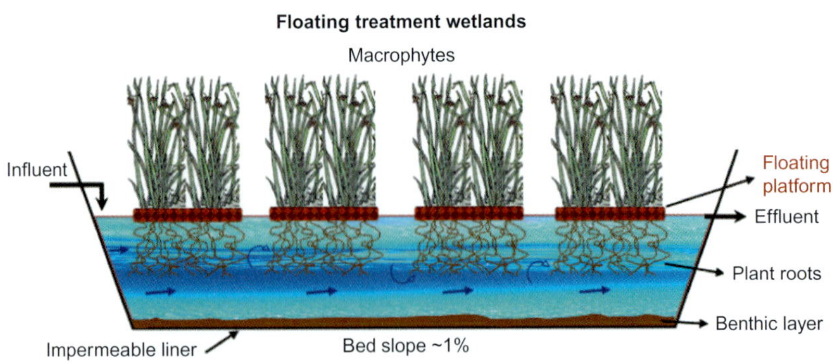

FIGURE 2.6 Floating treatment wetlands (FTWs; schematic representation).

2.5 FLOATING TREATMENT WETLANDS

FTWs (also known as Floating Islands) represent a new development in the field of constructed wetlands. Their main characteristics include parts from both CWs and pond systems (Van de Moortel et al., 2010). The basic difference compared to FWS, HSF, and VF systems is that the plants are not established in porous media or soil but on an element that floats over water (Stewart et al., 2008; Van de Moortel et al., 2010; Tanner and Headley, 2011) (Figure 2.6). The platform is usually made of plastic in order to be resistant to various hydraulic and environmental conditions and allows for the growth of emergent macrophytes. These plant species develop an extensive and dense root system below the platform and inside the water. The system floats on a free water surface, and therefore, it is not affected by varying water level in rivers, lakes, canals, etc. (De Stefani et al., 2011). Similar to the other CW types, the role of plants remains more or less the same; the root system provides the necessary attachment area for biofilm creation and operates as a physical filter, while the plants uptake nutrients for their growth directly from the water column (Tanner and Headley, 2011; Xin et al., 2012).

FTWs have been implemented for the treatment of stormwater (Tanner and Headley, 2011; Headley and Tanner, 2012), domestic wastewater (Van de Moortel et al., 2010), agricultural and municipal wastewater (Stewart et al., 2008; De Stefani et al., 2011), river and lake water (Hubbard, 2010), and eutrophicated water (Xin et al., 2012; Zhao et al., 2012), among other applications.

Chapter 3

VFCW Types

3.1 HYDRAULIC MODE OF OPERATION

Based on the water flow direction along the vertical axis and the level and duration of saturation, VF systems can be distinguished in the following variants:

3.1.1 VFCWs with Intermittent Loading (Downflow)

This is probably the most common operation mode, especially in Europe (Figure 3.1). This concept of operation was developed by Seidel (1966). The wastewater is applied in large volumes onto the bed surface in a very short period of time, thus flooding the entire CW surface (Vymazal et al., 2006; Stefanakis and Tsihrintzis, 2009b, 2012a). Usually, an appropriate device or network of perforated tubes provides the uniform distribution of the wastewater onto the entire bed surface. The rapid introduction of wastewater onto the surface creates temporary water ponding of 3-5 cm and temporary saturated conditions in the top layers. The trapped air is forced by the supernatant water to move downward. The water is then drained vertically by gravity through the porous media layers of the bed in an unsaturated flow mode. As the water drains, air from the atmosphere enters the system and fills the void space of the media (Maier et al., 2009), replacing the water volume that drains. In this way, the bed aeration is enhanced and the microbial activity is stimulated (Cooper, 1999; Vymazal, 2007). After the full drainage of the water, a resting period is applied in order to fully restore aerobic conditions within the bed, and to allow for the oxidation of the accumulated OM to prevent clogging of the bed (Cooper et al., 1996; Vymazal et al., 2006; Vymazal, 2007; Kadlec and Wallace, 2009). This feeding strategy also favors the nitrification process (Cooper et al., 1996; Green et al., 1997a; Laber et al., 1997; Tietz et al., 2007a; Vymazal, 2007). Usually, the number of wastewater doses per day is not very high (3-10), in order to allow for the complete drainage of the loaded water between two batches.

☆"To view the full reference list for the book, click here"

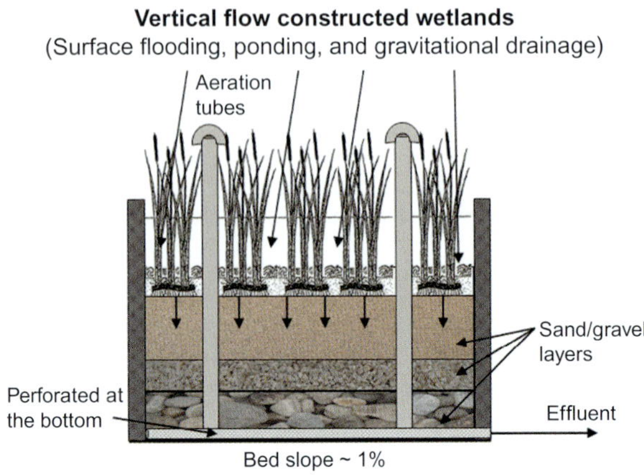

FIGURE 3.1 VFCWs with intermittent loading (schematic representation).

3.1.1.1 The French System

A special variant of vertical flow constructed wetlands (VFCWs) has been developed in France over the last two decades (Boutin et al., 1997). The so-called French system is an alternative arrangement of intermittent loading VFCW beds. The main difference of this system is the absence of a fine-grained material top layer in the bed and the placement of an extended layer with coarse-grained gravel. A typical configuration of the French system includes a primary and a secondary treatment stage (Boutin et al., 1997; Boutin and Liénard, 2003; Molle et al., 2005). A coarse bar screen is usually installed before the main treatment stages to remove large particles (>2 cm; Boutin et al., 1997). Usually, the first stage includes three parallel beds. The main advantage of the French system is that raw wastewater is fed directly onto the beds of the first stage, which means that a primary settling tank is avoided (Molle et al., 2005) and each bed receives the full organic load and the SS content, resulting in a decreased total surface area needed. The feeding strategy is the same with the common VFCW type with intermittent loading and comprises feeding and resting periods. This high load application represents, however, a risk for fast clogging of the bed, taking into account that the major part of SS is retained in the first stage beds and a sludge layer is gradually created (Molle et al., 2006). This risk is minimized by the application of proper resting periods which allow for aerobic conditions restoration. Moreover, sludge accumulation on the bed surface means that management and handling cost for primary sludge are reduced (Boutin and Liénard, 2003). French systems have been used in series with horizontal flow CWs (hybrid systems; Molle et al., 2008).

3.1.2 Recirculating VFCWs

A problem with intermittently loaded VF systems has to do with the relatively short contact time between the wastewater and the porous media, as it drains vertically through the bed matrix. To counterbalance this issue, partial effluent

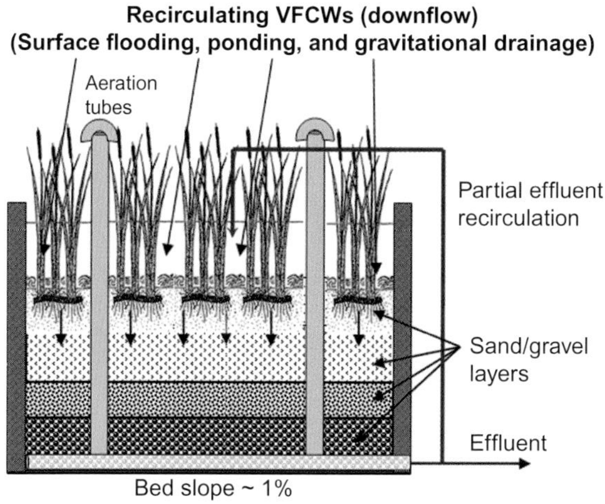

FIGURE 3.2 Recirculating VFCWs (schematic representation).

recirculation has been proposed (Sun et al., 2003; Figure 3.2). Reintroducing part of the effluent volume onto the bed surface provides more oxygen for the enhancement of the aerobic microbial activities, while it increases the contact time and the respective interactions between the wastewater and the biofilm attached to the plant roots and sand/gravel particles (Sun et al., 2003; Sklarz et al., 2009). The result of this is dilution of the inflow water and increased removal of OM. When recirculation of the well-nitrified VFCW effluent takes place into the pretreatment stage (e.g., a sedimentation tank), the available carbon and the generally anoxic conditions can also enhance denitrification, which results in an improved removal of total nitrogen (Laber et al., 1997; Arias et al., 2005).

These systems have been proved to be a good and reliable solution for the treatment of domestic wastewater as small onsite treatment systems for single households or small communities (Laber et al., 1997; Arias et al., 2005; Sklarz et al., 2009; García-Pérez et al., 2011; Zapater et al., 2011; Porst-Boucle and Molle, 2012). VFCWs with effluent recirculation have also been used to treat other wastewater types, e.g., swine wastewater (He et al., 2006a), landfill leachate (Lavrova and Koumanova, 2010), graywater treatment and reuse (Gross et al., 2007), agricultural wastewater (Sun et al., 2003), and livestock wastewater (He et al., 2006b), among others.

3.1.3 Tidal Flow CWs

The operation strategy of these systems relies on the regular filling of the bed with wastewater (creating saturated conditions) followed by draining (unsaturated conditions) (Sun et al., 1999a, 2006; Zhao et al., 2004a; Kadlec and Wallace, 2009; Figure 3.3). During the filing of the bed, air existing within the pores of the substrate is forced to escape the bed and the system gradually becomes saturated with wastewater. After a certain period of time, during which

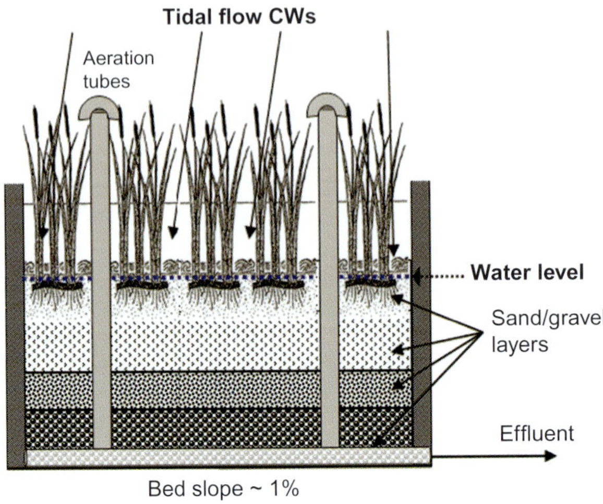

FIGURE 3.3 Tidal flow CWs (TFCWs; schematic representation).

the bed remains completely submerged, the wastewater drainage begins. Then, fresh air from the atmosphere is sucked into the bed, since the draining wastewater operates as a passive air pump (Sun et al., 1999b, 2006). The advantages of this feeding strategy is the increased contact time between the wastewater and the CW components (plant roots, porous media, biofilm) that results in improved pollutant removals. The incoming air favors the creation of aerobic conditions within the bed (increased oxygen transport) and enhances the microbial activity (increased oxygen consumption rates), thus improves the aerobic OM decomposition and nitrification (Sun et al., 2005; Wu et al., 2011a). Maintaining aerobic conditions within the bed through air introduction and application of resting periods is also crucial for bed clogging prevention (Sun et al., 2007).

Various studies have been implemented aiming to investigate the importance of design and operation parameters, such as the fill and drain cycle time, the drainage depth, the substrate depth, the hydraulic and organic loading rates, and the presence of plants. The saturated times tested vary from 20 min (Sun et al., 1999a,b), 45 min (Austin, 2006), 1 h (Zhao et al., 2004a; Sun et al., 2005), 1-3 h (Zhao et al., 2004b; Sun et al., 2007; Wu et al., 2011a), 4 h (Zhao et al., 2011a), 3-10.5 h (Cui et al., 2012), and reach even 7 days (De Feo, 2007). These values indicate that there is not a widely accepted saturation period, although a shorter time of a few hours seems to be adequate.

Usually, the TFCW system comprise several parallel beds, so that the fill and drain mode can be alternately applied. Overall results imply an improvement in system performance under the tidal flow operation compared to the intermittent loading strategy. Due to their treatment advantages, TFCWs have been mainly tested for special wastewater types with high pollutant loads, e.g., agricultural (pig farm slurry) (Sun et al., 1999a,b, 2006, 2007;

Zhao et al., 2004a,b), a mixture of landfill leachate and activated sludge (De Feo, 2007), livestock (Zhao et al., 2011a), mixture of dried cheese whey, urea fertilizer, and well water (De Feo, 2007), septic tank effluent (Cui et al., 2012), as also urban stream water with Fe and Cu (Scholz and Xu, 2002a) and domestic wastewater (Tunçsiper et al., 2009). Positive results have also been reported for VFCWs which combine tidal flow and effluent recirculation (Sun et al., 1999a,b, 2005; Zhao et al., 2004a; Tunçsiper et al., 2009; Cui et al., 2012).

3.1.4 Saturated Vertical Upflow CWs

In these systems, wastewater introduction takes place at the bottom of the bed with an appropriate network of pipes (Figure 3.4). The wastewater gradually moves upward through the porous media layers and the plant root zone. The effluent is collected at the bed surface or a few cm below the surface (Breen, 1990, 1997; Ghosh and Gopal, 2010). The advantages of this operational mode is the longer residence time that can be applied, which respectively allows for longer contact time between the wastewater and the system components (porous media, plant roots, biofilm) and, thus, improved nutrient removal rates (Farahbakshazad and Morrison, 1997, 2000, 2003; Moreno et al., 2002). It is reported that the increased contact time between wastewater and roots significantly upgrades the role of the plants in nitrogen removal by direct uptake (Breen, 1990; Heritage et al., 1995; Farahbakshazad and Morrison, 2000, 2003; Moreno et al., 2002; Fuchs et al., 2007). An interesting finding is that the vertical oxygen profile within the bed presents decreased oxygen concentration with depth, i.e., aerobic conditions tend to dominate at the top layers

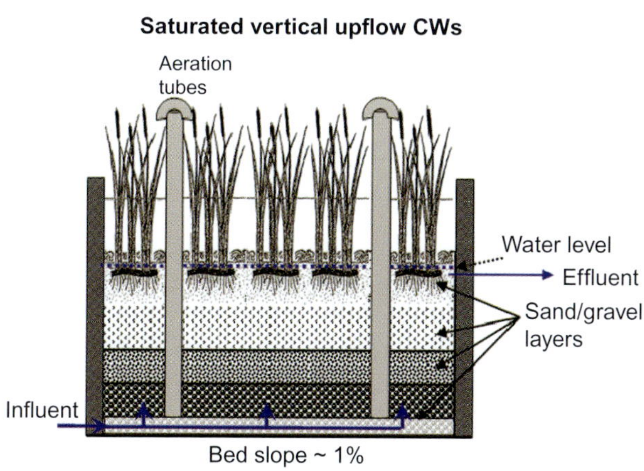

FIGURE 3.4 Saturated upflow vertical CWs (schematic representation).

of the bed (Ong et al., 2010). This is related to another finding according to which about 70% of the root mass is concentrated in the upper layers (Breen and Chick, 1995).

This water movement from the bottom anaerobic region to the upper aerobic layer indicates the simultaneous presence of aerobic and anaerobic processes (Imfeld et al., 2009). This fact, coupled with methane production at the bottom anaerobic layer, makes upflow VFCWs attractive for the biodegradation (dechlorination) of chlorinated solvents and the removal of pollutants present in contaminated groundwater (e.g., volatile organic compounds, hydrophobic organic compounds) (Kassenga et al., 2003; Tan et al., 2004; Amon et al., 2007; Imfeld et al., 2009; Roen, 2011; Zhou et al., 2013). This unique combination of aerobic and anaerobic zones has also been exploited for the treatment of textile wastewater, for the removal of color and acid orange 7 (AO7; an anionic acidic dye), besides organic compounds (Ong et al., 2009).

The hydraulic residence time (HRT) appears to be a crucial parameter for the system performance, since it determines the contact time between the wastewater and the rootzone (Breen, 1997; Ghosh and Gopal, 2010). Various HRT values have been tested in upflow VFCWs. For agricultural runoff treatment, the HRT applied was 10 h (Farahbakshazad and Morrison, 2000), for secondary effluent treatment 5 d (Breen, 1990), for tertiary treatment 1-4 d (Ghosh and Gopal, 2010), for groundwater treatment 6.9-17.5 d (Roen, 2011), for diluted urine 4-10 h (Farahbakshazad and Morrison, 1997), and for septic tank effluent treatment 2-3 d (Xuan, 2009). The relatively high HRT that can be applied also makes upflow VFCWs attractive for the investigation of possible ways to increase phosphorus removal, e.g., with use of special porous media (Lantzke et al., 1998; Brooks et al., 2000). Effluent recirculation has been found to enhance system performance concerning nutrients removal, as experiments in Sweden (Moreno et al., 2002; Farahbakshazad and Morrison, 2003) and Brazil showed (Salati et al., 1999; Farahbakshazad et al., 2000).

3.1.5 Saturated Vertical Downflow CWs

These systems are more or less similar to the upflow beds. The main difference is that the wastewater is introduced on the top of the bed with uniform distribution (Figure 3.5). With an appropriate hydraulic arrangement at the outlet, the water level within the bed is kept a few cm (5-20 cm) below the surface (Visesmanee et al., 2008; Dan et al., 2011). The HRT applied (1-8 d) allows for the removal of various pollutants from the wastewater, due to the increased contact time between the wastewater and the system components.

This type appears as an attractive option for acidic effluents, e.g., acid mine drainage treatment, classified into the passive treatment systems (Mitsch and Wise, 1998; Norton et al., 1998; Demchak et al., 2001; Masferrer, 2002; Thomas and

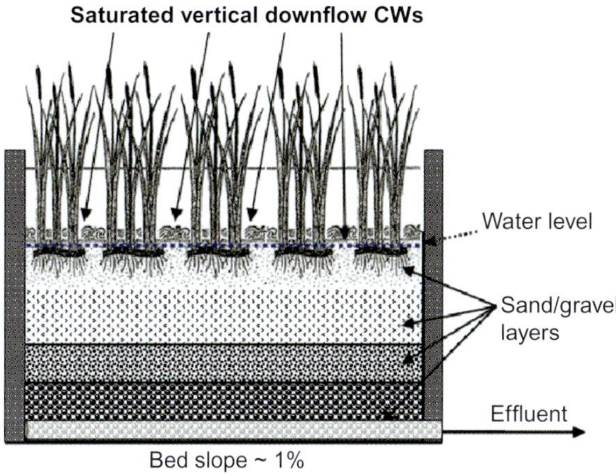

FIGURE 3.5 Saturated downflow vertical CWs (schematic representation).

Romanek, 2002; Ford, 2003; Collins et al., 2004; Johnson and Hallberg, 2005a; Wolf et al., 2007). These wetland systems are also known as anaerobic wetlands or "compost bioreactors" if they are unplanted (Johnson and Hallberg, 2005a,b). Vertical downflow wetlands for acid mine drainage treatment are also called "successive alkalinity producing systems" and two modes can be distinguished: oxidizing aerobic wetlands and reducing anaerobic wetlands (Ford, 2003; Kosolapov et al., 2004; Johnson and Hallberg, 2005a). While aerobic wetlands (mainly systems with surface flow) can treat net alkaline water (Johnson and Hallberg, 2005a), anaerobic wetlands (with subsurface flow and sometimes subsurface and surface flow simultaneously) also treat acidic water with high metal concentrations (e.g., Al, Fe). Anaerobic wetlands are planted beds (generally >30 cm), deeper than aerobic ones, usually filled with organic substrates for the promotion of reducing conditions and with limestone for acid neutralization, while aerobic wetlands usually contain soil, clays, or mine spoil (Ford, 2003; Kosolapov et al., 2004; Johnson and Hallberg, 2005b; Skousen and Ziemkiewica, 2005). A sufficient HRT provides adequate contact time for the water and improves the remediation, thus, anaerobic wetlands usually possess higher land area demands.

3.1.6 Integrated VFCWs

Integrated VFCWs (IVFCWs) are a relatively new type of CWs with subsurface flow. This type was first introduced in China by the European Union in 1996 as a joint international project (INCO-DC project) (Perfler et al., 1999; Wu et al.,

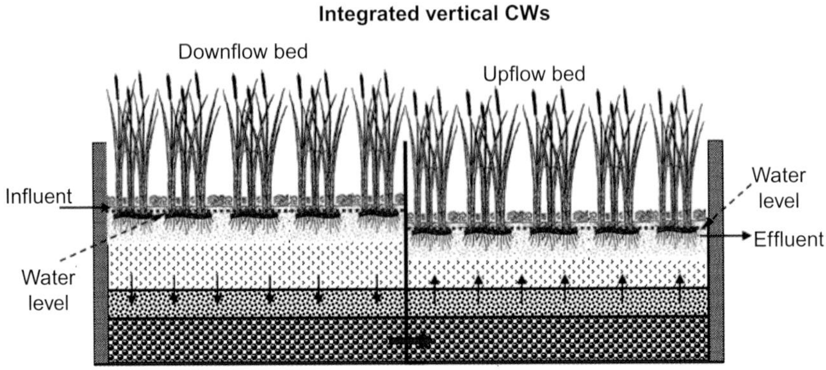

FIGURE 3.6 Integrated vertical flow constructed wetland; arrows show the waterflow direction (IVFCWs; schematic representation).

2004; Liu et al., 2008; Chang et al., 2012; Xie et al., 2012). This system is practically another type of hybrid CWs. IVFCWs comprise a downflow-planted vertical unit followed by an upflow-planted vertical unit (Figure 3.6). The two systems are separated by a partition wall (Chang et al., 2012; Xie et al., 2012). The downflow vertical system usually is 10 cm higher than the upflow one, so that water flows from the first to the second system without the use of any pump (Figure 3.6).

IVFCWs are considered as promising systems for nitrogen removal, as also for COD and TP removal (Wu et al., 2004; Kantawanichkul et al., 2008; Yuan et al., 2008; Wang et al., 2009a; Zhou et al., 2009; Zhao et al., 2011b; Chang et al., 2012; Zhang et al., 2012a). Aerobic processes (e.g., nitrification) appear in the upper root zone in the downflow bed, while anaerobic processes (e.g., denitrification, sulfate reduction, methanogenesis) dominate in the upflow bed (Yuan et al., 2008; Zhao et al., 2011b). Therefore, the combination of down- and upflow units provides good treatment efficiency. The longer wastewater retention time and the higher oxygen transfer due to the flow length enhance this performance.

These systems have also been employed with effluent recirculation (He et al., 2006b), in combination with a recirculation system with IVFCWs and ponds (Fei et al., 2011) and a submerged membrane bioreactor (MBR) unit (Xiao et al., 2010). The various applications of IVFCWs include the purification of polluted surface waters (rivers, lakes) for reuse options (Liang et al., 2003; Zhou et al., 2009; Tang et al., 2011; Chang et al., 2012; Zhang et al., 2012a), domestic wastewater treatment (Zhao et al., 2011b; Chang et al., 2012), secondary effluent treatment and sewage (Wu et al., 2004; Wang et al., 2009a; Xiao et al., 2010), livestock wastewater (e.g., pig farm effluent) (He et al., 2006b; Kantawanichkul et al., 2008), aquaculture ecosystem restoration and protection

(Fei et al., 2011), water polluted with organic compounds (acid esters) (Zhao et al., 2004c), saline wastewater (Wang et al., 2010a), and paper mill effluent (Li et al., 2009).

3.2 THE PROBLEM OF BED CLOGGING

Clogging of the bed is probably one of the most serious problems that subsurface flow CWs (both horizontal and vertical) may present during their operation. It takes place through the gradual accumulation of solids (e.g., organics) within the wetland substrate, which subsequently results in decreased hydraulic conductivity of the porous media, and thus, reduces the water flow that passes through the bed (Blazejewski and Murat-Blazejewska, 1997). Substrate clogging may be caused by a variety of biological, chemical, and physical processes that take place within the system (Knowles et al., 2011; Nivala et al., 2012a). The appearance of clogging phenomena affects the system performance in terms of pollutant removal (Blazejewski and Murat-Blazejewska, 1997; Langergraber et al., 2003; Winter and Goetz, 2003; Zhao et al., 2009; Knowles et al., 2011; Nivala et al., 2012a).

In VFCWs, clogging may result in the creation of a surface pond above the porous media layers, since the influent volume cannot be completely drained between two feeding batches due to the decreased hydraulic conductivity of the substrate media. As a result, a portion of the influent wastewater that may be gathered onto the bed surface can bypass the bed (Knowles et al., 2011) and flow into the effluent ditch. Another clogging consequence is that the amount of oxygen that penetrates the bed is decreased and is not sufficient to cover the needs for the various aerobic processes (e.g., organic matter decomposition, nitrification) (Langergraber et al., 2003; Hua et al., 2010). This highly affects the treatment capability of the system, taking into account that the basis for the proper system operation is air transport from the atmosphere to the bed via diffusion and convection. In general, clogging affects not only the system efficiency but it can also reduce its useful operational lifetime.

3.2.1 Clogging Mechanisms and Contributing Factors

Generally, the mechanisms responsible for clogging are complicated and not fully understood. There are several studies reporting the various processes that influence the permeability of the bed. The recognized mechanisms that contribute to the clogging are (Platzer and Mauch, 1997; Langergraber et al., 2003; Winter and Goetz, 2003; Hua et al., 2010):

1. *Accumulation of organic and inorganic solids*: it refers to the gradual deposition of SS onto the bed surface which results in the creation of a surface mat (outer blockage) and the SS deposition onto the surface of the substrate grains in the pores (inner blockage). The level of volatile (biodegradable)

solids affects the blockage type. Organic solids can be decomposed by aerobic microorganisms. The creation of the surface mat limits the oxygen supply for the microorganism metabolism, which means that the increase of the outer blockage leads to respective increase of the inner blockage. If there is a pretreatment stage (e.g., a sedimentation tank or Imhoff tank) then the major portion of the incoming solids is removed, which is crucial for the effective and long-term operation of the system.

2. *Biomass production*: the development of the biofilm on the surface of the porous media grains may result in the clogging of the substrate pores. As the system is fed with nutrients contained in the influent wastewater, microorganisms metabolize them for their growth. The increase of microorganism population forms the biofilm layer. When the biofilm layers on two adjacent grains increase and come to touch each other, the pore is plugged and the substrate hydraulic conductivity is reduced to that extent determined by the biofilm. In order to prevent pore clogging, the biomass production rate has to be in equilibrium with their decay rate.
3. *Composition of clogging material*: the percentage of the biodegradable matter that is accumulated in the substrate pores affects the clogging process. Clogging occurs from both organic and inorganic matter accumulation; therefore, the composition and origin of the wastewater to be treated determines the clogging characteristics.
4. *Rootzone effect*: concerning VFCWs, many researchers report that the presence of plants is beneficial for the system operation and prevents clogging of the bed (Brix, 1994b, 1997; Bahlo and Wach, 1995; Coleman et al., 2001). The exact role of plants from a hydraulic point of view is still a controversial issue, especially for horizontal flow systems (Knowles et al., 2011). It is accepted that the presence of a dense root system affects the hydraulic characteristics of the substrate: the roots and the biofilm attached on them (and on the substrate pores) contribute to the pore clogging, but, at the same time, the decomposition of the dead root parts and the continuous new root development result in the formation of new secondary pores within the bed (Vymazal et al., 1998; Cooper, 2009). In VFCWs, it is reported that the movement of reed shoots creates cracks in the "schmutzdecke" (i.e., sludge and litter layer) which provides openings for the water to move downward (Brix, 1994b; Stottmeister et al., 2003; Cooper, 2009).
5. *Chemical processes*: precipitation (e.g., of $CaCO_3$), adsorption, and deposition may play a role in clogging development. Usually, these processes are not so intensive to create significant clogging problems. Possible problems may occur during the treatment of industrial wastewater with high concentrations of specific constituents (e.g., metal hydroxides, sulfur) (Kadlec and Wallace, 2009; Knowles et al., 2011).

Based on the clogging mechanisms, a number of factors that influence the creation of pore blockage can be determined (Blazejewski and Murat-Blazejewska, 1997; Platzer and Mauch, 1997; Langergraber et al., 2003; Knowles et al., 2011):

1. *Porous media characteristics*: the grain size distribution, the porosity and the shape of the filter media are important parameters since they determine the volume of the media pores and influence the hydraulic conductivity.
2. *Wastewater characteristics*: pollutant loads, organic load, and solid content. Knowing these parameters of the influent wastewater makes it easier to understand the behavior of the system and the clogging processes. The solid content and the organic portion (biodegradable) are directly connected with the accumulated matter in the pores and the creation of the biofilm.
3. *Operational strategy*: this includes the flow rates applied to the bed and the feeding regime. Extreme flow rates, which may deliver high solid concentration onto the VFCW bed surface, may lead to rapid bed clogging. Application of an appropriate number of intermittent loadings and sufficient resting periods in VFCWs is crucial for the longevity of the system, since the restoration of aerobic conditions and the sufficient aeration of the bed are prerequisites for the mineralization of the clogging material. Additionally, the optimum distribution of the wastewater onto the entire bed surface is important for the effective oxygen transfer into the bed (Stefanakis and Tsihrintzis, 2012a).
4. *Pretreatment*: an appropriate pretreatment stage can provide significant protection of the VFCW bed from shock solid load (Winter and Goetz, 2003). Usually, a primary settling tank (sedimentation or an Imhoff tank) is used to remove the greater portion of solids before the wastewater introduction to the VFCW bed. Alternately, other technologies can be used, e.g., anaerobic sludge blanket reactors (Singh et al., 2009).

Among the various mechanisms, the first three are considered as the most important. The accumulated total matter is of both organic and inorganic composition. The ratio between these two fractions is important because it determines the level of biodegradability of the total accumulated matter, and thus, directly affects the clogging development. This becomes even more important when the wastewater to be treated is strong and contains high organic matter concentration. Based on these, the mode of operation of a VFCW system with intermittent loadings and application of resting periods is crucial for the clogging prevention. It has been found that the infiltration rate of a clogged bed can be restored with the appropriate feeding strategy and operation, due to mineralization of the biodegradable organic matter by aerobic microbes during the resting periods (Platzer and Mauch, 1997). Inorganic material cannot be removed by mineralization, so it remains in the pores and acts as a blockage.

Although the significance of these mechanisms is generally accepted, there is still an ongoing debate on the issue of which is the dominant clogging mechanism. Solid accumulation and biofilm development are the main mechanisms, but the relation and/or interactions among them are still under investigation. It is reported that clogging due to biofilm growth accounts for only a small pore fraction and that solid accumulation is the main mechanism

(Langergraber et al., 2003; Winter and Goetz, 2003; Kadlec and Wallace, 2009). Indeed, a non-direct relation between the decrease of the infiltration rate and the increase of OM accumulation has been found during treatment of strong dairy wastewater, while the majority of the accumulated solids were non-biodegradable minerals (Tanner et al., 1998; Nguyen, 2000; Caselles-Osorio et al., 2007). On the contrary, another study reports that the biodegradable solid fraction (volatile) is not an important clogging factor, and that the infiltration rate is mainly dependent on the OM distribution in the media pores (Platzer and Mauch, 1997). In a study where solid accumulation and biofilm growth were simultaneously examined in a VFCW unit, results showed that solid accumulation had a greater contribution to the clogging process (Zhao et al., 2009). It is interesting that in an anti-sized system with reverse media gradation (coarse material at the top of the bed) the clogging occurred more slowly compared to the unit with the typical design (Zhao et al., 2005). This unusual media configuration provided more available pore volume for solids trapping and allowed for better aeration of the substrate, indicating that the grain size of the filter media can influence the clogging process.

It is also mentioned that the biofilm has a synergetic role, meaning that it accelerates the clogging process. In another study, the clogging process investigation showed that it proceeds in three stages: solids trapping within the beds; formation of the blanket-like deposition layer; and final formation of the clogging layer (Hua et al., 2010). The infiltration rate appeared to decrease as the clogging process proceeds through these three stages. A recent study investigated the biofilm development and biological clogging in porous media on the basis of bacterial growth and hydraulic conductivity variations for different flow rates and substrate concentrations (Kim et al., 2010). This study confirmed that the created biofilm accelerates the clogging process at high flow rates and showed that the flow rate affects the biofilm morphology.

The use of gravel instead of soils (with lower hydraulic conductivity), which was a common practice in the past, has minimized the possibilities for clogging occurrence (Cooper, 2009). On the whole, clogging phenomena are likely to occur in VFCWs. Bad design and improper management, hydraulic overloading, peak inflow solid concentrations during storm incidents, inadequate porous media selection, and unequal wastewater distribution onto wetland surface have been pointed as typical reasons which can lead to pore clogging (Rousseau et al., 2004a). The most common practice to prevent clogging is the loading of the beds with pre-treated wastewater (mechanically or in a primary settling stage) in order to reduce the content of particulate organic matter and the application of reasonable loading rates (Rousseau et al., 2004a, 2008; Langergraber et al., 2009). In case of fully clogged substrate, the bed is excavated and the clogged material removed and replaced with new clean material or with the old material after cleaning. For clogging reversing, rinsing with hydrogen peroxide has been proposed for the oxidation of the accumulated organic matter (Nivala and Rousseau, 2009) as well as backwashing cleaning (Fei et al., 2010).

Chapter 4

VFCW Components

4.1 VEGETATION

Plants are among the most important components in CW systems, including VFCWs. Their presence in CWs is probably the main reason that CWs are called a "green technology." Plant species used in CWs are usually the same species that exist in natural wetlands. Wetland plants (mainly vascular plants, also known as macrophytes) grow in semisaturated or fully saturated water conditions. In order to be suitable for use in CWs, the selected macrophytes should meet the following criteria (Tanner, 1996):

- They should be well adapted to the local ecological conditions. This is essential so that they will not possess any risks regarding possible disease development to the local flora and to nearby ecosystems.
- They should be viable in the local climatic conditions and withstand possible pest, insect, and disease appearances.
- They should be tolerant against a variety of pollutants present in wastewater (e.g., organic matter, nitrogen, phosphorus, heavy metals, etc.) with simultaneous high removal capacity either by direct uptake or indirectly by providing the necessary conditions (e.g., enhanced oxygen transfer) for other removal mechanisms.
- They should be easily adjusted to the local CW environment and show a relatively fast growth and spread.
- They should be easily available in the local market to buy or, preferably, in the region for shoot transplanting.

It is crucial that local, indigenous species are used in a CW facility, since the import and establishment of exotic species, not naturally present in the area, may create several ecological risks, e.g., invasion and/or diseases. Furthermore, there is always the possibility that exotic species will not be well adapted to new climatic conditions. Plants which grow in nearby natural wetlands and can develop an extensive root system are preferred, since this means that they can

☆"To view the full reference list for the book, click here"

grow well in a saturated substrate. Finally, the selected species should be tolerant against periods with varying saturation level (high influent volume and short dry periods).

4.1.1 Wetland Plant Classification

Generally, the classification of wetland plants is based on the hydrological regime and water flow characteristics, as also on growth characteristics (Cronk and Fennessy, 2001). The most common plant categories are (Brix, 1994b; Haberl et al., 1995; USEPA, 1995, 1988; Kadlec and Knight, 1996; Vymazal et al., 1998; Cronk and Fennessy, 2001; Kadlec and Wallace, 2009):

1. *Emergent*: These plants grow in soils or porous substrates. Their root system and shoots develop below the water surface of the wetland, while leaves and stems grow above the surface. They find the necessary nutrient amounts for their growth in the soil and the water (Cronk and Fennessy, 2001). They are competitive species, compared to other plant types. Common species in this category belong to the monocotyledons family, which dominates in both fresh- and saltwater, mainly in temperate climates, e.g., *Phragmites* spp. (common reed), *Typhaceae* (cattail), *Juncaceae* (rushes), *Scirpus* spp. (bulrushes), *Glyceria* spp. (Mannagrasses), *Iris* spp. (blue and yellow flags), *Zizania aquatica* (wild rice), and *Cyperaceae* (sedges, e.g., *Carex*, *Cyperus*). Emergent plants like cattails, reeds, rushes, and bulrushes are more frequently found in CWs for wastewater treatment. Woody emergent plants include tree and shrub species.
2. *Submerged*: These plants can be rooted in the bottom substrate, while there are also rootless species that float on the water surface. Their photosynthetic parts are usually below the water surface, while they are quite flexible to avoid damages due to water flow. Rooted species takeup nutrients mainly from the substrate zone and secondarily from the water. This category includes families of *Callitrichaceae* (water starwort), *Ceratophyllaceae* (hornwort), *Haloragaceae* (water milfoil), *Potamogetonaceae* (pondweeds), and *Lentibulariaceae* (bladderworts), while the largest family is the *Hydrocharitaceae* (frogbit) (Cronk and Fennessy, 2001). Submerged plant types are considered sensitive to anaerobic conditions, present a strong diurnal effect, and can be shaded out by algae (USEPA, 1988).
3. *Floating-Leaved*: These plants are rooted in the substrate while their leaves float on the water surface. Petioles and/or stems connect the bottom parts with the leaves. This species shade the water column and, therefore, can dominate against submerged species for sunlight absorption. This category includes *Nymphaea* spp. and *Nuphar* spp. (water lilies), *Nelumbo lutea* (water lotus), and *Hydrocotyle vulgaris* (pennyworth).
4. *Floating*: These species are also known as floating attached plants (Cronk and Fennessy, 2001); they float on the water surface. Floating plants usually

show high growing rates. This category includes *Eichhornia crassipes* (water hyacinth), *Pistia stratiotes* (water lettuce), and *Lemna* spp. and *Spirodella* spp. (duckweed).

4.1.2 Plant Species in VFCWs

In CWs with subsurface flow (either horizontal or vertical), typically emergent macrophytes are used, while in FWS CWs all types of plants may be applied. Generally, the most frequently used plant species are *Phragmites*, *Typha*, *Scirpus,* and *Schoenoplectus*, i.e., common reeds, cattails, and bulrushes. In Europe, the most common plants species is the common reed (*Phragmites australis*; Figure 4.1) followed by cattails (*Typha latifolia*; Figure 4.2) (Haberl et al., 1995; Vymazal et al., 1998). For this reason, CWs are also called "reed beds" (Vymazal, 2001). For example, Danish guidelines for VFCWs construction include the use of common reeds (Brix and Arias, 2005). It is also reported that *P. australis* is mostly used in Ireland in both HSF and VF systems (Babatunde et al., 2007), in the Netherlands (Verhoeven and Meuleman, 1995), in Belgium (Rousseau et al., 2004a), in the Czech Republic (Vymazal, 2013), in Estonia (Mander and Mauring, 1997), in Greece (Akratos and Tsihrintzis, 2007; Tsihrintzis et al., 2007; Stefanakis and Tsihrintzis, 2009b, 2012a; Gikas and Tsihrintzis, 2010, 2012; Kotti et al., 2010; Tsihrintzis and Gikas, 2010;

FIGURE 4.1 Cluster of common reeds (*Phragmites australis*). Picture taken during spring months (growing period) under Mediterranean climate (North Greece).

FIGURE 4.2 Cluster of cattails (*Typha latifolia*). Picture taken during spring months (growing period) under Mediterranean climate (North Greece).

Gikas et al., 2011), and in France (Molle et al., 2005, 2006, 2008; Torrens et al., 2009a). The same species is also common in systems in North America, but in some parts of the USA, use of this plant is forbidden to avoid possible invasion in natural wetlands; cattails or bulrush are used instead (USEPA, 2000a). *P. australis* is usually the dominant species in an area, but it can coexist with *Typha* (cattails), *Scirpus* (bulrush), or *Spartina alterniflora*.

Plant growth starts in the first days of spring months, when the first new shoots are regenerated (Figure 4.3). During the last spring month and summer period, plants are fully developed (Figure 4.4). In winter period, when temperatures are lower, plants are yellowing and the above ground part is practically inactivated and dead (Figure 4.5). In CW systems, it is a common practice to cut the above-ground dried plants in late autumn months (Figure 4.6) at the level of only 10-15 cm height, although there is still a debate on the actual benefits of this action.

Table 4.1 presents some basic information about the most common plant species used in VFCWs. The main characteristics of these species are (USEPA, 1995, 1988; Cronk and Fennessy, 2001):

- *Common reeds* (*P. australis*): It is also called *Phragmites communis*. It is the most widely distributed angiosperm and it is present in temperate climates and also in tropical regions. It is an invasive species and can grow even in saline water. Its population is decreasing in Europe and there are

FIGURE 4.3 Pictures of new shoots appearance during first days of spring months in VFCWs for wastewater treatment (a and b) and sludge dewatering (c).

FIGURE 4.4 Fully developed plants during spring and summer months in: (a) and (b) pilot-scale VFCW beds and

(*Continued*)

ongoing efforts to restore its coverage, while in North America its expansion is controlled due to its limited wildlife value. It is a tall perennial grass and can reach a height of up to 4 m. Its rhizome expands vertically at a higher depth, compared to cattails, which results in good oxygen transfer to the substrate.

- *Cattails* (*Typha* spp.): Another common plant species. Cattails are spread worldwide, which implies their ability to grow in various climatic conditions.

FIGURE 4.4, Cont'd (c) full-scale VFCW and (d) on-site HSF CW beds under Mediterranean climate (Greece).

In winter, its metabolism decelerates to prevent oxygen shock; thus, it demands lower oxygen amount. It is capable of transferring relatively high oxygen amount to the substrate. It presents fast shoot regeneration during the first growth months after winter period. *T. latifolia* dominates in shallow water compared to *Typha angustifolia*, which—on the other hand—expands in relatively deeper water to counterbalance the competition. Both *Typha* species are capable of assimilating N and P and producing high rates of annual biomass. This species also creates a dense and extended root system (Figure 4.7).

FIGURE 4.5 Pictures of plants during late autumn months (yellow above ground parts) from (a) a full-scale VFCW and (b) an onsite HSF CW systems under Mediterranean climate (Greece).

- *Bulrushes* (*Scirpus* spp.): These plants also present worldwide expansion. They can grow in inland and coastal waters, even under saline conditions. The recommended water depth for their growth should not exceed 30 cm. They develop deep roots. Bulrush also presents high nutrient uptake during the growing season. Common species of this genus are *Scirpus californicus* and *Scirpus validus*.
- *Rushes* (*Juncus* spp.): It is also present in saline waters, but in smaller numbers. Rushes are perennial, grass-like herbs, which also show relatively high nutrient uptake rates. It can tolerate both dry and wet periods. *Juncus effusus*, a common species of this genus, rapidly develops deep roots and presents a high level of resistance against pests and winter conditions.

Of course, there are also several other plant species that have been used in pilot- or full-scale VFCW systems. Table 4.2 presents the scientific names of various plant species used in VFCWs found in the literature.

FIGURE 4.6 Pictures taken after plant cutting in pilot-scale VFCW units for (a and b) wastewater treatment and (c) sludge dewatering in late autumn months under Mediterranean climate (North Greece).

TABLE 4.1 Information About Common Emergent Plant Species Used in VFCWs (Cronk and Fennessy, 2001; USEPA, 1988)

Common Name	Scientific Name	Distribution	Temperature (°C)		Max. Salinity Tolerance (ppt)	Effective pH Range	Growth Rate (cm/d)	Min. Survival in Anoxia (d)	Root-to-Shoot Ratio
			Desirable	*Seed Germination*					
Common reed	*Phragmites australis*	Worldwide	12-23	10-30	30	2-8	10.8	>28	1.8-9.9
Cattails	*Typha* spp.	Worldwide	10-30	12-24	30	3-8.5		>28	0.4-0.6
Rush	*Juncus* spp.	Worldwide	16-26		20	5-7.5		4-7	
Bulrush	*Scirpus* spp.	Worldwide	18-27		20	4-9		>28	2.3-3.9
Sedge	*Carex* spp.	Worldwide	14-32			5-7.5		4	

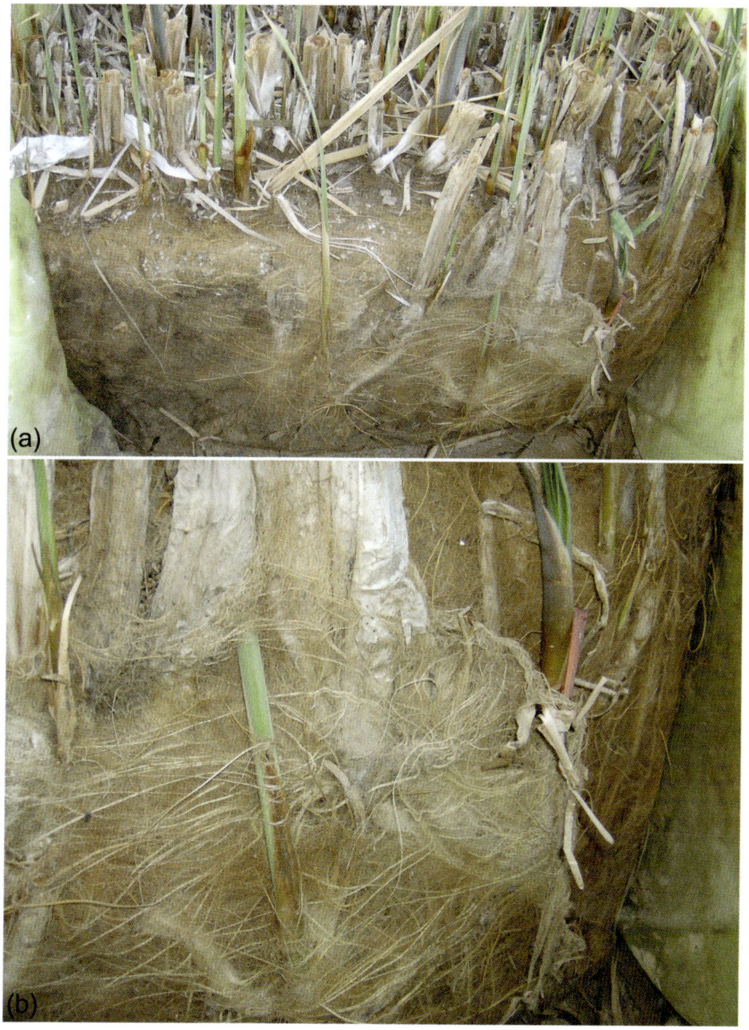

FIGURE 4.7 Pictures (a and b) of the root zone of cattails (*Typha latifolia*) taken from a FWS CW system during spring months (growing period) under Mediterranean climate (North Greece).

4.1.3 The Role of Plants

The significance of plant presence in CWs is today recognized and the positive effects on the system operation and performance have been—more or less—proved (Stottmeister et al., 2003). Although it was a controversial issue in the past, today it is generally accepted that CW plants provide a series of benefits and contribute to the creation of the necessary conditions which directly or indirectly affect the system efficiency.

TABLE 4.2 Various Plant Species Used in VFCWs

Scientific Names of Plant Species Used in VFCWs	
Phragmites australis	*Phragmites japonica*
Typha latifolia	*Typha angustifolia*
Typha orientalis	*Juncus effusus*
Scirpus lacustris	*Scirpus validus*
Scirpus radicans	*Scirpus triqueter*
Phalaris arundinacea	*Cyperus involucratus*
Cyperus flabelliformis	*Cyperus alternifolius*
Iris setosa	*Iris ensata*
Iris pseudacorus	*Acorus calamus*
Villarsia exaltata	*Schoenoplectus validus*
Conocarpus erectus	*Baumea articulata*
Zizania latifolia	*Zizania aquatica*
Thalia dealbata	*Paspalum pennisetum*
Arundo donax	*Miscanthus sinensis giganteus*
Oryza sativa	*Canna indica*
Pennisetum purpureum	*Vetiveria zizanioides*
Spathiphyllum wallisii	*Plantago asiatica*
Pistia stratiotes	*Dracaena fragrans*
Dracaena sanderiana	*Lactuca sativa*

- *Physical effects*: The presence of plants causes a number of physical effects (Brix, 1994b, 1997). The deep, complex, and extended root system within the substrate contributes to the water velocity deceleration, which respectively increases the contact time between the wastewater and the substrate media and the roots, as it moves vertically through the VFCW bed. This may result in enhanced removal of nutrients (USEPA, 1995; Vymazal et al., 1998). However, due to the gravitational vertical movement of the wastewater, the contact time is practically limited. The plant parts (roots, rhizome, and stems) also stabilize the substrate and bind the media grains. Additionally, the plant presence, combined with the intermittent loading regime usually applied in VFCWs, prevent clogging of the substrate media (see Chapter 3). The movement of plant stems by the wind prevents the blocking of the substrate surface and the "schmutzdecke" (sludge/litter layer) by creating cracks which allow for the downward flow of the wastewater (Bahlo and Wach, 1995). Moreover, tall and dense reed stems shade the substrate surface and act as an insulation layer for the bed, especially in winter. The same effect is also achieved by the litter layer on the bed surface.

- *Hydraulic conductivity*: The movement of the stems and the respective crack creation is also beneficial for the vertical permeability of the bed, and thus, the preservation of the hydraulic conductivity (Brix, 1994b, 1997; Haberl et al., 1995; Sundaravadivel and Vigneswaran, 2001). The extensive root system within the substrate, coupled with the pores of the media, manage to maintain a wastewater flow along the roots, even in winter, when the bed surface could be frozen or covered by ice. The vertical percolation of the wastewater across the media contributes this way to the good system performance, since the maintenance of the bed permeability in VFCWs is crucial for the proper operation.
- *Biofilm development*: The extensive and dense root system that gradually develops within the substrate layer functions as an attractive attachment area for the microbial population (Brix, 1997; Vymazal et al., 1998). This thin biofilm layer, which develops along the roots and onto the surface of the media grains, is important for the efficiency of the system, since it affects and activates various microbial transformation processes of the pollutants present in the wastewater.
- *Oxygen supply*: The presence of plants ensures the enhanced aeration of the bed. It is known that plants are capable of absorbing oxygen from the atmosphere through their leaves and transferring it to the deeper layers of the substrate via release from their roots (Brix, 1994b; Vymazal et al., 1998). This oxygen provided by the roots is then consumed by the aerobic microorganisms in the biofilm and enables various aerobic processes (e.g., nitrification, aerobic degradation of OM). It is reported that this oxygen transfer mechanism via plant roots accounts for the majority of the oxygen amount in the rhizosphere (Brix, 1994b). On the other hand, in VFCWs with intermittent loading, large oxygen amounts are provided through the feeding regime of creating flood onto the bed surface. Therefore, the level of importance of each transfer mechanism is still questionable. However, it is mentioned that regarding OM decomposition, plant contribution is not as important compared to microorganism activity (Stottmeister et al., 2003). In a large experiment with several VFCW beds, planted beds with *P. australis* and *T. latifolia* were found to be more efficient in OM and N removal by 6% and 10% compared to unplanted bed, indicating that the presence of plants does indeed improve the system performance (Stefanakis and Tsihrintzis, 2012a). On the other hand, the differences between the two species were not significant.

- *Direct constituent uptake*: Another function of plants in all CW types is the direct uptake of various constituents present in wastewater (Brix, 1994b; Vymazal et al., 1998; Vymazal, 2007). For example, plants utilize nutrients in wastewater (nitrogen and phosphorus) for their growth. The uptake takes place through the roots. Nutrient assimilation reaches higher levels during the growing period (spring months) when the plant needs are respectively higher. However, the amount of nutrient uptake by plants accounts for only a small percentage of the total removed amount in the system (Tanner, 1996; Stefanakis and Tsihrintzis, 2012a).
- *Organic compound release*: Besides oxygen, plant roots also release other substances. Although the nature of these constituents is not yet clearly identified, experiments have shown that plant roots release a range of organic compounds and antibiotics (Brix, 1997). The organic carbon that is excreted this way may be utilized as a carbon source for denitrifying microorganisms, but the range of this process is not exactly determined yet.

The extent of the importance of the plants is still an issue under investigation (Lee and Scholz, 2007). Many studies in the literature report significant difference between planted and unplanted units (Tanner, 2001; Akratos and Tsihrintzis, 2007; Stefanakis and Tsihrintzis, 2012a; Stefanakis et al., 2014), especially for nitrogen removal. Today, more focus is given on the possible differences that may appear between different plant species (Heritage et al., 1995; Akratos and Tsihrintzis, 2007; Iamchaturapatr et al., 2007; Yang et al., 2007; Brisson and Chazarenc, 2009; Kantawanichkul et al., 2009; Stefanakis and Tsihrintzis, 2009b, 2012a; Kotti et al., 2010).

4.2 SUBSTRATE MATERIAL

The selection of the substrate in a VFCW system represents a very important design parameter which might significantly affect the performance of the bed. The characteristics of the filter media which define its permeability should be carefully selected. Clogging problems are not uncommon when the filter material does not possess an adequate permeability for the applied hydraulic and organic load. Therefore, the substrate can be related to several drawbacks that may appear in VFCWs, regarding the improper hydraulic operation of the bed and the low efficiency in pollutant removal.

In the past, soil was almost exclusively used as filter media in VFCWs (Arias et al., 2001). This practice, however, often resulted in clogging problems due to the relatively low hydraulic conductivity of this material. Today, most systems contain gravel layers of different kinds and origins as filter media, usually with a top layer of sand. The presence of gravel offers a series of benefits, which could be summarized as follows:

- It supports the growth of the planted macrophytes.
- It stabilizes the bed (interaction effects with developed plant roots).

- It provides filtration effects.
- It ensures a high permeability, i.e., hydraulic conductivity, for the unhindered downward passage of the wastewater (assuming an appropriate pore volume), thus diminishing the appearance of possible clogging problems.
- It enhances the treatment efficiency, acting as a sink of various biotic and abiotic elements.
- It provides an attractive attachment surface area for various microorganisms (biofilm creation) which are involved in the pollutant removal processes.
- It supports several transformation and removal processes, which can be enhanced by using specialized materials (e.g., zeolite, bauxite, etc.; Stefanakis and Tsihrintzis, 2012a).

As mentioned in Chapter 3, clogging is one of the main problems common in constructed wetland systems. The key factor for the proper operation and performance is the maintenance of an adequate hydraulic conductivity. During the past years and with the increased use of coarse and fine gravel materials in VFCWs, such problems are becoming increasingly rare. As mentioned in the previous chapter, the synergetic effects of sufficient pore volume and gradual development of plant roots manage to maintain the hydraulic conductivity of the bed. Typical range values for the effective sizes (d_{50}) for substrate media found in the literature are: sand 0.2-0.6 mm, fine gravel 6-16 mm, medium gravel 24-32 mm, and coarse gravel (cobbles) 60-130 mm.

Recently, focus is given on the identification of filter materials that could enhance P removal in VFCWs. Generally, the efficiency of VFCWs in P retention is limited compared to the other pollutants (Lüderitz and Gerlach, 2002; Rustige et al., 2003; Brix and Arias, 2005; Prochaska and Zouboulis, 2006; Stefanakis and Tsihrintzis, 2012a). The reason is the relatively short contact time between the wastewater and the system components (media grains, plant roots) as the vertical wastewater movement proceeds by gravity in VFCW beds with intermittent loading. Taking into account that P adsorption to the substrate is regarded as the main removal mechanism, the research around filter media today is targeting mainly at the investigation of various filter materials with special properties for the enhancement of P adsorption (Sakadevan and Bavor, 1998; Drizo et al., 1999; Arias et al., 2001; Brix and Arias, 2005; Wetholm, 2006; Stefanakis and Tsihrintzis, 2009b, 2012a,b; Stefanakis et al., 2009a; Vohla et al., 2011; Gikas and Tsihrintzis, 2012).

The media that have been tested or proposed for use in CWs in general and, particularly in VFCWs (Table 4.3), can be classified as:

1. *Natural materials*: Minerals, rocks, soils, marine sediments. These are naturally occurring materials that can be used without any or only with a slight pretreatment to enhance their P adsorption capacity if necessary (Wood and McAtamney, 1996; Sakadevan and Bavor, 1998; Drizo et al., 1999; Arias et al.,

TABLE 4.3 Various Filter Materials Used in CWs

I. Natural Materials (Wood and McAtamney, 1996; Sakadevan and Bavor, 1998; Drizo et al., 1999; Arias et al., 2001; Brix et al., 2001; Lüderitz and Gerlach, 2002; Brix and Arias, 2005; Prochaska and Zouboulis, 2006; Wetholm, 2006; Xu et al., 2006; Vohla et al., 2011; Stefanakis and Tsihrintzis, 2012b)	
Igneous gravels[a]	Carbonate gravels[a]
Dolomite[a]	Natural zeolite[a]
Bauxite[a]	Limestone (sedimentary rocks)[a]
Opoka (marine sediment)	Maerl (marine sediment)
Laterite[a]	Wollastonite
Apatite (sedimentary)	Apatite (igneous)
Calcinated Alunite	Marl
Oyster shell	Spodosols
Peat[a]	Polonite
Sands[a]	Shell sand[a]
Crushed marble[a]	Hornblende[a]
Bentonite	Clay[a]
Shale[a]	Soils[a]
II. Synthetic Materials (Drizo et al., 1999; Wetholm, 2006; Albuquerque et al., 2009; Białowiec et al., 2011; Vohla et al., 2011)	
Synthetic zeolites	Filtralite®[a]
LECA (light-weight expanded clay aggregate)[a]	Filtralite-P[a]
LWA (light-weight aggregates)	Calcite
Vermiculite	Cat litter (burnt diatomaceous earth)
Activated carbon[a]	
LESA (light-weight expanded shale aggregate)[a]	
III. Industrial by-products (Sakadevan and Bavor, 1998; Drizo et al., 1999; Wetholm, 2006; Xu et al., 2006; Zhao et al., 2008; Vohla et al., 2011)	
Slag (from steel industry)[a]	Blast or steel furnace slag[a]
Fly ash[a]	Coal ash
Iron ore	Burnt oil shale[a]
Quartz sand	Ochre
Charcoal[a]	Dewatered alum sludge[a]

[a]Indicates substrates tested in VFCWs.

2001; Brix et al., 2001; Lüderitz and Gerlach, 2002; Brix and Arias, 2005; Prochaska and Zouboulis, 2006; Wetholm, 2006; Xu et al., 2006; Vohla et al., 2011; Stefanakis and Tsihrintzis, 2012b).

2. *Synthetic (man-made) materials*: These are materials that are produced in the laboratory or by various treatment processes, e.g., heating, or known natural materials or mixtures (Drizo et al., 1999; Wetholm, 2006; Albuquerque et al., 2009; Białowiec et al., 2011; Vohla et al., 2011).
3. *Industrial by-products*: Materials of this category usually are residuals produced during industrial processes (Sakadevan and Bavor, 1998; Drizo et al., 1999; Wetholm, 2006; Xu et al., 2006; Zhao et al., 2008; Vohla et al., 2011).

Chapter 5

Treatment Processes in VFCWs

5.1 GENERAL POLLUTANT REMOVAL MECHANISMS

All CW systems have been proved to be capable of removing a variety of pollutant present in wastewater, namely, organic matter (BOD_5 and chemical oxygen demand - COD), suspended solids, nitrogen, phosphorus, heavy metals, pathogenic microorganisms, and micro-organic compounds. The removal of the various pollutants proceeds through a series of mechanisms and processes taking place within the CW bed. These processes can be physical, chemical, and biological, and occur as a result of the various and complex interactions between the elements of the CW system (water, substrate media, vegetation, litter layer, microorganisms). The various pollutant removal mechanisms may appear independently from each other and/or act simultaneously, creating synergetic effects. In general, the mechanisms that remove pollutants from wastewater during treatment in CWs are the following:

- *Microbial activity* within the biofilm which is developed onto the surface of the filter media grains and along the plant roots. This biofilm contains a rich and dense microbial flora which enables microbial-mediated removal/transformation mechanisms.
- *Filtration* of the largest incoming particles and retention within the media pores.
- *Sedimentation* of the influent solids as the wastewater moves through the wetland substrate.
- *Direct uptake* of various pollutants (e.g., nutrients) and water by plants which use them for their growth needs.
- *Adsorption* of constituents (mainly nutrients) onto the filter media grain surface.
- *Natural die-off* of pathogens due to prolonged wastewater residence time within the system.
- *Direct exposure to UV* radiation and possible *excretion of antibiotics* by the plant roots which eliminate pathogens.
- *Predatory activities* by bacteriophages, *Bdellovibrio* like organisms, protozoa, metazoa, etc.

☆"To view the full reference list for the book, click here"

TABLE 5.1 Pollutant Removal Processes in CWs

Pollutant	Removal/Transformation Process		
	Physical	*Chemical*	*Biological*
Organic matter (as BOD_5 and COD)	Filtration and settling (particulate OM)[a]	Oxidation	Bacterial degradation (soluble OM)[a] Microbial consumption[a]
Suspended solids (as TSS)	Filtration[a], sedimentation[a]		Bacterial decomposition[a]
Nitrogen	Volatilization	Ion exchange	Nitrification[a]/denitrification, microbial consumption[a], plant uptake
Phosphorus	Filtration[a]	Adsorption[a], precipitation	Plant uptake, microbial consumption
Pathogens	Filtration[a]	UV-degradation, adsorption	Predation[a], natural die-off[a]
Heavy metals	Settling	Adsorption[a], precipitation[a]	Biodegradation, phytodegradation, phytovolatilization, plant uptake

[a]Indicates processes that are intense in VFCWs.

Table 5.1 contains the classification of the various processes (physical, chemical, and biological) together with the main pollutants being removed by each process (Kadlec and Knight, 1996; Vymazal et al., 1998).

Each removal mechanism is influenced by different parameters, which have to do with the type of the CW (flow direction, level/time of saturation), the wastewater characteristics to be treated, and the climatic conditions. One of the most crucial parameters is the hydraulic residence time (HRT) which represents the time that the wastewater remains within the wetland body and is in contact with the various CW elements. Since the natural processes are relatively slow, a sufficient HRT is usually necessary for the effective treatment. Another important parameter is of course the plant species used. As mentioned in Chapter 4, plant species well adopted to the local environment and those creating an extensive and dense root system are highly preferable. High rates of plant biomass production are also desirable for nutrient removal by harvesting. Moreover, the selected filter media and, specifically, the properties it possesses, can affect the extent of various treatment processes. Many of the microbiological processes are also susceptible to temperature and pH variations.

In the following sections, the removal processes for each pollutant are presented separately. Focus is given on those processes that are particularly intense in VFCWs.

5.2 ORGANIC MATTER

Organic matter (OM) is a major pollutant, especially in municipal wastewater, and is expressed by the biodegradable part (BOD_5) and the total OM (COD). Particulate OM is removed by settling and physical filtration, followed by hydrolysis. The soluble/colloidal part of the OM is decomposed both aerobically and anaerobically (Vymazal et al., 1998; García et al., 2010; Figure 5.1).

Aerobic degradation of soluble OM is accomplished by the groups of chemoheterotrophic and chemoautotrophic microorganisms, as shown in Table 5.2.

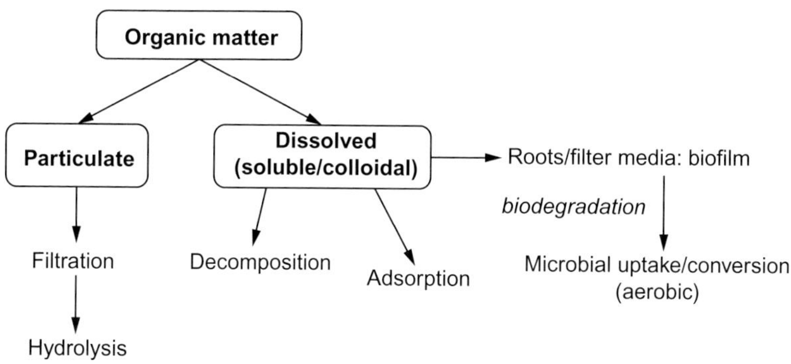

FIGURE 5.1 Organic matter transformation and removal processes in VFCWs.

TABLE 5.2 Bacteria Groups Responsible for Various Pollutant Removal Processes in VFCWs

Microorganisms		Energy Source	C Source	Pollutant Removal
Autotrophic	Photoautotrophic	Solar energy	CO_2	NH_4^+-N
	Chemoautotrophic	Inorganic oxidation-reduction	CO_2	
Heterotrophic	Photoheterotrophic	Solar energy	Organic C	BOD_5
	Chemoheterotrophic	Inorganic oxidation-reduction	Organic C	

The first group proceeds faster and accounts for the major BOD reduction, according to the following reaction:

$$CH_2O + O_2 \rightarrow CO_2 + H_2O \quad (5.1)$$

The microbial aerobic decomposition of OM (e.g., glucose: $C_6H_{12}O_6$, proteins, lipids), which demands high oxygen supply, is a major removal process in VFCWs. The typical feeding regime of intermittent loadings with the creation of a flood onto the bed surface and the subsequent gravitational wastewater drainage provides high rates of oxygen availability for aerobic microbial processes (Vymazal, 2007; Stefanakis and Tsihrintzis, 2012a). Thus, VFCWs can be very effective in BOD_5 removal. Autotrophic bacteria can also aerobically degrade organic compounds containing nitrogen (nitrifying bacteria, see Section 5.4). OM biodegradation takes place in the biofilm (attached microbial population) along the plant roots and stems and the surface of the substrate grains.

Stepwise anaerobic degradation is carried out in the absence of oxygen by acid- or methane-forming bacteria and proceeds slower than aerobic degradation. This process is not so extended in VFCWs (García et al., 2010). The first step (fermentation) is driven by facultative bacteria, as follows:

$$C_6H_{12}O_6 \rightarrow 3CH_3COOH + H_2 \quad [\text{Acetic acid}] \quad (5.2)$$

$$C_6H_{12}O_6 \rightarrow 2CH_2CHOHCOOH + H_2 \quad [\text{Lactic acid}] \quad (5.3)$$

$$C_6H_{12}O_6 \rightarrow 2CO_2 + 2C_2H_5OH \quad [\text{Ethanol}] \quad (5.4)$$

In the second step, the first-step products are converted by anaerobic bacteria, as follows:

$$CH_3COOH + H_2SO_4 \rightarrow 2CO_2 + 2H_2O + H_2S \quad (5.5)$$

$$CH_3COOH + 4H_2 \rightarrow 2CH_4 + 2H_2O \quad (5.6)$$

The OM degradation also depends on the OM composition and the HRT applied. Readily biodegradable organics are rapidly oxidized in VFCWs, under sufficient oxygen amounts, by aerobic bacteria which utilize oxygen as an electron acceptor (aerobic respiration), while refractory substances are partially degraded since they demand higher contact time. It is also reported that the major OM part is removed in the first 10-20 cm of the VFCW bed (usually within the sand layer), since the top layer is dominated by aerobic conditions and contains high microbial density (Kadlec and Wallace, 2009; Tietz et al., 2008; Stefanakis and Tsihrintzis, 2012a). Practically, there will always be a minimum BOD_5 effluent concentration (3-5 mg/L), due to OM addition to the system by the dead plant material (plant litter), as also organic substances (e.g., acids, sugars, vitamins, etc.)

released by plant roots in small concentrations (rhizodeposition) during the degradation process (Stottmeister et al., 2003).

5.3 SUSPENDED SOLIDS

Inorganic and suspended solids are removed by physical processes. The main removal mechanism for total suspended solids (TSS) in VFCWs is gravitational settling (sedimentation) and filtration. As the wastewater drains vertically, it passes through the pores of the substrate media and the velocity of the water current is decreased. The solids are trapped within the pores either mechanically or by adhesion (Kadlec and Wallace, 2009). As mentioned in Section 3.2 in Chapter 3, gradual accumulation of solids is probably the main parameter affecting substrate clogging. Solids accumulate on top of the bed (usually above the sand layer), creating a sludge/litter layer (also known by its German term "schmutzdecke"), as also within the substrate pores (physical blocking) and onto the surface of the media grains. In VFCWs, intermittent loading and application of resting periods between loadings allow for the good aeration of the bed and the oxidation of the accumulated organic solids, which also prevents the bed clogging.

5.4 NITROGEN

Nitrogen (N) is a pollutant of great interest for removal from wastewater. N can negatively affect the quality of effluent receiving surface water, alter the dissolved oxygen (DO) balance to insufficient levels for the living organisms in water (due to aerobic N transformation processes), and contribute to the appearance of the eutrophication phenomenon. Typical N forms in wastewater are (Kadlec and Knight, 1996; Vymazal, 2007; Kadlec and Wallace, 2009; Faulwetter et al., 2009; García et al., 2010; Saeed and Sun, 2012):

- organic forms: urea [$CO(NH_2)_2$], amino acids (-NH_2 and -COOH), uric acid ($C_5H_4N_4O_3$), purine, and pyrimidines,
- inorganic forms: ions like ammonium (NH_4^+), nitrate (NO_3^-), and nitrite (NO_2^-), and gases like nitrous oxide (N_2O), nitrogen (N_2), nitric oxide (NO_2), and free ammonia (NH_3).

Usually, analytical methods include the determination of ammonia, nitrate, nitrite, total Kjeldahl nitrogen (TKN = organic + ammonia N), organic N (TKN − ammonia N), oxidized N (nitrate + nitrite), inorganic N (oxidized N + ammonia N), and total nitrogen (TN = TKN + oxidized N). Typical composition of domestic wastewater consists of ~60% ammonia N and 40% organic N.

Generally, various biological and physicochemical mechanisms are responsible for N transformations and removal in CWs, while specifically in VFCWs,

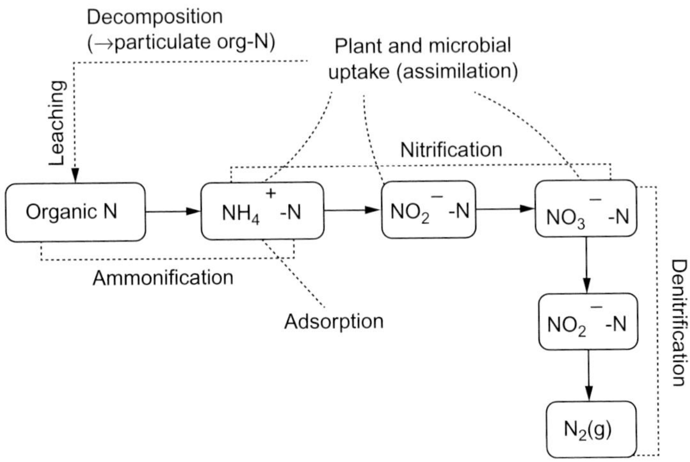

FIGURE 5.2 Nitrogen transformation and removal processes in VFCWs.

N removal mostly relies on the effectiveness of the microbial processes nitrification, denitrification and ammonification, as shown in Figure 5.2.

5.4.1 Ammonification

Ammonification represents the first step in the N transformation chain. The organic nitrogen contained in the influent wastewater is converted to ammonia, as Equation (5.7) depicts (Vymazal et al., 2006; Vymazal, 2007; Faulwetter et al., 2009; Saeed and Sun, 2012)

$$\text{Amino acids} \rightarrow \text{Imino acids} \rightarrow \text{Keto acids} \rightarrow NH_3 \tag{5.7}$$

$$\text{e.g., amino acid: } RCH(NH_2)COOH + H_2O \rightarrow NH_3 + CO_2 \tag{5.8}$$

$$\text{urea: } CO(NH_2)_2 + H_2O \rightarrow 2NH_3 + CO_2 \tag{5.9}$$

Usually, most of the organic N is transformed to ammonia by microbes. This process takes place in both aerobic and anaerobic areas of the bed, but proceeds rapidly in the oxygen-rich layers. Ammonification takes place faster than nitrification in terms of kinetics (Kadlec and Knight, 1996; Vymazal et al., 2006; Vymazal, 2007). This process is affected by temperature, pH, C/N ratio, nutrient content, and soil conditions. Optimum pH area is between 6.5 and 8.5 and temperature between 40 and 60 °C, while it is reported that the ammonification rate doubles with 10 °C temperature increase (Kadlec and Knight, 1996; Vymazal et al., 2006; Vymazal, 2007; Saeed and Sun, 2012). Ammonification has not been investigated at the same level, compared to other processes, while various rates have been reported in the literature up to 0.53 g N/m^2/d (Tanner et al., 2002; García et al., 2010).

In VFCWs, the ammonification rate decreases with substrate depth, thus higher rates are observed in the upper aerobic zone. Practically, N is not removed by this process but converted to ammonia, which is then subjected to other removal processes. Due to well-aerated conditions, the ammonification rate is at high levels in VFCWs compared to other CW types.

5.4.2 Nitrification

Nitrification is probably the most important removal process in VFCW systems (Cooper, 1999; Tanner et al., 2002; Vymazal, 2007; Faulwetter et al., 2009; García et al., 2010). It represents the second step in the N transformation chain (Figure 5.2), during which ammonia N is converted to nitrate by bacteria. First, ammonia N is oxidized to nitrite under aerobic conditions by chemolithotrophic *Nitrosomonas*, *Nitrosococcus*, *Nitrosolobus*, and *Nitrosospira* bacteria, and then to nitrate by facultative *Nitrospira*, *Nitrospina*, *Nitrococcus*, and *Nitrobacter* (Vymazal et al., 1998; Tanner et al., 2002; Vymazal, 2007; Kadlec and Wallace, 2009; Faulwetter et al., 2009; Lee et al., 2009; Saeed and Sun, 2012). Ammonia oxidation provides energy to the nitrifying bacteria, while CO_2 is utilized as carbon source. The two process stages are (Metcalf and Eddy, 2003) as follows:

$$NH_4^+ + 1.5O_2 \rightarrow 2H^+ + NO_2^- + H_2O \quad [\text{stage 1}] \tag{5.10}$$

$$NO_2^- + 0.5O_2 \rightarrow NO_3^- \quad [\text{stage2}] \tag{5.11}$$

The overall reaction can be written as:

$$NH_4^+ + 2O_2 \rightarrow NO_3^- + 2H^+ + H_2O \tag{5.12}$$

As Equation (5.12) shows, during respiration activities of the nitrifiers, oxygen is consumed and protons are produced, which results in a pH decrease in the wastewater. Nitrification is an oxygen-consuming process. Oxygen levels affect directly the extent of the process. Complete ammonia N oxidation requires ~4.6 mg O_2/mg N and DO concentration of 1 mg/L (optimum value 3-4 mg DO/L), while the conversion consumes 8.64 mg HCO_3^-/mg NH_4^+-N (Vymazal, 2007; Faulwetter et al., 2009; Saeed and Sun, 2012). Parameters that affect nitrification are temperature, pH value, water alkalinity, inorganic carbon source, moisture, microbial population, ammonia N concentration, and DO (Cooper, 1999; Vymazal, 2007; Lee et al., 2009). It is generally accepted that the optimum temperature range for nitrification is from 25 to 35 °C, while the process is practically inhibited for temperatures below 4-5 °C (Kuschk et al., 2003; Vymazal, 2007). Optimum pH values are around 7.5 to slightly alkalic values close to 7.8 (Cooper et al., 1997; Vymazal, 2007).

In VFCWs, nitrification is believed to be the most important mechanism for ammonia transformation and removal, compared to other mechanisms, due to the good aeration conditions created in this CW type (see Chapter 3; Cooper et al., 1997; Sun et al., 1999a; Kayser et al., 2002; Stefanakis and Tsihrintzis, 2012a). Experiments have shown increased nitrate concentration in VFCW effluent, indicating intensive nitrification (Platzer, 1999; Stefanakis and Tsihrintzis, 2012a). Reported oxygen consumption rate in VFCWs can reach values of 29.7 g $O_2/m^2/d$ (Sun et al., 2003). Passive air pumps have been used to enhance nitrification in VFCWs, through the increase of the oxygen supply to the system (Green et al., 1997a; 1998). However, it is also reported that the simultaneous presence of high OM content in the wastewater possibly creates antagonistic conditions to nitrification due to oxygen consumption for OM decomposition (Sun et al., 1999a,b; Zhao et al., 2004a; Stefanakis and Tsihrintzis, 2012a).

5.4.3 Denitrification

The next step in the transformation chain of ammonia N includes the reduction of the produced nitrite and nitrate to nitric oxide, nitrous oxide, and gas N by denitrifying bacteria (Figure 5.2). Although denitrification represents a significant N and OM removal process, due to usually low nitrate concentrations in municipal wastewater, it is considered as a combined process with nitrification for total N removal. In this process, facultative heterotrophic bacteria utilize N oxides or oxygen as electron acceptors and organic material as electron donors (Vymazal et al., 1998, 2006; Kadlec and Wallace, 2009; Saeed and Sun, 2012). Autotrophic bacteria, which utilize inorganic compounds as energy source and CO_2 as carbon source, can also act as denitrifiers (Lee et al., 2009). Heterotrophic bacteria like *Pseudomonas*, *Bacillus*, *Micrococcus*, and *Spirillium* (among others) convert nitrate under anaerobic or anoxic conditions, according to Equation (5.13):

$$NO_3^- \rightarrow NO_2^- \rightarrow NO \rightarrow N_2O \rightarrow N_2 \tag{5.13}$$

The process is affected by the redox potential, the availability of organic carbon and nitrate, the DO concentration, the moisture content, the pH value, and the temperature (Vymazal, 2007; Faulwetter et al., 2009; Lee et al., 2009). Plant litter can serve as source of organic carbon in all wetland systems. Optimum temperature range is between 60 and 75 °C, while the process proceeds very slowly for temperature values below 5 °C, and optimum pH values are 6-8. Denitrification takes place under low oxygen conditions (DO < 0.3 mg/L); therefore, it is very limited in VFCW beds, which usually are aerobic systems (Platzer, 1999; Kayser et al., 2002; Faulwetter et al., 2009; Saeed and Sun, 2012; Stefanakis and Tsihrintzis, 2012a), while it is more abundant in horizontal flow systems (Kayser et al., 2002; Akratos and Tsihrintzis, 2007; Kotti et al., 2010).

5.4.4 Plant Uptake

The presence of plants contributes to the removal of nitrogen in CW systems both directly and indirectly. Plants absorb and utilize nutrients present in wastewater (e.g., nitrogen and phosphorus) to support their growth (Figure 5.2). The extent of this uptake is dependent on various parameters, including the bed configuration, the loading rate, the wastewater type, the plant type, and environmental conditions (Saeed and Sun, 2012). Also, the N storage capacity of the plants has to do with the nutrient concentrations in plant tissues and the density (reed stands/m^2) and height of the plant biomass (Vymazal et al., 1998; García et al., 2010), which means that plant species with rapid growth, high potential of nutrient storage in tissues, and ability to reach high standing stems are preferable. The nutrient content assimilated in plant tissues is removed from the system by annual harvesting of the produced biomass. However, in this way, only a small portion of the total N load is practically removed (up to 4-5%) in VFCWs (Brix, 1994b; 1997; Kantawanichkul et al., 2009; Stefanakis and Tsihrintzis, 2012a), while higher rates (even up to 40%) have been reported for horizontal flow systems (Shamir et al., 2001; Tanner, 2001; Lee et al., 2004; Kadlec and Wallace, 2009; Meers et al., 2008), probably due to the permanent saturation of the bed. Nitrate and ammonia N are the two forms of N that are absorbed by plants, with the latter being more favored (Stefanakis and Tsihrintzis, 2012a), although in wastewater with high nitrate concentrations this form acts as the major nutrient source (Vymazal, 2007; Kadlec and Wallace, 2009). N uptake by plants mainly takes place during the growing season (spring and summer months) (Stefanakis and Tsihrintzis, 2012a), when the plant needs for nutrients are higher, while during winter the uptake rate is reduced. If harvesting of the produced biomass is neglected, there is the possibility of nutrient release back to the system at the end of the growing season (leaching), as also by the mineralization of the dead plant material above the porous media layer.

As mentioned in Chapter 4, the presence of plants allows for the creation of the biofilm along the plant roots (microbes attachment area), while the oxygen transfer by plant roots into the deeper zones of the substrate enhances the activity and growth of microorganisms in the rhizosphere. In this way, the nitrification process is enhanced by the plant presence.

5.4.5 Adsorption

Ammonia N can be adsorbed on to the surface of the substrate material (Figure 5.2). However, ion exchange of ammonia N depends on the characteristics of the porous media used. Specific porous media with high cation exchange capacity have been tested as substrates in VFCWs in order to enhance N removal, like zeolite and gravel (Yalcuk and Ugurlu, 2009; Stefanakis and Tsihrintzis, 2012a), blast furnace artificial slag or coal burn artificial slag (Cui et al., 2010), high-density polyethylene and shale (Austin, 2006), among others (Saeed and Sun, 2012). In VFCWs with intermittent loading, the relatively short contact time between the wastewater and the porous media, due to relatively rapid gravitational flow, does not allow for intense cation exchange between the wastewater and the media (Stefanakis and Tsihrintzis, 2012a); thus, the nitrogen amount removed by this process is not significant. Practically, the adsorbed amount of ammonia N onto media grains is oxidized to nitrate (nitrification) during the resting period between two loadings, taking into account the aerobic conditions that dominate in VFCWs.

5.4.6 Other Nitrogen Processes

5.4.6.1 *Ammonia Volatilization*

In this process, unionized ammonia (NH_3) is lost from the water surface to the atmosphere through diffusive and advective forces, i.e., ammonia N is in equilibrium between the gaseous and hydroxyl forms (Vymazal et al., 1998; Vymazal, 2007; Saeed and Sun, 2012).

The process is pH dependent and losses are significant for pH values higher than 8.0, while for pH > 9.3, the conversion of ammonium to NH_3 gas increases (Bialowiec et al., 1991; Vymazal, 2007; Saeed and Sun, 2012). In VFCWs, as in all subsurface flow CWs, the extent of ammonia volatilization is practically of minimum importance, especially for pH values below 8.0.

5.4.6.2 *N_2 Fixation*

Fixation process is the synthesis of ammonia, amino acids, and proteins from gaseous nitrogen by heterotrophic soil bacteria, actinomycetes, and cyanobacteria (Kadlec and Knight, 1996; Vymazal, 2007; Kadlec and Wallace, 2009). It occurs in floodwater, on soil surface, in aerobic and anaerobic soils, in plant roots, and on plant stem and leaf surface. In wetlands with high influent nitrogen concentration, fixation is insignificant compared to the other nitrogen conversion processes, since it consumes high amounts of cellular energy (Kadlec and Knight, 1996). Thus, in VFCWs, this process does not play an important role, although the process has not been investigated thoroughly yet.

5.4.6.3 *ANAMMOX*

ANAMMOX (anaerobic ammonium oxidation) is a relatively recently discovered process for N removal, which converts ammonium to gaseous N_2 (van de Graaf et al., 1995; Faulwetter et al., 2009) under anaerobic conditions. Both nitrite and nitrate can serve as an inorganic electron acceptor for ANAMMOX, although nitrite appears of greater importance (Jetten, 2001). Compared to coupled nitrification/denitrification processes, the ANAMMOX process demands lower energy and oxygen and does not require an external carbon source or additional aeration (Lee et al., 2009; Saeed and Sun, 2012). This process has been identified in CWs (Tanner et al., 2002; Kadlec et al., 2005;

Shipin et al., 2005; Dong and Sun, 2007; Paredes et al., 2007; Tao and Wang, 2009; Saeed and Sun, 2012). The overall reaction is the following (Tao and Wang, 2009; Saeed and Sun, 2012):

$$NH_4^+ + 1.32NO_2^- + 0.066HCO_3^- + 0.13H^+ \rightarrow 1.02N_2 + 0.066CH_2O_{0.5}N_{0.15} + 0.26NO_3^- + 2.03H_2O \quad (5.14)$$

With the ANAMMOX process, about 85% of ammonia is converted to gaseous N_2, 15% to nitrate, and <0.1% to N_2O. The fact that the main final product is gaseous N_2 promotes the use of ANAMMOX in combination with other processes, e.g., denitrification, for the complete removal of ammonia in VFCWs (Dong and Sun, 2007). However, there are still certain uncertainties concerning the existence of ANNAMOX bacteria in CWs, their growth conditions, and the extent of the process (Vymazal, 2007; Faulwetter et al., 2009), although some studies point out to this direction (Tao and Wang, 2009).

5.4.6.4 CANON

Another relatively new pathway of N transformation is the CANON (completely autotrophic nitrogen removal over nitrite) process, which is a combination of partial nitrification and ANAMMOX processes (Sun and Austin, 2007a). CANON is an aerobic single-stage processes. The equation of this process is the following (Third et al., 2001):

$$NH_4^+ + 0.85O_2 \rightarrow 0.434N_2 + 0.133NO_3^- + 1.3H_2O + 1.4H^+ \quad (5.15)$$

ANAMMOX bacteria cofunction with autotrophic nitrosobacteria for the oxidation of ammonia to nitrite, which then reacts with the remaining ammonium and produces gaseous N_2 (Lee et al., 2009). It may also appear in VFCW systems, which are well aerated, for the treatment of wastewater with low OM content and high nitrogen load (Sun and Austin, 2007a,b). As for ANAMMOX, both processes need to be further investigated to identify the optimal operating parameter and driving bacteria in CWs.

5.5 PHOSPHORUS

Phosphorus (P) is a macronutrient of special importance for all biological organisms. It is present in various wastewater types. High P content, together with nitrogen, in surface water contributes to eutrophication which favors algal and plant growth and results in the depletion of the oxygen amount in the water column with negative impacts on the aquatic organisms. Major sources of P are untreated or insufficiently treated wastewater; agricultural practices; domestic, urban, and industrial runoff. P, like nitrogen, is a useful nutrient for plants, which utilize it for their growth. P in wastewater appears in various organic and inorganic forms. Free orthophosphate (OP: PO_4^{-3}-P) is a common form in wastewater and exists in ionic equilibrium (Vymazal, 2001):

$$H_3PO_4 \leftrightarrow H_2PO_4^{-} \leftrightarrow HPO_4^{-2} \leftrightarrow PO_4^{-3} \quad (5.16)$$

Under pH values between 5 and 9, $H_2PO_4^-$ and HPO_4^{-2} are abundant. Polyphosphates (cyclic, linearly condensed) are another inorganic form, while organically bound compounds can also be found, like phospholipids, nucleic acids, and phosphorylated sugars, among others (Vymazal, 2001, 2007; Kadlec and Wallace, 2009). Dissolved phosphates are the most reactive and result in hydration change with pH variations (Reddy et al., 1999).

In wetland environments, many transformations of P forms take place (Figure 5.3). Dissolved organic P and insoluble organic and inorganic P have to be converted into soluble inorganic forms, in order to be subjected to biological consumption. Plants take up soluble reactive P to cover their growth needs and convert it to tissue P, while P is also sorbed to wetland substrate (Vymazal, 2007). The main transformation/removal mechanisms of P in CWs include: adsorption/desorption, precipitation, plant/microbial uptake, and mineralization. As mentioned previously, VFCWs are capable of removing high amounts of nitrogen and OM, as a result of various processes. On the other hand, P removal is more problematic compared to other pollutants, i.e., OM or nitrogen (Lüderitz and Gerlach, 2002; Brix and Arias, 2005; Rousseau et al., 2008; Stefanakis and Tsihrintzis, 2012a).

5.5.1 Adsorption and Precipitation

P adsorption onto the substrate media probably is believed to play the most important role in the removal of P from wastewater in CWs (Drizo et al., 1999; Vymazal, 2007). The physicochemical characteristics of the substrate define the adsorption capacity, and thus, the retention of P (Kadlec and Knight, 1996; Drizo et al., 1999; Sakadevan and Bavor, 1998). The content of clay and other mineral components affects P adsorption (Vymazal, 2007). The presence of high content of aluminum (Al), iron (Fe), and calcium (Ca) oxides in the filter media increases the adsorption and precipitation reactions of P (Sakadevan and Bavor, 1998; Drizo et al., 1999; Arias et al., 2001; Brix et al., 2001; Stefanakis et al., 2009; Vohla et al., 2011; Stefanakis and Tsihrintzis, 2012a,b).

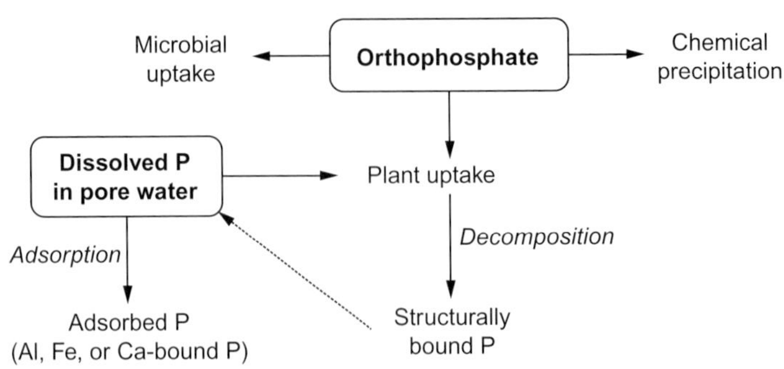

FIGURE 5.3 Phosphorus transformation and removal processes in VFCWs.

During adsorption, soluble inorganic P is accumulated onto the surface of the minerals of the filter media, without penetrating below the surface. Phosphate ions can then react with Al, Fe, Ca, or Mg cations to form new solid precipitates, resulting in the formation of amorphous or crystalline solids (Vymazal, 2007).

Adsorption and precipitation are considered as coupled processes and, usually, no differentiation is reported among them in most of the studies (Del Bubba et al., 2003). This has enabled the search for suitable media for P adsorption, which is still ongoing. Gravel is the most common filter media used in CWs, with relatively low content of minerals. Among others, filter media tested in CWs include sands, slags, zeolite, bauxite, dolomite, shale, limestone, LECA, fly ash (Zhu et al., 1997; Drizo et al., 1999; Arias et al., 2001; Brix et al., 2001; Del Bubba et al., 2003; Rustige et al., 2003; Prochaska and Zouboulis, 2006; Wetholm, 2006; Xu et al., 2006; Stefanakis et al., 2009; Vohla et al., 2011; Stefanakis and Tsihrintzis, 2012a,b). All these porous media have specific characteristics which potentially make them better P adsorbents compared to common gravel materials with low P-sorption capacity. Moreover, other parameters, like pH and E_h, can also affect the adsorption process (Reddy and D'Angelo, 1997; Zhu et al., 1997; Drizo et al., 1999; Lüderitz and Gerlach, 2002; Stefanakis and Tsihrintzis, 2012b). Also, the grain size of the filter media could affect adsorption, since fine materials provide higher available specific surface area for adsorption but are more susceptible to clogging due to lower hydraulic conductivity (Vymazal, 2007; García et al., 2010; Stefanakis and Tsihrintzis, 2012a,b).

In VFCWs, respectively, low P removal rates have been reported (Verhoeven and Meuleman, 1999; Lüderitz and Gerlach, 2002; Esser et al., 2004; Rousseau et al., 2004a; Brix and Arias, 2005; Prochaska and Zouboulis, 2006; Tang et al., 2008; Stefanakis and Tsihrintzis, 2012a). However, the use of filter media with special properties (e.g., increased content of mineral oxides) in VFCWs has not been proved to enhance the adsorption of P (Lüderitz and Gerlach, 2002; Stefanakis and Tsihrintzis, 2012a). The fact is that the relatively short contact time between the wastewater and the media grains (gravitational flow) in VFCWs with intermittent loading does not allow for intense adsorption and precipitation reactions. This also shows that the feeding regime and the frequency of the loadings also affect P removal. It is reported that in VFCWs under the fill and drain regime (see Section 3.1.3 in Chapter 3), P removal increased with HRT, due to increased contact time (Lantzke et al., 1999; Tang et al., 2009). This indicates that the HRT is a crucial parameter for the effective P removal (Drizo et al., 2006). Although the substrate is the main P sink in CW systems, adsorption is a finite process that depends on the media adsorption sites and usually declines with time (Meuleman et al., 2003; Vohla et al., 2007; Stefanakis and Tsihrintzis, 2012b). However, it has been found that reduced redox levels for prolonged time period can possibly increase P retention in VFCWs, due to regeneration of the adsorption capacity through building of new amorphous Fe structures (Lantzke et al., 1999; Rustige et al., 2003).

5.5.2 Plant Uptake

As for nitrogen, plants utilize P present in wastewater to cover their growth needs (Greenway and Woolley, 1999; Merlin et al., 2002; Engloner, 2009). P uptake by plants takes place at a slower rate compared to adsorption to the substrate (Lantzke et al., 1999; Lüderitz and Gerlach, 2002), mainly through their root system (Vymazal et al., 1998). Plant uptake proceeds faster and is more intense during the growing period, especially in the first months after the winter period when the new shoots are regenerated (Reddy et al., 1999). Generally, the amount of P removed by plants and subsequent harvesting of the produced aboveground plant biomass is negligible compared to the total removed amount, taking also into account that the main portion of the assimilated P is located in the belowground biomass (Vymazal, 2005a; Kadlec and Wallace, 2009; García et al., 2010; Stefanakis and Tsihrintzis, 2012a). However, biomass harvesting is beneficial for the system, since release of P back to the wetland through litter decomposition is avoided (Vymazal et al., 1998). Plant uptake could be more significant in tropical regions due to year-round plant growth.

5.5.3 Microbial Uptake

Bacteria and other microorganisms (e.g., algae, fungi) can rapidly assimilate and store P, a mechanism which is partly reversible (Vymazal et al., 1998; García et al., 2010). The initially assimilated P amount by microorganisms is released back to the system when they die off and decay. Microorganisms take up organic P and transform it to inorganic P through catabolic activities, while during their growth P is stored in the microbial biomass (Reddy et al., 1999). Thus, only a small portion of the assimilated P is permanently removed. Generally, there are not many studies investigating P uptake by microbes and available information about quantitative consumption in VFCWs is limited. It is reported that in an HSF CW system, the assimilated P in the microbial biomass can reach values of up to 15% (Silvan et al., 2003), while the P uptake rate is higher in the upper parts of the substrate where aerobic conditions dominate and the microbial population is dense (Edwards et al., 2006). But, again these values are typical for systems with high HRT, which is not the case in common VFCWs.

5.6 HEAVY METALS

Unlike other pollutants and other toxic organic compounds, HM are not degradable, and thus, show the trend to accumulate in the environment. In the long run, part of HM can reach human beings through the food chain, causing serious and chronic health damages. The term "heavy metals" refers to those elements with a molecular weight greater than that of calcium in the periodic table of elements. The main HM of interest in water/wastewater treatment are lead (Pb), zinc (Zn),

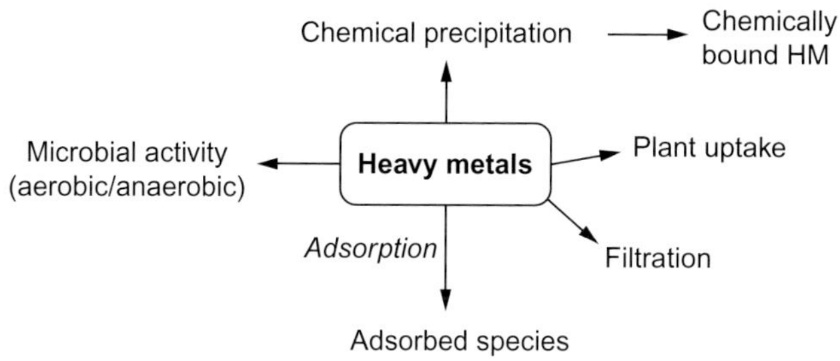

FIGURE 5.4 Heavy metal transformation and removal processes in VFCWs.

cadmium (Cd), nickel (Ni), copper (Cu), chromium (Cr), mercury (Hg), iron (Fe), and arsenic (As). HM present in CWs are in dissolved, colloidal, or particulate form, while the free metal ions are available for biological conversion (Cooper et al., 1996; Yeh, 2008).

CWs have been applied for the retention of HM from water containing high levels of them, e.g., industrial wastewater, sewer outflow, mine drainage, landfill leachate, etc. Processes that take place in CWs include adsorption and precipitation, filtration, microbial interactions, and plant uptake (Vymazal, 1998; Sobolewski, 1999; Stottmeister et al., 2003, 2006; Yeh, 2008; García et al., 2010). Among them, the main removal mechanisms are: (a) adsorption to organic substances, (b) coprecipitation with iron oxyhydroxides, and (c) precipitation as metal sulfides (Figure 5.4).

5.6.1 Adsorption and Precipitation

HM removal through adsorption and binding processes occurs within the wetland components, i.e., filter media, vegetation, and water. Dissolved metals are capable of binding to organic substances on plant roots or filter media grains (Vymazal et al., 2008). The positive charge of HM is attracted by the negative-charged surface of the adsorption material, which results in the binding of HM with suspended particles and the subsequent settlement on the substrate (Vymazal et al., 1998; Sundaravadivel and Vigneswaran, 2001). Adsorbed HM on filter media grains can be exchanged by other cations, depending on the cation exchange capacity of the substrate and the water pH (Vymazal et al., 1998; Ouki and Kavannagh, 1999; Scholz and Xu, 2002b; Sheta et al., 2003; Santona et al., 2006; Sheoran and Sheoran, 2006).

Complexation of metals with organic ligands is considered as an important parameter for reducing metal toxicity (Yeh, 2008). Organic compounds like humic acids containing carboxyl and phenolic groups drive these interactions and result in the formation of stable metal complexes (Sheoran and Sheoran, 2006). The litter layer can be used as a source of OM. However, possible biological OM decomposition may result in the desorption of metals. Additionally, HM precipitate as insoluble salts, e.g., carbonates, bicarbonates, sulfides, and hydroxides, depending on the solubility and concentration of the metal ions and the wastewater pH (Kadlec and Knight, 1996; Walker and Hurl, 2002; Lee et al., 2005; Sheoran and Sheoran, 2006; Yeh, 2008; García et al., 2010). The formed salts are insoluble and after their precipitation they become fixed within the substrate. These insoluble salts are formed by the oxidation and hydrolysis of Fe, Al, and Mn, which represent the main removal processes for these metals. Among them, Al and Mn are less dependent on pH (Sheoran and Sheoran, 2006). Metal binding with Fe and Mn oxides is not a long-term removal mechanism, since these oxides are very dependent on DO variations and may be released back to the system. Moreover, under anaerobic conditions and the presence of a sufficient carbon source, sulfate is reduced by sulfate-reducing bacteria to sulfides, which precipitate metals as metal sulfide solids (Sheoran and Sheoran, 2006; Yeh, 2008):

$$2CH_2O + SO_4{}^{-2} \rightarrow H_2S + 2HCO_3 \tag{5.17}$$

$$M^{+2} + H_2S + 2HCO_3 \rightarrow MS + 2H_2O + 2CO_2 \tag{5.18}$$

where in these chemical reactions, CH_2O represents the OM and M stands for metals. This process results in water quality improvement since the water mineral acidity is reduced. This kind of precipitation is considered as the main long-term removal process in CWs.

Generally, the mechanisms of HM removal in CWs (also in VFWCs) are more or less known, but there are still uncertainties concerning the importance and the extent of each process in HM retention. The majority of the published literature focuses on the performance of horizontal flow systems. However, taking into account that VFCWs with intermittent loading are dominated by aerobic conditions, the processes of precipitation and coprecipitation with hydroxides should be more common in VFCWs, while precipitation with sulfide formation takes place in HSF CWs (typically, anaerobic conditions). VFCW beds with greater depths perform better, due to more available effective surface for adsorption and enhanced anaerobic conditions which favor sulfate reduction and the final metal sulfide precipitation (Yadav et al., 2010). However, Lesage et al. (2007) report the detection of sulfides in a VFCW bed, which was attributed to the formation of a sludge layer and stagnant water above the media layer and the creation of anaerobic conditions (reduction of Fe and Mn hydroxides). VFCWs with a respectively high HRT (fill and drain mode) allow for higher contact time between wastewater and filter media, which enhances the adsorption process (Yadav et al., 2010). Moreover, in VFCWs, the top sand layer has been found to play a significant role in HM adsorption compared to the bottom layers (Cheng et al., 2002). On the whole, it is not completely clear yet, whether design (media, plant species) or operational parameters regulate the efficiency of the process, although it is reported that the HRT applied, the influent load and the influent pH play an important role (Lee et al., 2005).

5.6.2 Filtration and Sedimentation

CWs also function as active filters for HM. As the wastewater passes through the pores of the filter media and the dense and extensive plant root system, HM are trapped and blocked and, thus, retained inside the system. Additionally, the presence of plants contributes to the retention of the precipitated metal hydroxide particles (Cooper et al., 1996; Vymazal et al., 1998). The filtration efficiency can be affected by the water flow velocity, the particle settling velocity, the pore volume of the substrate media, and the wetland length (Sheoran and Sheoran, 2006). Sedimentation is another physical process for HM removal. It does not occur alone, but it follows other processes like precipitation and the formation of larger particles which can sink and be trapped within the media grains, especially in HSF and FWS CW systems (Walker and Hurl, 2002).

5.6.3 Plant Uptake

Wetland plants are capable of assimilating HM from wastewater, mainly in the dissolved form (Brix, 1994a,b; Kadlec and Knight, 1996; Vymazal et al., 1998; Sobolewski, 1999; Cheng et al., 2002; Sheoran and Sheoran, 2006; Yeh, 2008; Engloner, 2009; García et al., 2010; Marchand et al., 2010; Stefanakis and Tsihrintzis, 2012c). There are several species that can extract and accumulate metals and/or metalloids. As for the other assimilated constituents (nitrogen and phosphorus), the HM amount that can be assimilated by plants is not significant (up to 5%) compared to the total amount removed in CWs (Stottmeister et al., 2003; Lee and Scholz, 2007; Yeh, 2008; García et al., 2010; Stefanakis and Tsihrintzis, 2012c). Common wetland species (e.g., *Phragmites australis*, *Typha latifolia*, *Glyceria fluitans*, *Juncus effusus*, *Scirpus spp.*) have proved to be able to grow in metal-rich environments (Lee et al., 2005; Sheoran and Sheoran, 2006; Stottmeister et al., 2006; Marchand et al., 2010). HM are taken up by roots and distributed to the other parts of the plants. The main portion of the assimilated HM amount is stored in the roots and only a small part is translocated to the other plant parts, e.g., shoots and leaves (Fediuc and Erdei, 2002; Vymazal and Krása, 2003; Deng et al., 2004; Bragato et al., 2006, 2009; He and Yongfeng, 2009; Khan et al., 2009; Bonanno and Giudice, 2010; Stefanakis and Tsihrintzis, 2012c); this has been observed in all CW types, including VFCWs, while small differentiations occur between the accumulation and translocations rates of various metals (Cheng et al., 2002; Lesage et al., 2007; Yadav et al., 2010).

Generally, harvesting of the aboveground plant biomass contributes only a small percent to the total HM removal in VFCWs (Cheng et al., 2002; Kosolapov et al., 2004; Lee and Scholz, 2007; Marchand et al., 2010). It is interesting that the HM mass removal by harvesting in VFCWs appears to be higher than in HSF CWs (Lesage et al., 2007). However, the role of plants in HM removal is more important in other processes, e.g., filtration, sedimentation, and adsorption, and in the provision of OM and oxygen to the microbial population involved in metal complex formation (Kosolapov et al., 2004; Yeh, 2008).

The excretion of protons and other exudates by plant roots and the supply of organic carbon through the decay of dead plant material affect microbial activity, and thus, metal removal in CWs (Stottmeister et al., 2003; García et al., 2010). This rhizodeposition is more important in the treatment of wastewater with low content of organic carbon (e.g., industrial effluents), while its role is limited in urban wastewater treatment. Additionally, it is reported that increased evapotranspiration rates by plants enhance the reduction of Cr(VI) to Cr(III) due to the respective increase of the chemical constituents in water (Zazo et al., 2008). Also, the supply of carbon by plant litter decomposition supports the precipitation of Fe oxides and Zn sorption (Batty and Younger, 2007). However, there are still several processes to investigate and certain functions to clarify concerning the exact role of plants.

5.6.4 Microbial Activities

As already mentioned in Chapters 1–4, a dense microbial population is developed along the extensive network of plant roots. The root surface is covered by a variety of bacteria, which utilize various substrates for their growth. Such substrates are excreted by roots, e.g., sugars, which are then exchanged for N or P. There are microorganisms present in the biofilm, which can interact with metals and metalloids, and can affect their mobility and speciation through a series of mechanisms (Kosolapov et al., 2004; Marchand et al., 2010).

Sulfate reduction and co-precipitation with metals is considered as a main removal mechanism from wastewater. Sulfate-reducing bacteria, such as *Desulfovibrio* spp., precipitate metals (e.g., Zn, Cu, Fe) and form insoluble precipitates (Sobolewski, 1999; Sheoran and Sheoran, 2006; Faulwetter et al., 2009; Marchand et al., 2010). This mechanism may represent a long-term sink for HM removal, because anoxic precipitated metals remain in the sediments or media grains (Kosolapov et al., 2004). Parameters, like pH values >5.5 and redox potential below -100 mV, positively influence the process (Faulwetter et al., 2009; García et al., 2010). Sulfate reducers are present at high numbers in wastewater, which means they demand high amounts of carbon and energy sources. Therefore, in certain industrial wastewater with low carbon content the process is limited (Kosolapov et al., 2004; Lloyd et al., 2004). Low temperatures may also negatively affect the process (Stein and Hook, 2005). Under aerobic conditions, sulfur oxidizing bacteria, such as *Thiobacillus* spp. or *Thiomonas*, promote the oxidation of sulfides to sulfites and then to sulfates, and it is believed that they are involved in the oxidation of ferrous iron (Hallberg and Johnson, 2005). The exact role and the transformations of sulfur in CWs are not yet well understood and the microorganisms involved in the various processes have not been identified in detail. Obtained results imply the good efficiency in HM removal and the importance of sulfate-reducing bacteria activity for the carbon and metal cycles in CWs.

5.7 PATHOGEN REMOVAL

Although the main target when designing a CW system is the removal of OM, SS, and N constituents, these systems have been proved capable of providing additional sufficient removal of microbiological pollution. Pathogenic germs present in wastewater possess a potential health risk if they are not sufficiently removed from the treated wastewater, especially when it is reused for agricultural or irrigation purposes. Migration and subsequent pollution of surface- and groundwater may cause a series of waterborne diseases. The most important pathogenic bacteria include *Salmonella* sp., *Shigella* sp., *Vibrio cholerae*, *Yersinia enterocolitica*, *Yersinia pseudotuberculosis*, *Leptospira* sp., *Francisella tularensis*, *Dyspepsia coli*, *Escherichia coli*, and *Pseudomonas* (Craun, 1985; Matthess and Pekdeger, 1985). Common methods for hygienization include chlorination, ozonation, UV illumination, and filtration, but most of them possess high construction/supply and maintenance/operation costs (Mezzanotte et al., 2007).

CWs represent a very efficient and economical biofilter for pathogen removal (Kivaisi, 2001), capable of providing an acceptable effluent quality based on the recommendations of the World Health Organization (WHO, 1989). Pathogenic bacteria in CWs are removed through various physical, chemical, and biological mechanisms and processes, which include (Green et al., 1997b; Decamp and Warren, 2000; Karim et al., 2004; Stevik et al., 2004; Vymazal, 2005b; Wand et al., 2007; Kadlec and Wallace, 2009):

- *Physical*: filtration, sedimentation;
- *Chemical*: oxidation, UV radiation by sunlight, exposure to plant biocides, adsorption to OM;
- *Biological*: antimicrobial activity of root exudates, predation by nematodes and protozoa, activity of lytic bacteria or viruses, retention in biofilms and natural die-off.

Analytical measurements for the determination of microbiological pollution usually include the usage of indicator organisms, which are both easy and relatively inexpensive to monitor and show satisfying correlation with the overall population of microbes. The most common indicator organisms come from the coliform bacteria group, measured as total or fecal coliforms, as also the fecal streptococcus group. *E. coli* is among the most frequently used indicators of human fecal contamination. Other genera used are enterococci, *Clostridium perfringens*, *Citrobacter*, *Enterobacter*, or *Klebsiella*.

Generally, it is still not clear which is the dominant mechanism of pathogen removal in CWs nor an adequate quantification, as also the relative extent of each mechanism. Although there are many studies reporting the removal efficiency of various CW types, relatively limited work has been undertaken to identify the dominant mechanisms. Besides, most of the published work focus on the fecal bacteria indicator removal and little information is available for other microorganisms, e.g., specific bacteria, protozoa, viruses, or helminths. Thus, although several removal mechanisms are mentioned in various studies and considered responsible for bacteria removal, most of them are not yet fully understood, and their extent and effectiveness have not been surely proved (García et al., 2010).

5.7.1 Abiotic Mechanisms: Sedimentation, Filtration, and Adsorption

Pathogens are removed as the wastewater flows through the CW bed via physical mechanisms, like sedimentation and filtration, and chemical mechanisms, like adsorption and oxidation.

Sedimentation is considered a main mechanism for pathogen removal (Green et al., 1997b; Stott et al., 2003; Karim et al., 2004; Sleytr et al., 2007). Sediments and media grains in CWs are capable of accumulating high concentrations of pathogens, indicating that bottom gravel layers could act as a sink for pathogenic germs (Karim et al., 2004). Filtration and microbial attachment onto plant root surfaces also contribute to the removal (Gerba et al., 1999). These processes are closely related to the size and density of the particles and the associated bacteria (Boutilier et al., 2009). Sedimentation importance increases with the size of the associated organic/inorganic particles on which pathogens are adsorbed.

In VFCWs, filtration is believed to play a more significant role in pathogen removal (Arias et al., 2003a). However, a pre-treatment stage (usually a sedimentation tank) often exists in VFCW systems where large organic particles are removed; this implies that the majority of pathogenic organisms will associate with smaller particles (Boutilier et al., 2009). In this case, it is reported that smaller particles (<80 μm) remain in suspension which limits the role of sedimentation and increases the importance of other mechanisms like inactivation or natural die-off. Generally, microorganisms (bacteria, viruses) that are associated with relatively large particles settle faster compared to those associated to smaller colloidal particles which remain suspended for a longer time (García et al., 2010). *Cryptosporidium* and *Giardia* settle more rapidly, since they are larger than other bacteria and viruses.

Adsorption of microorganisms also takes place to plant roots and filter media grain/sediments and the associated biofilm (Brix, 1997; Gerba et al., 1999; Sleytr et al., 2007). Adsorption of coliform bacteria (*E. coli*) and viruses is affected by the size and type of the attachment particles, the presence of

OM, porous media characteristics, and the particle electrochemical charge (Stevik et al., 2004; Jamieson et al., 2005). However, in VFCWs treating domestic wastewater, it is reported that adsorption is not the main removal mechanism (Wand et al., 2007).

5.7.2 Biotic Mechanisms: Predation, Retention in Biofilm, and Plant Presence

5.7.2.1 Predation Activity

Among the various pathogen removal mechanisms, probably the least investigated are the biotic ones and, especially, predation by other microorganisms and biolytic processes. Predation is attracting increasing interest when dealing with bacterial populations. It is known that most of bacterial pathogens are food for other microorganisms, like protozoa, bacteriophages, and *Bdellovibrio*-like organisms (BLOs) (Kuschk et al., 2012; Wand et al., 2007). Protozoa and bacteriophages are common in wastewater and can reach concentrations up to 10^4-10^6 cells/mL and 10^8-10^9 virus-like particles/mL (Shapiro et al., 2010), respectively. The predation activity probably depends on the target bacteria-prey characteristics (e.g., concentration and species), the specific characteristics of the predator itself (morphology, physiology, feeding strategy, etc.), and various physicochemical parameters (e.g., temperature, redox conditions) (Momba, 2010; Shapiro et al., 2010). Prey characteristics for grazing susceptibility and resistance also need further investigation. However, the range of their significance or the main principles that regulate their activity is not well known yet. It is only reported that predation activity is the main removal mechanism in VFCWs (Wand et al., 2007).

Protozoa are known predators in various wastewater treatment processes and are used as biotic indicators due to their ability to reduce the bacteria concentration and suspended particles in the treatment process (Madoni, 1994; Pauli et al., 2001; Bomo et al., 2003; Wand et al., 2007). Protozoa grazing (mainly by ciliates) has been investigated in HSF CWs with good results for pathogen removal (Decamp and Warren, 1998; Stott et al., 2001; Wand et al., 2007). Ciliates can prey upon *E. coli* at a higher rate in planted HSF CWs than in unplanted ones (Decamp and Warren, 1998; Decamp et al., 1999). Protozoa grazing shows a rate of selectivity over prey size, cell wall chemistry, nutritional value, activity, and the ability of bacteria to produce inhibitory compounds (Boenigk and Arndt, 2000). Generally, grazing parameters are not yet clearly understood, while the correlation with bacteria removal and quantification of the removal rate has not been made in CW systems. So far, experiments in pilot VFCWs indicate that predation by protozoa and other biolytic processes is the main pathogen removal process (Wand et al., 2007), while it has been found that removal activities take place mostly within the upper layer (up to 20 cm) of the bed.

Bacteriophages are viruses that attack bacteria and are considered to be the most abundant and diverse biological entity on Earth; about 10 phages exist for every bacterial cell in most studied ecosystems (Withey et al., 2005; Shapiro and Kushmaro, 2011). Limited knowledge on their function comes from other wastewater treatment techniques, mainly activated sludge systems (Redman et al., 1997; Schijven et al., 2003; Arraj et al., 2005; Withey et al., 2005; Zhang and Farahbakhsh, 2007; Goldman et al., 2009; Shapiro and Kushmaro, 2011). Most of the studies present their use as indicators of survival and release of pathogenic viruses into the environment. It is reported that in activated sludge the number of bacteriophages is the highest compared to any other environment (Shapiro and Kushmaro, 2011). Bacteriophages, together with BLOs, are the least investigated groups in CWs and are used as pollution indicators (Thurston et al., 2001; Yousefi et al., 2004; Abdulla et al., 2007), e.g., to measure their significance in *E. coli* removal (Withey et al., 2005). As for the other predators, their significance in pathogen removal in CWs is still to be evaluated. At the present, a positive correlation between the bacteriophages and the classic bacteria indicators is implied (Yousefi et al., 2004; Abdulla et al., 2007).

Bdellovibrio spp. and BLOs require prey bacteria to grow, complete their life cycle, and propagate; they have been characterized as the "*world's smallest hunters*" (Williams and Piñeiro, 2007). Predation is the main mechanism for them to cover their energy needs. They consume Gram-negative bacteria (like *E. coli*), while Gram-positive bacteria are not subjected to predation (Fratamico and Whiting, 1995; Núñez et al., 2005; Williams and Piñeiro, 2007). Similar to phages, BLOs are dependent on a prey cell for replication, but are less specific than phages and can use various prey. Some recent studies have tried to explain the reactions by which *Bdellovibrio* recognize and attach to susceptible prey bacteria, based on the presence of lipopolysaccharide and sugar in the prey core (Schelling and Conti, 1986; Martin, 2002), but it is not known how much of bacterial mortality should be attributed to them. The extent of their role in the biolytic activity in treatment processes remains still unidentified (Yair et al., 2003; Stevik et al., 2004; Williams and Piñeiro, 2007; Wand et al., 2007; Shapiro et al., 2010).

On the whole, predation activity investigation is limited and focuses only on ciliates and, more recently, on phages. BLO activity in wastewater treatment systems (including CWs) for pathogen removal, as also possible interactions among the different predators, is largely unknown.

5.7.2.2 Biofilm Retention

The thin biofilm layer on plant roots and onto the surface of filter media grains acts as a filter and attracts and traps various microbial species (Brix, 1997; Vymazal, 2001; Larsen and Greenway, 2004). Associated pathogens to particles and suspended solids can be retained by the biofilm, while most of the microbial activity in CWs is located in this thin layer of biofilm (Brix, 1997). The development of a biofilm in sand layer has been found to improve bacteria removal (Sélas et al., 2002), while the presence of protozoa in the biofilm is correlated with higher removal efficiency (Chabaud et al., 2006).

5.7.2.3 Plant Presence

Generally, it is believed that the presence of plants improves the efficiency of CWs in pathogen removal, as many studies have shown (Rivera et al., 1995; Soto et al., 1999; Decamp and Warren, 2000; Thurston et al., 2001; Vymazal, 2005b), since the extended root system and the respective oxygen supply favor microbial growth and activity in the biofilm. However, other studies tested VFCW units and found no significant differences between planted and unplanted systems (Vacca et al., 2005; Sleytr et al., 2007; Wand et al., 2007). Specific factors, like plant species and wastewater characteristics, may also affect the performance (Tanner et al., 1995). Finally, the excretion of secondary antibacterial metabolites (e.g., tannic and gallic acids) by plant roots is supposed to be another pathogen removal mechanism (Neori et al., 2000), although this is still to be investigated.

5.7.3 Other Factors

It is suggested that other (environmental and/or design) factors and parameters possibly affect the removal microorganisms, such as temperature, the HRT applied (mainly for horizontal flow systems), solar irradiation, filter media characteristics, wastewater flow velocity, and the CW type (Stevik et al., 2004; Reinoso et al., 2008). Increased HRT seems to improve the system efficiency concerning pathogen removal, especially in horizontal flow beds (Green et al., 1997b; García et al., 2004; Vymazal, 2005b; Molleda et al., 2008; Reinoso et al., 2008). On the other hand, increased temperature values are usually correlated with increased bacteria concentration (Thurston et al., 2001; Molleda et al., 2008). This may be attributed to the increased plant growth and subsequent oxygen release by plant roots, and the increased accumulation of OM which stimulates microbial activity. Moreover, the size of the grains of the porous media also affects bacteria removal (Vacca et al., 2005). Common gravel material used in CWs has pore volume greater than the bacteria size, which favors their adsorption. Fine materials like sands, clay, and silt have pore sizes close to the bacteria cells and act as filters (bacteria straining; Stevik et al., 2004) and improve the inactivation of microbes (García et al., 2010). This mechanism is significant when the average bacteria cell size is greater than the size of 5% of the media grains.

5.8 ORGANIC MICROPOLLUTANTS

5.8.1 Phenolic and Other Aromatic Compounds

Phenols are organic compounds which can be present in various wastewaters, mainly of industrial origin, like oil refineries, chemical industry, pharmaceuticals, tanneries, pulp and paper mills, and petrochemicals (Kumaran and Paruchuri, 1997). Their formula includes a hydroxyl group (-OH) which is directly bonded to an aromatic hydrocarbon group (benzene ring). Phenol is the simplest compound known as carbolic acid (C_6H_5OH). Phenolic compounds include simple phenols and flavonoids or polyphenols, product of phenol polymerization (Paraskeva and Diamadopoulos, 2006). Depending on the source, there can be various phenol-related aromatic compounds, e.g., tannins, catechin, chlorophenol, and nitrophenol, *m*-cresol, polycyclic aromatic hydrocarbons (PAHs), benzoic acid, hydrocarbons, benzenes. Phenolic content in wastewater may be responsible for the appearance of phytotoxicity effects on plants, while it can also be toxic for various bacteria (Farré et al., 2001; Farré and Barceló, 2003; Nair et al., 2008). For this reason, phenolic compounds have also been used as antiseptics, i.e., for killing microorganisms and as broad spectrum pesticides, herbicides, and bactericides in various domestic, agricultural, and industrial applications (Annadurai et al., 2000). In fact, phenols are considered as the most toxic pollutants in many wastewaters, e.g., in olive mill effluent (Herouvim et al., 2011).

Constructed Wetlands have been used for the treatment of wastewater with respectively high phenol concentrations, such as olive mill wastewater (Del Bubba et al., 2004; Paraskeva and Diamadopoulos, 2006; Kapellakis et al., 2012; Herouvim et al., 2011), cork producing industry (Santos et al., 2012), swine facility wastewater (Wood et al., 2000), refinery effluent and chemical industry (Knight et al., 1999; Sands et al., 2000; Yang and Hu, 2005), pulp and paper mill industry wastewater (Abira et al., 2005; Choudhary et al., 2013), stormwater runoff (Malaviya and Singh, 2012), winery effluent (Arienzo et al., 2009; Welz et al., 2012), contaminated groundwater (Stefanakis et al., 2013), among others. However, the majority of the published literature focuses on HSF CWs (Vymazal, 2009) and simply reports the system efficiency (% phenol removal).

Generally, the removal mechanisms of phenols are more or less known and include biodegradation, sorption, plant uptake, and volatilization (Davies and Cottingham, 1994; Polprasert et al., 1996; Reddy and D'Angelo, 1997; Wood et al., 2000; Haberl et al., 2003; Abira et al., 2005; Imfeld et al., 2009; Wu et al., 2012; Stefanakis et al., 2013). However, the fate of phenolic compounds in wetlands as well as defining the dominant removal mechanism need further investigation (Stefanakis et al., 2013). Volatilization mainly depends on the volatility of the microorganic compounds; thus, it is considered as a not extensive degradation process. However, in CW systems, the various processes and interactions are not yet completely understood. There are also various parameters that may affect the extent of a removal process, e.g., temperature variations, OM content, nutrient content, and the available electron acceptor.

5.8.1.1 Biodegradation

Phenol degradation can take place both aerobically and anaerobically, depending on the oxygen availability and the specific compound characteristics. Since phenols and generally aromatic compounds containing high carbon content, they can be utilized by some microorganisms (bacteria, fungi, and yeasts) as a carbon and energy source through enzymatic attack (Kumaran and Paruchuri, 1997). Oxygen is used by microbes as an electron acceptor in aerobic environment. Aerobic degradation includes the formation of catechol with simultaneous oxygen consumption, which is then degraded by two alternative pathways depending on the degrader organism (van Shie and Young, 2000). This process is generally known and studied in various environments (van Shie and Young, 2000). Bacteria species that degrade phenols include *Pseudomonas* spp., *Bacillus* spp., and *Acinetobacter* spp., among others (Wang et al., 2007; Nair et al., 2008). The genus *Pseudomonas* is widely used for phenol biodegradation and has been found to be effective as phenol degrader in CWs (van Shie and Young, 2000; Kurzbaum et al., 2010a), as also the *Streptomyces* sp. (Kurzbaum et al., 2010b). Also, a pure culture of yeast *Candica tropicalis* was found capable of biodegrading phenols and *m*-cresol (Yan et al., 2006). Moreover, the addition of extracellular enzymes to the CW substrate has been proposed as an alternative way to increase enzymatic activity and, thus, enhance the biodegradation process (Shackle et al., 2006). Anaerobic degradation of phenols occurs under methanogenic, nitrate-reducing, sulfate-reducing, and iron-reducing conditions. In this case, other constituents (e.g., nitrate) serve as electron acceptors. Known anaerobic degraders are *Thauera aromatic* K172 and *Azoarcus* spp. (nitrate reducers), *Desulfobacterium phenolicum* (sulfate reducer), and *Geobacter metallireducens* (iron reducer). Present knowledge on anaerobic degradation is not so more detailed as aerobic one.

Based on the above and taking into account that VFCWs are dominated by aerobic conditions, it is believed that aerobic degradation is the main removal mechanism (Welz et al., 2012). Welz et al. (2012) point out that the level of influent loading is crucial for the maintenance of the system efficiency, since high influent loads (and, respectively, high phenol concentrations) may result in accumulation of toxic metabolites and the respective loss of biofilm. Herouvim et al. (2011) also report that microbial activity in planted units was enhanced compared to unplanted unit, indicating that increased oxygen levels promote aerobic degradation of phenols. Also, recent studies in HSF CWs indicate that biodegradation seems to be the dominant phenol removal mechanism (Stefanakis et al., 2013).

5.8.1.2 Plant Uptake

It is reported that phenols and other organic pollutants can be absorbed by various aquatic plants (Harvey et al., 2002; Dhir et al., 2009). The various processes (e.g., phytostabilization, phytoaccumulation, phytodegradation) which take place in the uptake and transformation of organic compounds and other contaminants have been investigated, although they are not yet clearly defined (Paterson et al., 1990; Imfeld et al., 2009; Nwoko, 2010). Phenol uptake by plants takes place by phytostabilization (Susarla et al., 2002; Nwoko, 2010). Generally, the assimilated portion by plants is negligible for common pollutants (e.g., nitrogen and phosphorus) compared to the amount removed by microbial processes (see Sections 5.4 and 5.5). For more specific pollutants, like organic compounds, however, the role of plants is believed to be more important (Paterson et al., 1990; Haberl et al., 2003).

As shown in a review article, all known and commonly used plant species (e.g., *P. australis*, *T. latifolia*, *Scirpus validus*, etc.) have been tested in the treatment of various types of wastewater containing specific organic pollutants like hydrocarbons, chlorinated organics, aromatics (Haberl et al., 2003). Germination tests on *P. australis* have shown that this plant species is not affected by phenol toxicity and is capable of taking up limited amounts of phenol and transform it to different metabolites (Hübner et al., 2000). However, it is not yet clearly defined whether the phenol removal in VFCWs should be attributed to plant-mediated processes or to the extended biofilm present along the plant roots and the subsequent enhancement of microbial biodegradation. It has to be mentioned that higher phenol removals are reported in planted VFCWs (Herouvim et al., 2011) and HSF CWs (Tee et al., 2009; Rossmann et al., 2012; Stefanakis et al., 2013), while the authors attributed this performance to the presence of the plant root system and the biofilm creation along plant roots, i.e., to the enhancement of the biodegradation process.

5.8.1.3 Adsorption

Sorption of phenols or other organic compounds also contributes to the removal of these contaminants. As for other pollutants, the characteristics of the filter media play a crucial role in the extent and importance of this process (Imfeld et al., 2009). Adsorption proceeds rapidly during the first operational period of a CW system, since there are many available sorption sites onto the media surface; then, it gradually drops and comes to a more limited condition (Rossmann et al., 2012). Phenol adsorption onto filter media depends on a variety of parameters like the physical nature of the adsorbent and the adsorbate, particulate and dissolved OM and pH (Reddy and D'Angelo, 1997; Ahmaruzzaman and Sharma, 2005; Wu et al., 2012). Also, interactions with metals may play a role in the sorption/desorption of phenols, either through competition for adsorption sites on humic acid, through kerogen and black carbon micropore blocking by metal oxides, and/or through complexation of phenolic acids with the inner and outer spheres of hydroxylated Fe and Al compounds on the surface of clay particles (Welz et al., 2012). One of the most common media used for phenol adsorption in wastewater treatment is activated carbon (Ahmaruzzaman and Sharma, 2005). Studies of phenol adsorption and, generally, organic compounds in VFCWs are limited. Some information is derived from studies on HSF CWs. A comparison of two filter media in HSF CWs (gravel and rice husk) showed improved phenol removal for the second material; however, this difference was attributed to the denser rhizome and the extended biofilm rather than its role as adsorbent (Tee et al., 2009). The use of zeolite has also been found to enhance phenol removal in CWs treating olive mill wastewater (Yalcuk, 2011). Generally, although phenol adsorption is recognized as a phenol removal process, further research is needed toward the determination of its extent and role in CWs.

5.8.2 Pharmaceuticals, Personal Care Products, and Endocrine Disruptors

These pollutant groups are often detected in ground- and surface water and are considered as emergent contaminants consumed in modern societies like drugs, antibiotics, tranquillizers, antiepileptics, hormones, X-ray contrast media, musk fragrances, etc. (Suárez et al., 2008). All these compounds are considered as high-risk contaminants. Conventional wastewater treatment plants are not usually quite effective in the treatment of these compounds (Joss et al., 2006; Matamoros et al., 2009; Hijosa-Valsero et al., 2010a). Over the last years, the interest on these compounds has increased and new analytical techniques are available which significantly lowered the detection and quantification limits. Constructed Wetlands appear today as a rising and attractive method for the removal of these compounds. Thus, the number of studies implemented toward the investigation of the removal processes and system efficiency are increasing.

Compounds tested in CWs (mainly in HSF systems) include analgesic/anti-inflammatory drugs, antiepileptic drugs, lipid regulators, astimulants and fragrances, such as carbamazepine, naproxen, diclofenac, ibuprofen, caffeine, salicylic acid, ketoprofen and clofibric acid, ketoprofen, clofibric acid, caffeine, acetaminophen, paracetamol, 1,7-dimethylxanthine, cotinine, sulfamethoxazole, triclosan, trimethoprim, furosemide, ambrettolide, celestolide, cashmeran, galaxolide, tonalide, and methyl dihydrojasmonate (Hijosa-Valsero et al., 2010a,b; Ranieri et al., 2011; Gorsalitz, 2012; Zhang et al., 2012b).

As for the other organic compounds, the main removal mechanisms of pharmaceuticals, personal care products (PCPs), and endocrine disruptors in VFCWs are biodegradation and sorption onto the substrate grains (Matamoros et al., 2007). Aerobic conditions that dominate in VFCW beds enhance the biodegradation of various organic compounds, since the presence of sufficient oxygen amount is a crucial factor for the aerobic biodegradation.

Generally, biodegradation is favored and proceeds faster under aerobic conditions (Massmann et al., 2008). Sorption is affected by the physicochemical properties of the specific target compound, the sorbent properties like organic or inorganic phase, the hydration status, the grain size, the surface coating, and the solution characteristics, i.e., pH, temperature, ionic strength, and the presence of cosolutes (Gao and Pedersen, 2005).

Matamoros and Bayona (2008) presented a review on the treatment efficiency of CWs with different flow path (horizontal and vertical flow systems) and concluded that VFCWs represent the best system for the removal of these compounds among the various CW types (Matamoros et al., 2007a,b; Hussain, 2011). The same authors also consider sorption to the substrate as an important removal mechanism for compounds with high or moderate hydrophobicity like musk, estrogens, and some alkyl phenol derivates. In another study, where VFCWs were used to treat the effluent of two other treatment technologies (activated sludge and MBR), biodegradation was reported as the dominant mechanism for PCPs (Reif et al., 2011). The same was concluded in a study where VF soil filters were used for the removal of xenobiotics, e.g., antioxidants or plasticizers (Janzen et al., 2009). Song et al. (2009) report that biodegradation and even plant uptake are of major importance for the removal of estrogens in VFCWs, compared to sorption. On the other hand, the removal of selected antibiotics and the elimination of three tetracycline resistance genes from swine wastewater in VFCWs were mainly attributed to the sorption process (Liu et al., 2013). Thus, it seems that despite the fact that the dominant mechanisms of various organic micropollutants are known, the role of each mechanism regarding the different contaminants it not clear yet.

Chapter 6

Domestic/Municipal Wastewater Treatment with VFCWs

6.1 BASIC DESIGN CONSIDERATIONS

The design of a VFCW system should take into account various factors, e.g., meteorological, topographical, operational, etc. Generally, these factors include:

- Local climatic conditions,
- Topography (site appropriateness),
- Geological structure of the area,
- Present and predicted wastewater quantity to be treated,
- Availability of necessary land,
- Legal limits for the effluent quality,
- Possible effluent water reuse options,
- Characteristics of receiving water body, and
- Expected costs.

Before the construction phase, it has to be ensured that the necessary land area at reasonable cost is available. As mentioned in Chapter 2, CWs possess higher surface area requirements compared to conventional treatment plants, since they are slow-rate systems. Among the various types of CWs, VFCWs have the advantage of demanding lower surface area. Usually, the area requirement for VFCWs is about half than the respective requirement for HSF or FWS CWs.

6.1.1 Unit Area Requirement

The main design parameter that is used in practice is the unit area, i.e., the ratio of area requirement per person (m^2/pe), where pe stands for "person equivalent." Until today, there is not a generally accepted value. Different values are used from country to country, while even scientists or researchers from the same country may propose different values. Of course, this parameter is

☆"To view the full reference list for the book, click here"

mainly affected by the climatic conditions, but in general it is based on the experience of each individual engineer or scientist. Since VFCWs are very common in Europe, most of information comes from European countries. Proposed and applied values for required unit area in different countries are shown in Table 6.1.

It is clear that there is a large variety in the design characteristics of VFCWs, especially concerning the wastewater quantity to be treated per meter square, which directly affects the surface area demand. To this, the presence or not of a pretreatment stage also affects the required area since it reduces the pollutant load (as described below). As the numbers mentioned in Table 6.1 indicate,

TABLE 6.1 Proposed and Applied Values for Required Unit Area per Person Equivalent (m^2/pe) for VFCWs in Various Countries

Country	Required Area (m^2/pe)	References
Denmark	3.2	Brix and Arias (2005)
UK	1.0-2.0 for the "VF reed bed system" and for the proposed compact VFCW	Cooper et al. (1997), Cooper and Green (1998), Cooper (1999, 2005), and Weedon (2010)
Austria	4.0-6.0, also lower values (0.75-4.0) are investigated	Laber et al. (1997), Schönerklee et al. (1997), Haberl et al. (1998), Langergraber et al. (2007), and Mitterer-Reichmann (2012)
Czech Republic	1.5	Vymazal and Kröpfelová (2011)
Germany	1.6	Olsson (2011)
	2.25	Kayser et al. (2002)
	3	von Felde and Kunst (1997)
	2.0-3.0	Bahlo and Wach (1995)
Belgium	3.8	Rousseau et al. (2004a)
France	2.0-2.5	Molle et al. (2005) and Paing and Voisin (2005)
Greece	1.0-1.5	Stefanakis and Tsihrintzis (2012a)
	3	Gikas and Tsihrintzis (2012)
Italy	1.3-3.2	Foladori et al. (2012)
Spain	1.0-3.2	Puigagut et al. (2007)
Canary Islands	1.5	Vera et al. (2013)

the tendency is to use values smaller than 3 m^2/pe in order to reduce the surface area of the system. In warmer regions, like the Mediterranean basin, the climatic conditions allow for even lower values to be applied compared to North America or North Europe. For example, in colder climates, values of up to 4 or 5 m^2/pe may be necessary to ensure the system effectiveness, since at low temperatures, the various treatment processes are generally affected by ambient temperature and tend to proceed slower, thus, a higher bed volume is needed. On the other hand, in warmer regions with respectively high temperature values, the treatment processes are more intensive and the necessary surface area can be reduced (1-2 m^2/pe).

6.1.2 Organic and Hydraulic Load

Although the parameter of m^2/pe is widely used for the design, it is not fully sufficient. In fact, it only gives a very good indication of the area demands for the system at the preliminary design level. Other design parameters that are also used are the organic loading rate (OLR; g BOD_5/m^2/d or g COD/m^2/d) and the hydraulic loading rate (HLR; m^3/m^2/d or m/d). Several studies and researchers investigated different levels of organic and hydraulic load in VFCWs (Table 6.2) in order to determine the appropriate value for optimum performance without creating problems in system operation.

From Table 6.2, it is clear that the VFCW system is still under investigation in various countries, either using pilot-scale or full-scale beds, in order to identify the boundaries of its strength at various levels of organic and hydraulic load. Of course, the system configuration is not the same in the above-mentioned studies, e.g., different materials are used, the number of substrate layers may also differ, and, thus, the bed thickness. As mentioned before, this fact again reveals the lack of a generally accepted setup and operation conditions for VFCWs. For example, as Table 6.2 shows, the highest loading rates are applied in France. The so-called French system does not include an upper sand layer nor pretreatment stage (i.e., sedimentation or Imhoff tank), which allows for higher loads to be applied directly to the beds.

A first indication that can be drawn from Table 6.2 is that the average value of organic load is around 20-30 g BOD_5/m^2/d or 80 g COD/m^2/d and the hydraulic load close to 0.05 m^3/m^2/d. This of course does not mean that different values should not be applied. Generally, research focuses on applying as high as possible OLR and HLR with a simultaneous search for design modifications and configurations that will allow for high treatment efficiencies, even at the highest levels of loading rates. As before, climatic conditions play again a crucial role in the choice of an appropriate value. In cold climates, lower loading rates should be applied, e.g., OLR up to 30 g COD/m^2/d and HLR 0.05-0.10 m^3/m^2/d. In warmer climates, these values can be, respectively, increased up to 80 g COD/m^2/d and 0.15 m^3/m^2/d.

TABLE 6.2 Organic and Hydraulic Loads Applied to VFCWs in Various Countries

References	Country	OLR		HLR ($m^3/m^2/d$ or m/d)	Comments
		($BOD_5/m^2/d$)	*($COD/m^2/d$)*		
Weedon (2003)	UK	13.7	26.1	0.119	Compact VFCW (single stage)
Weedon (2010)	UK	56		0.05	
Platzer (1999)	Germany			0.25	VFCWs for nitrification
von Felde and Kunst (1997)	Germany		Up to 70	0.05, 0.10, 0.15	
Kayser et al. (2002)	Germany			up to 0.08- 0.180	
Luederitz et al. (2001)	Germany	10.01, 21.44	20.6, 35.69	0.030, 0.044	
Winter and Goetz (2003)	Germany		20		
Langergraber et al. (2007a, 2008a,b)	Austria		20, 27, 40		
Prochaska et al. (2007)	Greece			0.08-0.17	
Arias et al. (2003b)	Denmark			0.520, 1.370	
Brix and Arias (2005)	Denmark	3.75			Domestic wastewater
Korkusuz et al. (2004)	Turkey	6.5		0.100	Domestic wastewater
Vymazal and Kröpfelová (2011)	Czech Republic		80-193	0.038, 0.181, 0.295	
Gikas and Tsihrintzis (2012)	Greece	2.9		0.006	Domestic wastewater

Coulibaly et al. (2008)	Ivory Coast			0.010
Jia et al. (2010)	China			0.010
Olsson (2011)	Germany	13-27		0.087-0.096
Cheng et al. (2011)	China		2.5, 5, 10	0.025
Foladori et al. (2012)	Italy		37-87	0.055-0.123
Abou-Elela and Hellal (2012)	Egypt	2.62-7.65		0.044
Wu et al. (2011b)	China			0.8-1.2
Zurita et al. (2009)	Mexico	4.6		0.04
Vera et al. (2013)	Canary Islands	20		0.083
Prigent et al. (2013)	France		300	0.38
Paing and Voisin (2005)	France	55 g		0.14
Molle et al. (2004)	France		250±70	1-1.4
Stefanakis and Tsihrintzis (2012a)	Greece	89.9-180.9	107.2-219.6	0.195-0.439

6.1.3 Oxygen Transfer Capacity

Oxygen transfer capacity (OTC) is another important parameter in the design and sizing of VFCWs. The main advantage of VFCW systems is the ability to transfer high amounts of oxygen inside the bed, when operating under the commonly used regime of intermittent loading. Therefore, the calculation of the OTC of a VFCW bed gives significant information about the system potential to remove organic matter (decomposition) and nitrogen (nitrification). Oxygen introduction into VFCW beds takes place through three possible ways: (a) dissolved oxygen already present in the wastewater, (b) convection due to intermittent loadings, and (c) diffusion. Usually, the first pathway is not so important compared to the others, while atmospheric diffusion can be crucial (Schwager and Boller, 1997; Platzer, 1999; Kayser and Kunst, 2005). Oxygen from the atmosphere has been found to be the most important oxygen source in VFCWs, especially in the upper layers of the bed (Ye et al., 2012). Moreover, the ability of plants to transfer oxygen and release it inside the bed through their roots is still one controversial issue. Although it is accepted that this transfer takes place, the oxygen amount transferred this way is not sufficient for the total oxygen needs. Among the most common plant species used in CWs (*Phragmites* spp., *Scirpus* spp., *Typha* spp., *Carex* sp.), reported values for plant-mediated oxygen transfer are within the range of 0.005-12 g $O_2/m^2/d$ (Nivala et al., 2012b).

Platzer (1999) proposed the following Equation (6.1) for the calculation of the total oxygen demand (OD) in VFCWs:

$$OD\,(g/d) = (0.7\,gO_2/gCOD_{in}) + (4.3\,gO_2/gTKN_{in}) - (2.9\,gO_2/gTON_{in}) \tag{6.1}$$

The three terms in this equation represent the oxygen demand for organic matter (COD) decomposition, for nitrification, and the oxygen recovery from denitrification. However, this equation does not take into account the dissolved oxygen in the wastewater nor the biological assimilation of nitrogen by plants and microorganisms. Cooper (2005) used the following equation to calculate the OTC:

$$OTC = (\text{Flow rate}) \times [(BOD_{5in} - BOD_{5out}) + 4.3(NH_4{}^+\text{-}N_{in} - NH_4{}^+\text{-}N_{out})]/(\text{Total bed area}) \tag{6.2}$$

where flow rate is in m^3/d, BOD_5 and NH_4-N concentrations in mg/L, area in m^2, and OTC in $g/m^2/d$.

Generally, actual data concerning the OTC in VFCWs are limited. Reported values in g $O_2/m^2/d$ show a range of variation, as follows: 23-64 (Cooper, 1999), 35.4 (Weedon, 2003), 55 (Kayser and Kunst, 2005), 56 (Schwager and Boller, 1997),

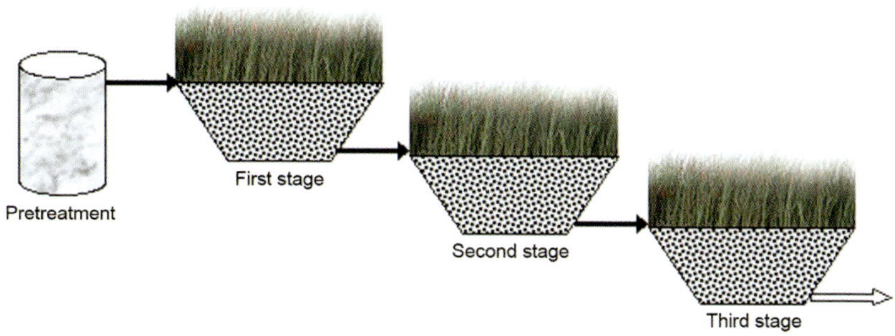

FIGURE 6.1 Typical layout of a full-scale VFCW facility: pretreatment stage followed by two to three VFCW stages.

60-80 (Kantawanichkul et al., 2009), 28-100 in various systems (Cooper, 2005), 147-156 (Ye et al., 2012), and 7.9-58.6 (Nivala et al., 2012b).

6.2 FACILITY LAYOUT, DESIGN, AND OPERATION

A full-scale VFCW system usually consists of several similar beds in parallel operation for each treatment stage (usually, two or three treatment stages in series are used; Figure 6.1). Before the VFCW bed, a pretreatment stage is usually present, which serves as the primary treatment stage. The second treatment stage contains several parallel units depending on the wastewater volume and load to be treated (Cooper and Green, 1998). The number of parallel beds in the third stage is usually smaller since they receive the treated effluent from the second stage. Respectively, smaller is the area requirement for the beds of the third stage. The third stage may be an HSF or a FWS CW, or even a different natural treatment system (e.g., a stabilization pond). This configuration is usually applied for large full-scale facilities.

Figure 6.2 presents an example of the plan view of a full-scale VFCW facility in Northern Greece (Kyprinos, Evros; Gikas et al., 2010). The system shown in Figure 6.2 could be characterized as a hybrid CW system. It is designed for 1,200 pe. This facility consists of three covered concrete anaerobic tanks to avoid odor problems (pretreatment stage), three first stage VFCW beds (1.2 m^2/pe), two second stage VFCW beds (0.8 m^2/pe), and a third stage of HSF CW bed (0.5 m^2/pe) for denitrification. The last stage of the system is chlorination/dechlorination in a specially designed tank, just before the final discharge of the effluent to the receiver. Another VFCW bed is constructed for dewatering of the sludge from the anaerobic tanks. All beds are covered at the bottom with high density polyethylene (HDPE) geomembrane, to avoid drainage to the underground aquifer. Both sides of the membrane are covered with geotextile, to protect it from possible damages due to the direct contact with the gravel material (obtained from a quarry). All beds are planted with *Phragmites australis*. The total surface of the facility is 9.5 acres.

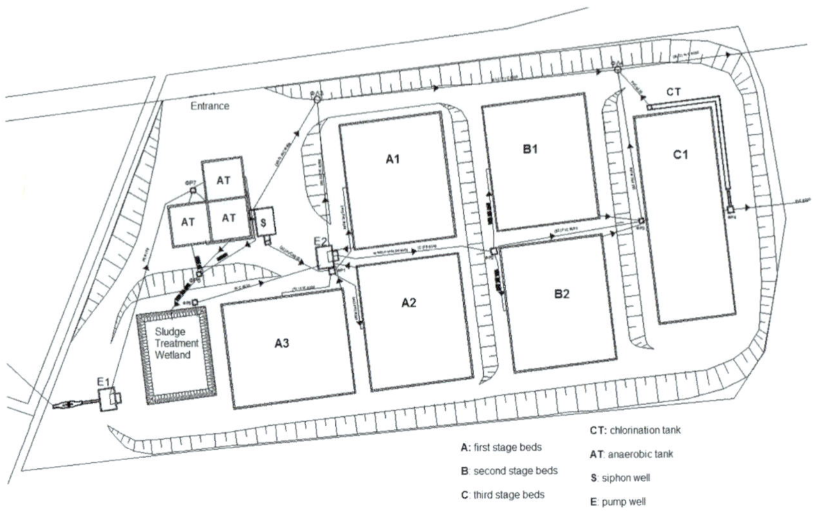

FIGURE 6.2 Plan view of a full-scale VFCW facility in northern Greece (Kyprinos, Evros), designed for 1200 pe: anaerobic tanks (AT), first stage VFCWs (A), second stage VFCWs (B), third stage HSF CWs (C), chlorination/dechlorination (CT), Sludge Treatment Wetland, siphon (S), pump well (E) (Gikas et al., 2010).

FIGURE 6.3 Picture of a VFCW bed in Columbia planted with *Cyperus papyrus*.

Figures 6.3–6.5 present a VFCW system located in Columbia for wastewater treatment of 500 pe. Anaerobic treatment is applied as primary treatment stage, two VFCW beds (100 m^2) planted with *Cyperus papyrus* are used as second stage and two HSF CW beds (400 m^2) planted with *Heliconia psittacorum*

FIGURE 6.4 Picture of a hybrid VFCW bed in Columbia planted with *Heliconia psittacorum*. Here the third stage HSF CW bed after the VFCW bed.

FIGURE 6.5 Picture of a VFCW bed in Columbia planted with *Cyperus papyrus*.

are used as third stage. Figure 6.6 shows a VFCW bed in Germany. Usually, each VFCW bed contains feeding pipes on the bed surface for the wastewater application (as shown in Figure 6.3) and drainage pipes placed at the bottom in order to collect the treated wastewater that drains vertically and collects at the bed. Moreover, the bottom is covered by a geomembrane and geotextile material to prevent leaching of the wastewater to the aquifer. Other options for

FIGURE 6.6 Picture of a VFCW bed in Harz Moutains, Saxony-Anhalt, Germany.

sealing the bed are a clay layer or bottom coverage with concrete. This bottom sealing is crucial, especially when the soil is permeable, and the groundwater table is elevated or it is pumped for irrigation or used as drinking water source. The bottom of the bed is constructed with a minimum slope in order to assist the gravitational flow and the collection of the wastewater. In most case, a slope of up to 2% is used but usually 1% is adequate.

This type of facilities, i.e., with two or three treatment stages and several parallel beds in each stage, is known as the "first generation" of VFCW beds (Cooper and Green, 1998; Cooper, 1999) and it is very common in many countries. In recent years, a new configuration was developed with only a single bed known as the "second generation of VFCWs" (Figure 6.7a) or "compact VF reed beds" (Figure 6.7b), which is a common setup for onsite applications implemented in many countries like Austria, Denmark, the UK, among others (Laber et al., 1997; Schönerklee et al., 1997; Haberl et al., 1998; Weedon, 2003, 2010; Brix et al., 2007; Langergraber et al., 2007a). In Austria, these single beds were initially operated with relatively low hydraulic load, i.e., unit areas of 5 m^2/pe compared to 1-2 m^2/pe used in the first generation VFCW beds, in order to ensure the good system performance. Soon, it was shown that higher loads (2 m^2/pe) can be effectively applied to the single bed, and more than 40 beds like this have been constructed in the UK (Weedon, 2010). Single beds with a pretreatment stage are promoted for onsite domestic wastewater treatment (Zakova and Zak, 2003; Brix and Arias, 2005; Gikas and Tsihrintzis, 2012).

A modification that has been proposed and applied in VFCWs is the partial effluent recirculation (Figure 6.7a). For this, an additional pump is used to transfer part of the effluent volume back to the pretreatment stage (usually a sedimentation tank, as described below) where it is mixed with the influent

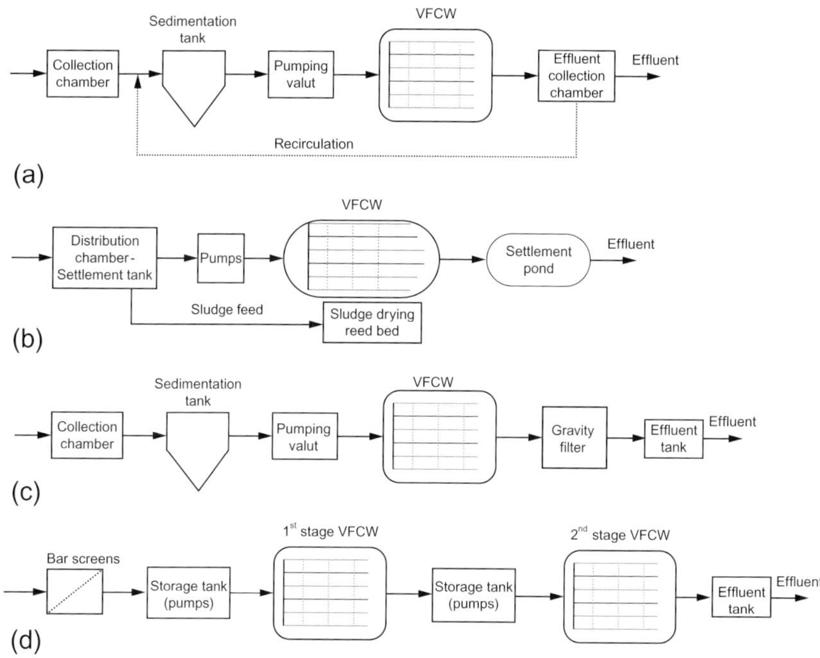

FIGURE 6.7 Schematic representation of the layout of various configurations: (a) single-bed VFCW system, (b) compact VFCW system, (c) single-bed VFCW with gravity filter for effluent treatment, and (d) French system.

wastewater (Laber et al., 1997; Bahlo, 2000; Arias et al., 2005; Brix and Arias, 2005; Gross et al., 2007; Sklarz et al., 2009; García-Pérez et al., 2011; Zapater et al., 2011). In this way, the risk of bed clogging is reduced, the treated volume is increased and the influent BOD_5 concentration is diluted, while denitrification is also enhanced. VFCWs with effluent recirculation are, therefore, considered as an alternative configuration for even more effective treatment performance, especially for onsite facilities.

6.2.1 Pretreatment

The pretreatment stage has been proved to be a crucial and necessary element in the design of a VFCW facility. The presence of a pretreatment stage upstream of the VFCW bed contributes to the retention of suspended solids, organic matter, and large particles (e.g., toilet paper, leaves, hair, wood pieces, etc.), thus minimizing the risk of bed clogging by solids accumulation (Vymazal, 2002). Therefore, wastewater pretreatment not only protects but also allows for the good system performance. The type and method of pretreatment is usually based on the type and quantity of the wastewater to be treated. Common methods used in CW facilities include septic tanks, Imhoff tanks, primary sedimentations tanks, anaerobic digesters, and stabilization ponds, as Table 6.3 shows.

TABLE 6.3 Wastewater Pre-Treatment Methods Associated with VFCWs in Various Countries

Pretreatment		Country	References
Anaerobic digester	BOD_5 removal = 66.2%	Germany	Luederitz et al. (2001)
Two unaerated ponds	430 m^3, 290 m^2 540 m^3, 415 m^2	Germany	Luederitz et al. (2001)
Settling pit and buffering tank		Germany	Haberl et al. (1995)
Settling tank		Germany	Kern and Idler (1999)
Screen and sedimentation (design for 150 and 300 pe)		Germany	Platzer (1999)
Multicompartment septic tank		Germany	Von Felde and Kunst (1997)
Settlement pond (1700 m^2, depth 1.5 m), facultative lagoon (2100 m^2, depth 1.0 m)		Germany	Kayser et al. (2002)
Septic tank and coarse screen		Germany	Olsson (2011)
Sedimentation pond (300 m^3)		Germany	Bruch et al. (2011)
Three-chamber septic tank		Austria	Langergraber et al. (2007a)
Settling tank		Austria	Laber et al. (1997)
Three-chamber sedimentation tank (septic tank)		Denmark	Arias et al. (2003b) and Brix and Arias (2005)
Sedimentation ditch		The Netherlands	Meuleman et al. (2003)
Septic tank		China	Cui et al. (2006)
Septic tank		China	Cui et al. (2010)
Septic tanks (single household) Imhoff tank (municipal wastewater) Bar, screens, grit chamber		Czech Republic	Vymazal (2002)
Mechanical pretreatment		Czech Republic	Vymazal and Kröpfelová (2011)
Septic tank		UK	Cooper (1999)
Settlement tanks		UK	Cooper et al. (1997)
Settlement tanks		UK	Weedon (2010)
Screen and sedimentation tank		Greece	Tsihrintzis et al., 2007

TABLE 6.3 Wastewater Pre-Treatment Methods Associated with VFCWs in Various Countries—cont'd

Pretreatment	Country	References
Settling tanks (two households facility)	Greece	Gikas and Tsihrintzis (2012)
Screen and Imhoff tank	Greece	Gikas et al., 2011
Bar screen	France	Boutin et al. (1997), Molle et al. (2005), and Paing and Voisin (2005)
Mechanical grid and Imhoff tank	Italy	Foladori et al. (2012)
Sedimentation tank	Italy	Morari and Giardini (2009)
Settling and Imhoff tanks	Spain	Puigagut et al. (2007)
Bar screen, Septic tank (50 m^3), Imhoff tank (15 m^3)	Canary Islands	Vera et al. (2013)
Sedimentation tank	Turkey	Korkusuz et al. (2004)
Coarse screen, oil removal, primary settling tank	Egypt	Abou-Elela and Hellal (2012)
Sedimentation tank	Mexico	Zurita et al. (2011)
Septic tank	Vietnam	Nguyen et al. (2007)
Settlement tank	Australia	Heritage et al. (1995)

Moreover, a preliminary mechanical treatment (Figure 6.8) is also applied to remove coarse materials with screens and bars (Vymazal, 2002; Haberl et al., 2003; Molle et al., 2005). Generally, septic and Imhoff tanks are preferred for small-scale facilities, while other methods like ponds or anaerobic digesters are used in full-scale systems.

The most common pretreatment method is the use of septic tanks. A septic tank is an underground tank comprising two or three chambers with walls made of bricks, concrete, PVC, or plastic materials (Tsihrintzis et al., 2007). Prefabricated tanks are also available in the market to serve small facilities. Tanks are the most common pretreatment method for onsite treatment of domestic wastewater (USEPA, 2000c; Gikas and Tsihrintzis, 2010, 2012). These systems are like anaerobic stabilization ponds, with a mean residence time of 1-4 days (depending on the climatic conditions), where solids are separated from the liquid phase by settling, creating a sludge layer (organic matter and solids) at the bottom which is stabilized by anaerobic digestion. Septic tanks are frequently used in developing countries (Kivaisi, 2001; Abdel-Shafya and El-Khateebab, 2013). They are used in onsite treatment systems for single households or blocks

FIGURE 6.8 Picture of a coarse screen (pretreatment stage) in N. Madytos (N. Greece) CW facility.

of apartments (usually up to 150-200 pe). A first treatment and removal of BOD_5 (up to 40%) and TSS (up to 70%) is accomplished in these systems (Bounds, 1997; USEPA, 2002; Metcalf and Eddy, 2003). Moreover, in case of single households, a separate tank for the removal of fats, oils, and grease from the kitchen effluent water may be used.

Imhoff tanks are another option for wastewater pretreatment in CW systems (Figure 6.9). They can be installed in larger facilities for small communities for the partial removal of organic matter prior the CW bed (Vymazal, 2002). Generally, their design includes three compartments for: (a) sedimentation, (b) sludge digestion, and (c) gas vent and scum collection (EPA, 1997; USEPA, 2002a,b). A pump is used for the removal of the sludge layer at the bottom. In Imhoff tanks, settleable solids from the fluid are trapped in the lower part of the tank, where digestion takes place (Metcalf and Eddy, 2003). These tanks have simple operation and advantage that no mechanical equipment is needed. Septic and Imhoff tanks represent the most widely used pretreatment methods for CWs, especially for small onsite systems up to full-scale facilities serving a few thousands person equivalent. For example, in Spain and in Italy, the majority of CW installations use these types of pretreatment (Masi et al., 2006; Puigagut et al., 2007).

FIGURE 6.9 Imhoff tanks (pretreatment stage) in N. Madytos (N. Greece) CW facility.

Anaerobic digesters appear as another pretreatment alternative, aiming to provide a more reliable and robust operation compared to septic and Imhoff tanks. UASB (upflow anaerobic sludge blanket) is a common anaerobic technology which has been used in combination with CWs (Álvarez et al., 2008; De la Varga et al., 2013). Compared to the other pretreatment methods, UASB reactors are more sophisticated technical systems, therefore, they are usually installed in relatively large CW facilities. They can be very effective in the removal of organic matter (mainly BOD_5), with relatively low sludge production, and produce an effluent with low TSS content, which can be very beneficial to minimize clogging in the following wetland system (De la Varga et al., 2013).

As mentioned, a differentiation of the common design with pretreatment exists in France (Figure 6.7d). The so-called French system does not include a pretreatment stage but only a preliminary treatment of the wastewater with a coarse bar screen for particles larger than 2-4 cm (Boutin et al., 1997; Boutin and Liénard, 2003; Molle et al., 2005; Prigent et al., 2013). In these systems, raw wastewater is applied directly to the surface of the first stage VFCW beds. A detail description of the French VFCW system is given later.

6.2.2 Feeding Strategy

The most common type of VFCW systems is that of intermittent wastewater loading. Usually, wastewater is loaded in large batches (varying from 3 to 12 per day) in order to create a flood onto the bed surface. This is made possible by using siphons or pumps. Then, the wastewater drains vertically by gravity through the substrate media and the plant root system. A drainage system placed at the bottom of the CW bed collects the treated effluent. The wastewater application proceeds in feeding cycles. Each cycle consists of a loading period followed by a resting period (Stefanakis and Tsihrintzis, 2012a).

Table 6.4 presents the various feeding regimes that have been used in several VFCW systems in many countries. It is clear that the general rule of a loading period followed by a resting period is used by all researchers/engineers when dealing with VFCWs. As mentioned before (see Chapter 3), this way of feeding, i.e., by applying intervals between the wastewater loadings and allowing for the wastewater to drain vertically, enhances the aerobic biodegradation of several

TABLE 6.4 Feeding Regimes Applied in VFCWs

Feeding Regime (Cycle of Loading and Resting Days)	References
Loading 2-3 times/week	Prochaska and Zouboulis (2006)
Daily doses	Weedon (2003)
3-12 times/d 1-2 times/d (for fine materials)	Von Felde and Kunst (1997)
Four loadings/d	Langergraber et al. (2007a,b)
1 day loading/3 days resting	Cooper et al. (1997)
2 days loadings / 4-6 days resting	Morris and Herbert (1997)
8-12 times/d	Brix and Arias (2005)
5 times/d	Heritage et al. (1995)
4 days loadings/10 days resting	Meuleman et al. (2003)
3 times/d	Vymazal and Kröpfelová (2011)
1-3 days loadings/1-2 days resting	Jia et al. (2010)
30-60 s/h or 2 h	Olsson (2011)
5 days loadings/2 days resting	Yan et al. (2012)
7 days loadings/7 days resting	Bruch et al. (2011)
Three loadings/d every 6.6 h	Foladori et al. (2012)
3 times/week	Ouattara et al. (2011)
8-12 h loadings / 24-36 h resting	Cui et al. (2009)
Loading every 3 h	Zurita et al. (2009)
2 days loadings/4-8 days resting	Stefanakis and Tsihrintzis (2012a)
3-4 days loadings/6-8 days resting	Boutin and Liénard (2003) and Molle et al. (2005)

substances (organic matter, nitrogen, etc.) and ensures higher OTC. However, it is also obvious that there is no widely accepted regime concerning the frequency of the batches applied during the loading period neither the duration of the resting period.

It is not yet clear which is the optimum feeding regime. In fact, most of the regimes presented in Table 6.4 have been proved to be more or less efficient. Frequent applications of small volumes seem to be one alternative (several doses per day). However, the risk taken in this case is that the resting intervals between the batches can be insufficient for the oxidation of the accumulated organic matter and the restoration of the aerobic conditions within the bed. On the other hand, there is the alternative of applying larger volumes which allows for a more prolonged resting time. Larger volumes, which are rapidly applied onto the bed surface, trap oxygen and force it to move downward through the bed matrix. Generally, up to 12 loadings per day can be applied, although a value around 4-6 seems more common. The loading period of 2-6 days should be followed by a double or triple number of resting days. This implies the existence of, respectively, two or three parallel beds in the first stage. The duration of the resting period defines the number of parallel beds needed, in order to ensure the continuous operation of the VFCW facility. An adequate number of parallel beds will, respectively, allow for an adequate duration of the resting period. A simple way to calculate the duration of the resting period is by using the following equation:

$$R = 2N - 2 \tag{6.3}$$

where R represents the duration of the resting period (days) and N is the number of the parallel units.

6.2.3 Wastewater Distribution and Collection

In VFCWs with intermittent loading, wastewater is applied onto the surface of the porous media layers. For the proper and effective system operation, it is crucial to ensure, as mentioned, that the wastewater is applied rapidly in order to create a flood onto the entire surface. This means that a large volume of wastewater should be applied in short time. For this, small tanks are usually used between different treatment stages to collect the wastewater; then, either siphons (Figure 6.10) or submerged pumps are used to empty it very fast onto the downstream VFCW bed. Siphons are preferred over pumps because of the minimal operation cost and maintenance. However, in order to use siphons, the downstream VFCW bed should be in a lower elevation.

The depth of the water column that is temporarily (usually for a few minutes) created on the bed surface should reach up to 5-10 cm, because it is important to ensure that the wastewater covers the entire bed surface. In addition, for this, a network of distribution pipes is designed with an appropriate diameter and hole size. Designing and establishing of appropriate wastewater

FIGURE 6.10 Siphon between treatment stages in N. Madytos (N. Greece) CW facility.

distribution pipes are crucial for the proper long-term operation of the facility. A commonly reported problem in VFCW operation is the uneven distribution of the wastewater on the bed surface (e.g., excess load applied on a small portion of the surface, usually around the pipe outlet). As a result, the entire CW area is not used, and system treatment performance is inferior to designed.

As mentioned, a network of perforated drainage pipes is also placed at the bottom of the bed (Figure 6.11). The distance between the pipes is usually less than 5 m but it depends on bed size. Drainage pipes are covered with coarse gravel material to protect them from clogging and allow for better drainage of the wastewater (Figure 6.12). The network of the drainage pipes is connected to the atmosphere via vertical aeration pipes, which are extending above the bed surface and allow for the better aeration of the deeper parts of the bed (Figure 6.13).

The drainage pipes are placed above the layer of geomembrane and geotextile material with which the bottom of the bed is usually covered to prevent wastewater percolation from reaching the local aquifer (Figures 6.14 and 6.15).

6.2.4 Bed Thickness and Porous Media Layers

VFCW beds consist of several layers of gravel and sand materials with finer materials at the surface and coarser materials at the bottom. Table 6.5 presents various values of bed thickness and number of media layers in different VFCW systems across many countries. It is clear that the general setup of a VFCW system includes a main layer of a porous medium which is the main treatment layer and occupies the largest part of the total bed depth. Usually, the main layer

FIGURE 6.11 Pictures of the pipe network in an onsite VFCW in N. Greece bed: (a and b) Drainage pipes at the bottom of the bed and vertical aeration tubes, (c) bed filled with gravel, and (d) feeding pipes on bed surface (Gikas and Tsihrintzis, 2012).

FIGURE 6.12 Drainage pipes placed above the layer of geomembrane and geotextile and covered with gravel in a VFCW bed in Koresteia, Kastoria (N. Greece; Gikas et al., 2006).

FIGURE 6.13 Aeration pipes extending above the bed surface in a VFCW facility in De Bouvron, France.

FIGURE 6.14 Geomembrane material used to cover the bottom of a VFCW bed in De Bouvron, France.

FIGURE 6.15 Geomembrane and geotextile material used to cover the bottom of a VFCW bed in Koresteia, Kastoria (N. Greece; Gikas et al., 2006).

contains sand material and has a depth of at least 40 cm. The bottom drainage layer of the system contains gravel material with coarse grain size. Usually, the drainage layer has a depth of 10-20 cm and contains the drainage pipelines that collect the treated wastewater and prevents the clogging of these pipes.

The addition of one more gravel layer on the top of the main treatment layer (even above the sand layer) is also possible to prevent the appearance of a free water surface if the system is clogged, while in colder climates, this additional layer also operates as an insulation layer against freezing. However, this top gravel layer is not necessary when the facility is controlled and properly designed and there is no access of the public to the treatment bed.

TABLE 6.5 Depth and Number of Media Layers in Various VFCW Systems in Different Countries

Number of Layers	Bed Depth (cm)	Country	References
1	50-180	Germany, Austria, and Switzerland	Rustige et al. (2003)
1	50	Japan	Zhou et al. (2008)
3	62 (3+42+17)	Greece	Prochaska and Zouboulis (2006)
3	75 (50+10+15)	Austria	Langergraber et al. (2007a)
3	100 (80+5+15)	UK	Weedon (2003, 2010)
1	100	Germany	Platzer (1999)
3	107 (2+98+7)	Germany	von Felde and Kunst (1997)
2	70 (20+50)	Germany	Kayser et al. (2002)
3	85 (5+60+20)	Germany	Luederitz et al. (2001)
2	80 (60+20)		
3	75 (50+10+15)	Austria	Langergraber et al. (2007a)
2	100 (80+20)	Denmark	Arias et al. (2003b)
3	140 (20+100 +20)	Denmark	Brix and Arias (2005)
4	161 (5+120+10 +26) 121 (5+80+10 +26)	Germany	Schönerklee et al. (1997)
3	70 (30+350+50)	UK	Cooper et al. (1997)
2	65 (25+40) 70 (40+30)	UK	Morris and Herbert (1997)
3	60 (15+30+15)	Turkey	Korkusuz et al. (2004)
1	100	Australia	Heritage et al. (1995)
1	80, 60	Austria	Laber et al. (1997)
1	100	Czech Republic	Vymazal and Kröpfelová (2011)
2	100 (40+60)		

TABLE 6.5 Depth and Number of Media Layers in Various VFCW Systems in Different Countries—cont'd

Number of Layers	Bed Depth (cm)	Country	References
4	50 (10+15+10+15)	Greece	Stefanakis and Tsihrintzis (2012a)
4	80 (30+10+25+15)		
3	90 (50+20+20)		
2	100 (30+70)	Greece	Gikas and Tsihrintzis (2012)
2	40 (10+30)	Ivory Coast	Coulibaly et al. (2008)
3	60 (5+40+15)	China	Jia et al. (2010)
2	45 (20+25)	China	Yan et al. (2012)
3	95 (10+65+20)	Germany	Olsson (2011)
3	60 (20+10+30)	Italy	Foladori et al. (2012)
2	85 (60+25)	Egypt	Abou-Elela and Hellal (2012)
4	150 (16+22+22+90)	Italy	Morari and Giardini (2009)
4	100 (12.5+12.5+25+ 50)	Poland	Białowiec et al. (2011)
2	40 (10+30)	Ivory Coast	Ouattara et al. (2011)
3	60 (15+35+10)	China	Li et al. (2013)
3	50 (15+25+10)		
2	35 (10+20+5)	China	Cui et al. (2010)
1	30	Mexico	Zurita et al. (2011)
2	80	Canary Islands, Spain	Vera et al. (2013)
2	45 (20+25)	China	Zhao et al. (2010)
2	100-120	France	Boutin and Liénard (2003)
3	70 (30+20+20)	France	Molle et al. (2005)

Finally, a freeboard of at least 15-20 cm should also be provided to allow for the temporary water accumulation during the feeding batches and for rainfall.

6.2.5 Post-Treatment

A post-treatment stage is usually not necessary in a full-scale VFCW facility, since the combined system of pretreatment and VFCW stages is capable of providing an effluent of high quality. However, a post-treatment stage may be present, especially when criteria for effluent reuse are to be met. Common disinfection techniques like chlorination, UV radiation, or subsurface chlorination may be used to ensure that the effluent meets specific criteria for reuse concerning the elimination of pathogenic microorganisms (Gross et al., 2007). Maturation ponds have also been used in certain systems for disinfection (Gikas et al., 2011).

The use of a post-treatment stage has also been proposed in order to improve the overall system efficiency, especially in terms of phosphorus removal. Although the system efficiency is more than satisfying in the removal of organic matter and nitrogen compounds, phosphorus removal is usually lower compared to the other compounds, due to relatively inadequate contact time between the wastewater and the filter media (Brix and Arias, 2005; Stefanakis and Tsihrintzis, 2012a). Therefore, one inexpensive and easy to build alternative that has been proposed is the use of separate filter units with special porous media as an additional treatment step of the VFCW effluent (Figure 6.7c; Arias et al., 2003b; Brix and Arias, 2005; Stefanakis and Tsihrintzis, 2012b). In this configuration, the choice of an appropriate porous media with specific characteristics that enhance the removal of phosphorus (e.g., high content of Al, Ca, or Fe minerals), or nitrogen (e.g., high cation exchange capacity (CEC)) is crucial for the effectiveness of the filter.

Porous media materials that have been tested include bauxite, zeolite, carbonate material and mixture of these materials (Stefanakis and Tsihrintzis, 2012b), calcite (Arias et al., 2003b), and lava sands with zeolite minerals (Bruch et al., 2011), showing very promising results. Increased contact time between the VFCW effluent and the porous media in the filter improves the overall system efficiency. Stefanakis and Tsihrintzis (2012b) proposed an HRT of 1 day in filter containing a 50-50% mixture of bauxite and zeolite. This configuration (pretreatment + VFCW + post-treatment) is also appropriate for onsite facilities, e.g., for single households, and represents a feasible and inexpensive solution. This setup, using zeolite in the filter, was tested for onsite wastewater treatment of two households (8 pe) and proved to be very efficient (Gikas and Tsihrintzis, 2012). Gikas et al. (2011) presented a full-scale VFCW system (3000 pe) in Greece, which consists of pretreatment (screen and Imhoff tank), three stages of VFCWs, and, as post-treatment, two maturation ponds. Results of this study showed that the presence of maturation ponds improved the quality of the effluent, mainly by further eliminating pathogenic bacteria population.

6.2.6 The French System

As previously mentioned, a special type of VFCW has been developed by IRSTEA (former: CEMAGREF) in France over the last 15-20 years, the so-called French System (Molle et al., 2005). The main characteristic of this system is the absence of a pretreatment stage (e.g., a sedimentation or Imhoff tank) and the use of a first stage French VFCW instead (Figure 6.7d). This means that raw wastewater is directly introduced onto the surface of the first stage French VFCW, while coarse screening is used to remove large solids (Boutin et al., 1997).

The French System usually comprises two stages of beds. The first stage contains 3 and the second 2 units, as described by Molle et al. (2005). All beds have a media depth of 80 cm. The first stage beds contain from bottom to top a drainage layer of gravel (20 cm, grain size 20-40 mm), a transition gravel layer (20 cm, grain size 5 mm), and a final layer of fine gravel (>30 cm, grain size 2-8 mm). A freeboard (ca 20 cm) is also left on top for sludge accumulation (about 1.5 cm/year), which has to be removed after 10-15 years of operation. First stage beds operate with intermittent wastewater feeding as most VFCWs. The feeding cycle consists of 3-4 days of feeding with raw (untreated) wastewater followed by 6-8 days of resting (Boutin and Liénard, 2003). The effluent of the first stage beds is applied onto the surface of the second stage filters.

The second stage beds enhance nitrification and OM and pathogen removal. They contain from bottom to top a drainage gravel layer (20 cm, grain size 20-40 mm), a transition gravel layer (20 cm, grain size 3-10 mm), and a final sand layer (>30 cm, 0.25 mm, $d_{10} < 0.40$ mm). All beds are planted with common reeds (*P. australis*). An alternative arrangement for the French System (Prigent et al., 2013) includes a depth of 120 cm with five media layers from bottom to top: a 20-cm gravel layer (10-20 mm), a 20-cm gravel layer (4-10 mm), a 20-cm layer with a special media (Mayennite®; 0.5-4 mm), a 30-cm gravel layer (4-10 mm), and a 30-cm final Mayennite® layer (2-4 mm). The first stage French System is designed for 1.2 m^2/pe, an organic load of 100 g COD/m^2/d and a hydraulic load of 120 L/m^2/d for each bed. Respective values for the second stage are 0.8 m^2/pe and 25 g COD/m^2/d (Molle et al., 2005). New hydraulic limits (1.80 and 0.90 m/d for the first 0-10 and 10-25 cm of the bed, respectively) have been proposed based on the accumulated material on top of the bed (Molle et al., 2006).

A plan view of a full-scale French VFCW system constructed in Greece is shown in Figure 6.16 (Gikas et al., 2006). This facility is designed for 600 pe and comprises three first stage VFCW beds (1.5 m^2/pe), two second stage VFCW beds (1.0 m^2/pe), and a third HSF CW (1.5 m^2/pe) for denitrification. As pretreatment stage, only a coarse screen is used (pore opening 2.0 cm). A pump is used to apply the wastewater onto the first stage beds, while the second stage is fed through a siphon and the third by gravity. The loading cycle consists in 3.5 days of feeding followed by 7 days of resting.

Figure 6.17 presents the first stage of VFCW system located in De Bouvron, France. This facility serves 2000 pe. The second stage of the system is an HSF

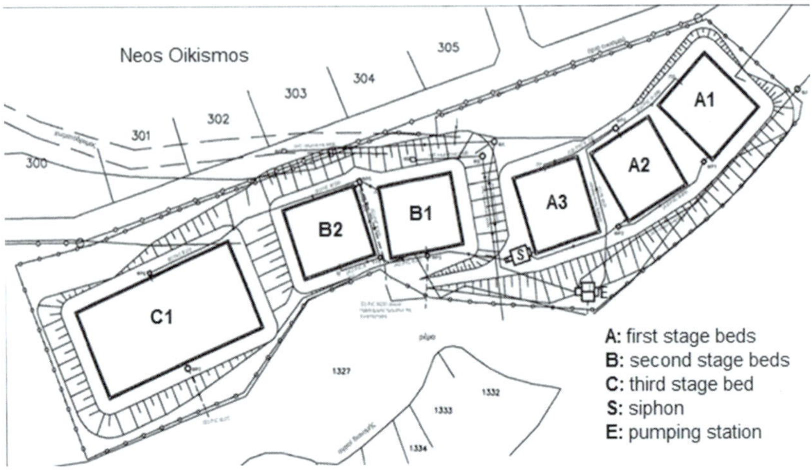

FIGURE 6.16 Plan view of a full-scale French VFCW facility in Neos Oikismos (600 pe), Kastoria, Greece: coarse screen, first stage VFCWs (A1, A2, A3), second stage VFCWs (B1, B2), third stage HSF CW (C1), siphon (S), pumping station (E) (Gikas et al., 2006).

FIGURE 6.17 A VFCW facility in De Bouvron, France for 2000 pe: the first stage VFCW bed.

bed (Figure 6.18) and the third a final VF. In this system, the HSF bed is placed prior the second VF bed and receives the effluent of the first VF bed, while the effluent of the second VF bed is also recirculated to the HSF bed for additional denitrification.

The main advantage of the French System is that it possesses lower construction cost, since there is no use of a pretreatment method (usually made from reinforced concrete which is quite expensive); furthermore, there is no daily sludge production, respectively. This makes these systems even more

FIGURE 6.18 A VFCW facility in De Bouvron, France for 2000 pe: the second stage HSF CW bed.

appropriate for small- or medium-size installations, e.g., small communities. On the other hand, area demands are usually higher compared to common VFCW design in other countries, especially in temperate and warm climates. Moreover, due to the fact that sludge is accumulated on the top of the first stage, for safety and hygienic reasons, caution should be employed when using this system for onsite treatment (e.g., single households).

6.3 PERFORMANCE

Generally, VFCWs with intermittent loading have been proved very efficient treatment systems for municipal wastewater. Reported performance from different countries under different climatic conditions shows that these systems provide a satisfying sanitation level for the major constituents like suspended solids, organic matter, nitrogen, and pathogenic microorganisms. However, as it is previously described, there is not a widely accepted setup or system configuration. Although most of the systems have common characteristics, e.g., existence of a sand layer, they also have many different system components and parameters, e.g., number of porous media layers, bed depth, feeding regime, etc. This makes the direct comparison of the efficiencies rather difficult. However, a general conclusion can be drawn. Most of the available data have to do with the presentation of influent-effluent concentrations and respective percentage removal rates. Based on these data, overall performance of VFCWs with different setups and configurations appears to reach high levels. The review of the data presented in the published papers included in Tables 6.1–6.5 can provide an overall mean removal efficiency of VFCW beds (upper and lower limits are presented in parentheses):

- 82% (52-99%) for TSS,
- 85% (48-99%) for BOD_5,
- 75.2% (44-95%) for COD,
- 74.8% (34-95%) for NH_4^+-N,
- 44.6% (20-94%) for TN,
- 62.6% (45-93%) for TKN,
- 62.8% (21-97%) for TP,
- 58.1% (26-92%) for OP, and
- 98.1% for total coliform (TC).

These values refer to a variety of VFCW designs (e.g., single beds and multi-stage facilities) and are only mentioned to give a global overview of the system capacity. Among the various systems evaluated for these mean values, there are high variations in performance as the upper and lower limit values indicate. This implies that the proper combination of the various design and operational parameters determines the efficiency. Also, it makes clear that currently there is not a widely accepted setup, but several design configurations are investigated, thus, there is a large dispersion depending on a series of parameters (e.g., design, operation, wastewater load, climate).

The VFCW type with downflow and intermittent wastewater loading is usually used for municipal wastewater treatment. As mentioned before (see Section 6.2), an integrated full-scale facility for municipal wastewater treatment usually includes in series a pretreatment stage, two treatment stages, and, sometimes, even a post-treatment stage. This setup has been proved to be capable of fulfilling the most stringent effluent limit values and it can reach removal rates of up to 95% for BOD_5, 90% for ammonia nitrogen, and 90% for TP.

6.3.1 Removal of Organic Matter and Nitrogen

Experimental results from both full- and pilot-scale facilities confirm that VFCWs with intermittent loading and alternating feeding and resting periods are capable of removing high amounts of BOD and ammonia, due to adequate bed oxygenation (Platzer and Mauch, 1997; Cooper, 1999).

Investigations in VFCWs constructed even 20 years ago showed the potential of these systems to achieve high-performance levels. Heritage et al. (1995) reports high removal for nutrients (nitrogen and phosphorus) and BOD_5 in planted VFCW beds. According to Cooper et al. (1999), VFCWs are already recognized as a better CW type for enhanced nitrification and OM degradation. A two-stage VFCW system was proved capable of removing about 95% and 90% of the influent BOD_5 and TSS, respectively. In these systems, nitrification was more limited (72% NH_4^+-N removal), which was attributed to the respectively low surface area per capita (1 m^2/pe). Another observation of this study was that nitrification was more intense in the first VFCW stage. A similar VFCW system with two stages also showed a very satisfying performance with effluent $BOD_5 < 3$ mg/L, TSS < 4 mg/L, and NH_4^+-N of 1 mg/L (Cooper et al., 1997). This system used a biofilter as pretreatment stage. The observed increased nitrate concentration in the effluent indicates that intense nitrification took place in the beds. Thus, in order to reduce nitrate concentration, hybrid systems (i.e., VF followed by HSF CWs) were investigated. The final stage of HSF CWs was used as a denitrification step (Cooper, 1999; Platzer, 1999).

Platzer (1999) also reported almost full nitrification in a VFCW bed for loadings below 6.5 g TKN/m^2/d with the final effluent NH_4^+-N concentration being always below 10 mg/L. Increased loads reduced the nitrification rate, which implied that for full nitrification, the calculation of the oxygen balance and the oxygen input should be taken into account when sizing the bed. Kayser et al. (2002) tested the nitrification rate in VFCW bed treating the wastewater from a village in Germany. With a mean NH_4^+-N influent concentration of 50 mg/L and a hydraulic load of up to 180 mm/d, they also observed very good nitrification rates and correlated them with wastewater temperature values. Thus, for temperatures above 10 °C, the nitrification rate was close to 90%. The same authors proposed a hydraulic load of 100 mm/d to achieve full nitrification.

Other modifications have been proposed to further improve the system efficiency, mainly in terms of enhancing denitrification and respective removal of the oxidized nitrogen. Laber et al. (1997) tested two similar systems with two different modifications. In the first, partial effluent recirculation to the settling tank took place, while in the second, an external carbon source was added. They also investigated different feeding regimes. It appeared that the addition of carbon to the system under the intermittent loading regime resulted to higher removal rates (82% for inorganic nitrogen and 78% for TN; Laber et al., 1997). With effluent recirculation, TN removal reached 72%. Improved results were also observed when using effluent recirculation by other authors (e.g., Brix et al., 2003; Arias et al., 2005; Gross et al., 2007). Arias et al. (2005) proposed an optimum recirculation ratio between 100% and 200% to avoid overload problems. Reported values for TN removal under the recirculation regime varied between 50% and 69% and above 90% for BOD_5 and COD (Arias et al., 2005; Gross et al., 2007). Platzer (1999) reported similar values for a recirculation rate of 200%. These results demonstrated that VFCWs are able to fulfill high treatment standards.

Weedon (2003, 2010) promoted the so-called Compact VFCW, which comprises one treatment stage (one VFCW bed) with no parallel beds. This system contains sand as the major media and the resting period applied is the interval between feeding batches. After 10 years of operation and with several units installed (mainly in the UK), the authors report similar high removal rates (90% for SS, BOD_5, and NH_4^+-N; Weedon, 2010) for the treatment of presettled domestic sewage at typical loading rates.

Rousseau et al. (2004a), in their review on CW performance in Flanders (Belgium), report that 34 VFCWs were operating at that time, mainly treating domestic wastewater. Efficiencies were again quite satisfying, with mean removal rates of 94% for COD, 98% for SS, 52% for TN, and 70% for TP. The authors also report that more than 95% of the effluent concentrations are in compliance with the effluent limit standards of the country. Moreover, problems observed in these facilities had to do with clogging of the beds due to hydraulic overloading and unequal wastewater distribution on the bed surface. These again indicate that if properly designed, VFCWs can reach high levels of performance.

Langergraber et al. (2007a) tested three experimental VFCW units in Austria under different organic loads and correlated the performance of the beds with the temperature variations. Results showed 97% removal of organic matter (BOD_5 and COD) for temperatures above 12 °C and 91-98% for BOD_5 and 80-94% for COD for temperatures below 12 °C, indicating dependence of the removal rate on temperature variations. The same was also observed for NH_4^+-N, whose effluent concentration was always below 10 mg/L (the national effluent standard). The authors concluded that for the lowest organic load (20 g COD/m^2/d) among the three tested, effluent quality was always below the legal limits throughout the year. Langergraber et al. (2008a,b) investigated a two-stage VFCW system with intermittent loading and they reported similar efficiency to a single-stage system for organic matter and nitrogen removal.

Morari and Giardini (2009) investigated the performance of two pilot-scale VFCWs in municipal wastewater treatment and the effluent suitability for reuse. The systems showed high removal rates, above 86%, for most constituents (COD, BOD_5, and N), but the effluent quality was not found to fulfill the criteria for reuse. The authors operated these systems without a pretreatment stage, which would improve the performance. Similar results (88%, 90%, 92%, 53%, and 57% for COD, BOD_5, TSS, TKN, and NH_4^+-N, respectively) are reported for a full-scale VFCW system in Egypt (Abou-Elela and Hellal, 2012), while an onsite VFCW bed followed by a zeolite filter for domestic wastewater treatment in Greece reached even higher removal rates (95.4% for COD and BOD_5 and 92.8% for ammonia; Gikas and Tsihrintzis, 2012). The overall good performance of VFCWs was also reported by Vera et al. (2013) for a system in Canary Islands (80% and 75% for COD and BOD_5 and 90% for TSS). VFCWs were also found capable of providing an effluent that meets the effluent criteria in Uganda concerning ammonia concentration (Elke and Lechner, 2012).

Olsson (2011) investigated the performance of six pilot-scale VFCWs with different characteristics, like the dosing regime, filter media, and plants. Removal rates were high for all units (85-99% for BOD_5, 76-99% for TSS, 77-94% for NH_4^+-N, 67-72 for org-N). Results of the same study showed that a feeding regime with more frequent, and thus, smaller wastewater doses provides an improved performance, and that the major portion of pollutants is removed in the upper parts of the bed. Similar conclusions are reported by Jia et al. (2010), who found that the intermittent loading enhanced the removal of ammonia nitrogen (>90%), but with lower TN removal rates.

Kadlec and Wallace (2009) reported that the majority of microbial activity is located in the first 20 cm of the bed, where more oxygen is also available (Tietz et al., 2008). Ye et al. (2012) investigated the vertical DO profile within a VFCW bed and observed a high DO concentration in the upper part followed by a reduction with depth. They also reported that atmospheric oxygen is the main oxygen source for the bed, while most of it is consumed within the first centimeter of the bed.

An interesting alternative to increase VFCW performance has to do with the use of earthworms. Xu et al. (2013) found that the addition of earthworms into planted beds enhanced nitrification potential by 105-268%, depending on the plants species used. A slight increase of the denitrification potential was also observed (5-15%). As a result of this, plant uptake of N was increased and the removal of N in the system was improved.

Stefanakis and Tsihrintzis (2009a, 2012a) performed a large experiment of 10 pilot-scale VFCW units which confirmed the high performance of VFCWs but also the potential of these systems to be effective even under high loading rates and respective pollutant concentrations. The VFCW beds of this study only simulated the first treatment stage of a full-scale facility and showed 78% mean removal rate for BOD_5 and COD, and 58% for TKN and NH_4^+-N. Three different VFCW designs were compared, i.e., different bed depths and porous media layers, based on the most common designs used in Europe, France, and North America. Among them, the tested French system with the coarser material had the lower performance among the three, with removal rates of 63% and 44.1% for OM and N, respectively.

Generally, various studies on the French system report relatively lower efficiencies compared to other designs. Boutin et al. (1997) report 82% COD and 53.7% TKN removals in the first treatment stage, with respective values for the entire system (two treatment stages) of 87.5% for COD, 92.5% for BOD_5, and 76% for TKN. Molle et al. (2005) present an overview of the two-staged French system performance with removal rates above 90% for COD, 94% for SS, and 85% for TKN. Respective values for the first treatment stage are 79%, 86%, and 58%. However, as it is already mentioned, the advantage of the French system is the absence of pretreatment stage. On the other hand, direct feeding of untreated wastewater onto the surface of the first stage beds demands the removal of the accumulated sludge layer after several years of operation, while area requirements are usually higher compared to other common designs.

Interesting conclusions can be drawn when comparing the removal efficiency as function of the influent loading rates applied. Figure 6.19 depicts the correlation that exists between the effluent BOD concentration (mg/L) and the respective BOD influent loading rate (g/m^2/d). The data of this graph refer to the summarized performance of 81 VFCW systems from different countries (Australia, Austria, Denmark, Egypt, France, Germany, Greece, Italy, Mexico, Spain, the Netherlands, and the UK). Each data point represents the performance of several systems from a specific country. A log-linear trendline

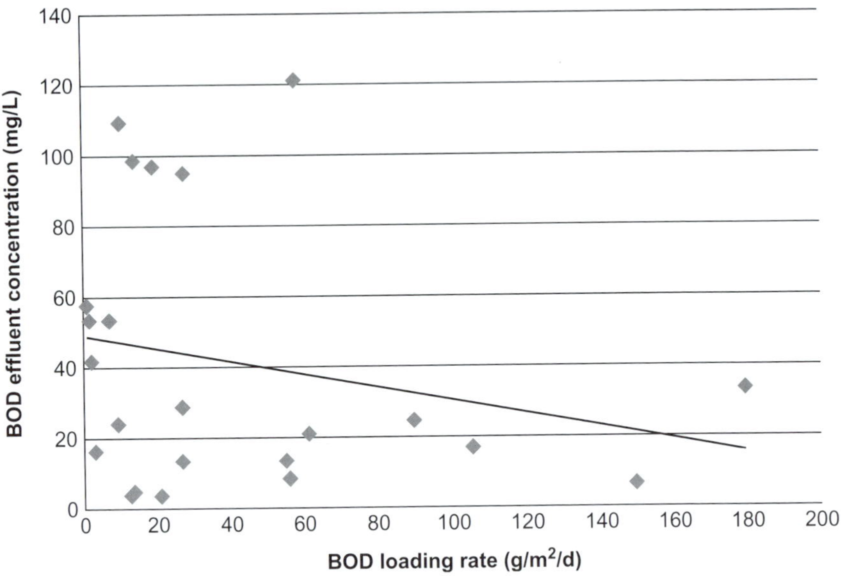

FIGURE 6.19 Effluent BOD concentration (mg/L) vs. respective BOD influent loading rates (g/m^2/d). Data obtained from 81 VFCW systems from different countries (Australia, Austria, Denmark, Egypt, France, Germany, Greece, Italy, Mexico, Spain, the Netherlands, and the UK).

is also fitted in order to give an indication of the tendency of the correlation between the BOD_5 effluent concentration and the BOD_5 loading rate.

This comparison does not take into account some parameters, such as differentiations in the system designs or different climatic conditions; however, it provides interesting information about the behavior and tolerance of VFCWs. This graph shows the wide range of loading rates that VFCW receive and reveals a rather unexpected tolerance of VFCWs to provide excellent removal of OM even at very high OLRs. Indeed, the majority of the tested systems gives a BOD_5 effluent concentration below 40 mg/L, while high performance is persistent even for systems of high loading rate. It is clear that VFCW systems (regardless of the variations of the system designs and the climate) possess a high potential for the removal of organic matter. The same conclusion was also reached by Kadlec and Wallace (2009), who compared the performance of 62 VFCWs. On the other hand, Kadlec and Wallace (2009) reported that the treatment performance is highly dependent on the influent loading rate, which is not in agreement with the tendency shown in Figure 6.19. These authors included a wide range of influent BOD concentrations (3-30, 30-100, 100-200, >200 mg/L). In the graph of Figure 6.19, mean influent BOD_5 concentration for all systems included was equal to 325 mg/L, i.e., the vast majority of the influent BOD_5 was far above 200 mg/L. Therefore, this graph indicates that—with the appropriate design—VFCW systems are capable of effectively treating high organic loads.

6.3.2 Removal of Phosphorus

It is generally known that the removal of phosphorus (P) in VFCWs is more problematic in the long-term compared to organic matter and nitrogen removal (Vymazal et al., 1998; Brix and Arias, 2005). Reported removal rates are lower compared to other constituents, usually in the area of 20-40% for most applications. This issue is closely related to the main P removal processes. As mentioned in Chapter 5, adsorption to the substrate media plays the most significant role in P removal, which means that the characteristics of the media used within a CW bed potentially could control P removal at a high level. Other factors that could regulate P removal is the hydraulic residence time (which practically determines the contact time between the media and the wastewater), plant uptake and assimilation, the characteristics of the influent wastewater (pollutant load) and the microbial activity. It is mentioned that in the long term, the only sustainable P removal mechanism in CW systems is direct plant uptake and subsequent harvesting of the reeds (Lantzke et al., 1998; Arias et al., 2001), meaning that this mechanism (continuous uptake followed by annual harvesting) is irreversible and the only one that takes place continuously for the entire life time of the system. But again, this process accounts only for a small percentage of the total P removal, while part of the assimilated P returns to the system via decomposition of dead plant material (Reddy et al., 1999; Luederitz and Gerlach, 2002; Kadlec and Wallace, 2009).

Schönerklee et al. (1997) reported a correlation of P removal rate and HLR, i.e., lower HLR results in higher performance. Mean removal rates were around 29% (3-65%). Authors suggest that the removal of P is affected not only by the available surface area but also by the concentration of specific P-binding metals. Similar performance is also reported by Cooper et al. (1997) and Cooper (1999) for a full-scale system, where the first two stages are VFCWs. In this system, OP removal in the VFCW beds reached 25.6% (19.4% and 7.7% in the first and second stages, respectively).

Luederitz et al. (2001) reported, on the other hand, a very good performance of two full-scale VFCW systems in Germany for TP removal (96.8% and 60.5%, respectively). However, they mention that P removal capacity was expected to start reducing after 5-7 years. This was also indicated by the reduced performance of the oldest second system (60.5%). The authors state that the initial high P removal capacity of VFCWs should be mainly attributed to plant assimilation (Lantzke et al., 1999). Gradual and fast P accumulation in plant biomass, as also gradual saturation of the filter media, resulted in the decreased performance of the systems concerning P removal. A similar situation (decrease of removal rate with time) is also reported by Weedon (2003) for a single bed system with a 80-cm sand layer. OP removal dropped from 91.5% to 44.7% after only 17 months of operation. The author attributed this reduction to the gradual saturation of the filter media. Verhoeven and Meuleman (1999) mentioned a similar removal rate (25%) for TP removal in the Netherlands, while

Meuleman et al. (2003) also reported an almost complete saturation of the filter media and low removal rate for P (24%) after 15 years of continuous operation, indicating the end of the useful life time of the system. Values of 20-30% for P removal are reported for VFCW beds alone in Danish and Greek systems for single households (Brix and Arias, 2005; Gikas and Tsihrintzis, 2012). The addition of pre- and post-treatment stages in full-scale systems, however, improves the efficiency of TP removal up to 62% (Tsihrintzis et al., 2007; Tsihrintzis and Gikas, 2010; Gikas and Tsihrintzis, 2012). Comparable performance (28%) was also achieved by a full-scale system in Canary Islands (Vera et al., 2013). Same order values (37% mean TP removal) are also mentioned by Stefanakis and Tsihrintzis (2012a), in an extended experiment with several pilot-scale VFCW units with intermittent loading and under different configurations.

Therefore, it is clear that the various configurations of common design setups may not be capable of removing P from the wastewater at an acceptable and satisfying rate, depending of course on the influent concentration which usually is relatively low. For this, some modifications have been proposed. One alternative is the addition of a coagulant at the pretreatment stage (usually in the sedimentation tank), in order to promote the chemical precipitation of P. This methodology is used in activated sludge facilities for the removal of P from the surplus sludge. Brix and Arias (2005) suggested the use of an aluminum compound (e.g., aluminum polychloride). Lauschmann et al. (2013) investigated the addition of sodium aluminate for P precipitation in the first chamber of the three-chamber tank which was used as mechanical pretreatment stage of a VFCW bed designed for 60 pe. These authors reported that satisfying results were obtained when the added dose exceeds 3.5 times the dose computed from stoichiometry. Although the desired effluent quality in terms of P concentration was achieved, some concerns are raised. It is certain that the use of chemicals alters the philosophy of CW systems and moderates their ecological character, while this option increases the operational cost for the facility (according to this study, up to 15%).

Since P removal is closely related with filter media used and its characteristics, research over the last years has focused on the determination and investigation of porous media with special characteristics and properties. The concept is to use these specific media either as substrate within the wetland body or as filter media in separate filter units treating the CW effluent. However, it has to be mentioned that specific actions for P removal improvement should be taken when the system aims at the removal of P and considers P as the primary target pollutant. It should be noted that, in most cases, the main target pollutants are organic matter and nitrogen or specific micropollutants. Thus, before the design and implementation of a VFCW system, this parameter needs to be clarified. In Europe, for example, the problem of eutrophication of surface water bodies, which appeared several decades ago, led to the alteration of the detergent production process and the non-use of P-containing additives. This, respectively, resulted in the decrease of P concentrations ending to treatment facilities. Present P-related problems in Europe mainly come from single households in rural areas where no proper wastewater treatment and subsequent disposal take place (Vymazal et al., 1998). On the other hand, treatment needs in developing countries are far more pressing, since in many countries, proper sanitation of the wastewater is still an issue and disposal of untreated wastewater is a daily practice. Thus, in these countries, P removal may not be the first problem that needs to be faced. Additionally, a crucial parameter that should always be in the mind of scientists and designers is that phosphorus represents one of the most essential nutrients in the world. It is predicted that current resources will last for 70 more years, while 80% of the extracted P is utilized to fertilize modern high-productivity farm soils (Evans, 2009). Therefore, besides meeting the effluent criteria for reuse of the treated effluent, saturated porous media can be considered as a valuable source of nutrients which could be further exploited; however, simultaneous measurement of the content of possible harmful compounds, e.g., heavy metals, should take place before the use of these materials.

6.3.3 Effects of Porous Media

Porous media represents one of the basic components of all CW systems. Specifically in VFCWs, the relatively limited performance concerning P removal (see Chapter 5) made it clear that new ways need to be found in order to improve this efficiency. Among the various processes that contribute to P removal in CW systems (adsorption, precipitation, plant uptake, microbial consumption), adsorption to the filter media is considered as the main removal process (Vymazal, 2007). Thus, research has mainly focused on enhancing this process. The main goal is to identify these alternative or special substrates that possess high P-sorbing capacity. Preferred filter media are those with sufficient content of Fe, Al, and Ca oxides, which favor P adsorption and precipitation processes. Moreover, the presence of these special media in VFCWs could positively affect the removal of other constituents as well, e.g., ammonia nitrogen. As already mentioned in Chapter 4, numerous materials of different origins and compositions (natural or synthetic materials and industrial by-products) have been proposed and investigated for potential use as substrates in VFCWs (see Table 4.3), some of them represent really innovative ideas (Westholm, 2006; Vohla et al., 2011), mainly focusing on the removal of P in wastewater. However, there are some issues that always will be raised.

The first issue has to do with the fact that adsorption and/or precipitation are finite processes, i.e., at some point in the future the materials may become saturated with P and their adsorption capacity will be reduced or even diminished. Therefore, it is always important to know the capacity of the proposed material in order to estimate its useful life time. Another issue has to do with the availability of the material. Often, new materials (either natural or synthetic) are tested and reported to possess a higher adsorption capacity. The problem here is the local availability of these materials. If the use of these "efficient" P-sorbing materials implies high transportation costs in addition to the purchase costs, then there is chance for this option not to go under serious consideration and to be often rejected. After all, the basic concept of CW systems is to use easy-to-find local materials and at reasonable cost, to avoid high expenses. Moreover, from a technical point of view, the use of special materials to enhance the P or other constituent removal should be clearly justified. It also has to be mentioned that the search for new specific materials is limited not only for domestic/municipal wastewater treatment applications but also for other types of wastewater, possibly with higher pollutant loads.

Stefanakis and Tsihrintzis (2012a) tested two materials (*zeolite*, *bauxite*) with special characteristics which could potentially improve the removal of ammonia nitrogen and P, respectively, in VFCW systems. Natural zeolite, e.g., clinoptilolite which is a common and abundant natural rock, possesses a high exchange capacity for ammonium cation (NH_4^+), and thus, appears as an appropriate media for ammonium removal from wastewater (Drizo et al., 1999; Stefanakis et al., 2009a; Stefanakis and Tsihrintzis, 2012a,b). Bauxite, on the other hand, is rich in hydrated aluminum and ferric oxides and low in alkali metals, alkaline earths, and silicates, which makes it a potential adsorbent for P (Drizo et al., 1999; Stefanakis and Tsihrintzis, 2012a,b). These authors found that when zeolite is placed as a porous media layer in intermittently loaded VFCWs, the relatively fast vertical gravitational drainage does not allow for a sufficient contact time between the wastewater and the media grains which would result in a respective removal rate increase. Thus, the use of these special materials in these systems was not suggested by these authors, since no significant difference was found between special materials and common gravel material. Luederitz and Gerlach (2002) also arrived at the same conclusion using mixtures of *sand* with *clay*; these authors also reported a decrease in P removal (from 44% to 27%) after 6 years of operation. Brooks et al. (2000) pointed out the importance of sufficient contact time between the wastewater and the porous media for increased P removal. These authors used *wollastonite* (a calcium metasilicate mineral abundant in New York, USA) in vertical flow columns and showed that there is a direct relationship between the residence time and the soluble P removed. Prochaska et al. (2007) reported promising results concerning P removal (39-64%) when sand is used as substrate in VFCW units; however, no improved P removal rate was reported for beds containing a 10:1 mixture of *dolomite and sand* as substrate compared to similar units containing *sand*.

A possible alternative is using these materials with a finer grain size, in order to reduce the drainage time and increase the reactive surface; however, this can negatively affect the hydraulic conductivity of the bed and result in a faster clogging of the bed. This means that the grain-size distribution is also a crucial parameter in the selection of the porous media. Therefore, a nominal high P-sorbing capacity of a media measured in the laboratory (e.g., batch column experiments) does not necessarily imply a respective performance in real operating conditions of a VFCW system, but needs to be combined with tests of proposed materials in operating VFCW beds in real conditions to get safer conclusions. On the other hand, in HSF CWs where sufficient residence time is usually applied, the use of these materials probably will result in improved efficiency (Luederitz et al., 2001).

Commonly used materials in VFCWs—and generally CWs—are sands and gravel which are locally available across the world. At first, there was no significant attention given to the choice of the media, e.g., to its sorbing capacity, since the primary purpose of the media was the stability of the bed and the assistance of the reed establishment and growth. Gradually, the P removal processes became more and more understood (Reddy et al., 1999) and it was clear that the selection of an appropriate media could represent an important parameter for improved P retention. At present, many different porous media have been suggested for use in CWs. Most of the studies found in the literature concerning porous media investigation, focus on the nominal P-sorbing capacity measured in the laboratory using, e.g., batch or column experiments. Thus, pilot- or full-scale applications of special porous media are limited, while the majority of them have to do with horizontal flow systems.

Sands of different origins and respectively different P-sorbing capacities have been investigated in Denmark (Arias et al., 2001). These experiments showed that the most important parameter is the Ca content which favors P precipitation. However, the predicted useful life time of Ca in full-scale facility is not expected to exceed 1 year more or less. Brix et al. (2001) included in their investigation not only different sands, but also LECA (light-expanded-clay-aggregates), diatomaceous earth, vermiculite, crushed marble, and calcite. The last were found to possess high P-sorbing capacities (up to 25 kg P/m^3), while the authors suggest that mixing these materials with sand or gravel could significantly improve the P removal capacity of the bed.

Pant et al. (2001) used Lockport *dolomite*, Queenston *shale*, and Fonthill *sand* as substrate media in a VFCW system in Canada and found that untreated sand could be a better media compared to the other two for P removal from wastewater. They also formulated the ascertainment that an increase in P-sorbing capacity does not necessarily ensure higher P removal rates, while many of the physicochemical parameters of the media (e.g., maximum P sorption, retention capacity) are subjected to alterations during the system operation. Similarly, Farahbakshazad and Morrison (2003) investigated the use of LECA as substrate in vertical planted upflow columns. Although this media showed improved equilibrium adsorption characteristics, no significant improvement of P removal compared to sand was observed under operating conditions. Another study tested the use of a mixture of *dolomite* and *sand* (1:10 w/w) as filter media in VFCW systems (Prochaska and Zouboulis, 2006). After 3 months of operation, P removal rate reached 45%, while the predicted time for media saturation was 5-7 months, depending on the loading rate. The P accumulation (6-18%) increased with the depth and the applied load. The same authors also reported the interesting observation that the majority of P is removed in the lower layers of the VFCW bed (Prochaska and Zouboulis, 2009).

Zeolite appears as one of the most tested porous media in VFCWs, alone or in a mixture with other materials, mainly for ammonia removal improvement. Huang et al. (2013) compared a VFCW system filled with natural zeolite with another system containing volcanic rock for piggery wastewater treatment and effluent recirculation; they reported better performance concerning ammonia removal (up to 94%) in the zeolite bed. An interesting observation of these authors has to do with the selectivity and preference of the zeolite over ammonium-oxidizing archaea and of the volcanic rock over ammonium-oxidizing bacteria. This preference toward ammonium-oxidizing prokaryotes is considered to affect ammonium removal in VFCWs. Increased (by 50%) ammonia nitrogen removal with zeolite presence was also reported by Wen et al. (2012) for a set of VF-HSF system, as also for a two-stage VF system (Canga et al., 2011). Zeolite was also used as substrate for olive mill wastewater treatment in Turkey and reported results for ammonium and OP were 37-49% and 95%, respectively (Yalcuk et al., 2010). Use of zeolite as single filter media was also proposed by Zhou et al. (2012), instead of a mixture of zeolite with dolomite, with reported ammonium removal rates improved by up to 36%. Nitrogen removal rates of 70% were also reported by Zhang et al. (2007) for zeolite and ceramic filter media. The same authors tested other materials and found that each could be used for specific pollutant removal, e.g., anthracite and steel slag for organic matter (70%) and P removal (60% and 90%, respectively). Bruch et al. (2011) also reported that mixing zeolite and lava sands (instead of common fluvial sands) could be a highly efficient filter media.

However, when the target pollutant to be removed is P, the use of other materials with different characteristics is usually preferred. A VFCW bed containing *marble stones*, *sand*, *lime*, and *gravel* was found to provide an adequate removal of P (60-90%) when treating primary effluent (Ayaz et al., 2012a). Wu et al. (2011b) investigated *anthracite* (a rock with a high carbon content), *steel slag* (an industrial by-product rich in Si and Ca oxides), and *vermiculite* (Si, Mg, and Al oxides) as substrate in VFCWs, alone and in various mixtures. Anthracite and steel slag were found effective in P removal (77% and 90% for TP, respectively) when treating domestic sewage, while anthracite showed an improved long-term useful operational lifetime.

Dewatered alum sludge is another material often used as substrate in VFCWs. Alum sludge from a conventional wastewater treatment plant has been used as substrate in VFCWs for dairy (Babatunde and Zhao, 2009), farmyard (Zhao et al., 2008), and animal (Zhao et al., 2011a) wastewater treatment. It is reported that this material possesses good hydraulic characteristics and shows high P removal (up to 90% of the applied load). Apparently, the high content of aluminum favors the adsorption of P, which makes this material appropriate for use in VFCWs. Lime and iron drinking water treatment residuals (i.e., *lime and iron sludge*) have also been evaluated as a potential substrate

(Leader et al., 2005). The first material has a high content of calcium carbonate and the second of iron and dissolved organic carbon. Both materials provided a mean P removal of 95% when treating a secondary municipal wastewater (TP influent 1.00 mg/L) and of 53% for anaerobically digested dairy wastewater (influent TP 48.5 mg/L). An additional advantage is that the tested materials, after their saturation with P, were used as agricultural soil amendments and fertilizers in agricultural lands.

Organic wood mulch has also been tested in both VFCWs and HSF CWs (Saeed and Sun, 2011a). Compared to gravel, the VFCW bed with organic mulch substrate showed an improved performance concerning TP removal (60%) and ammonia nitrogen (99.6%) and is considered as another potential substrate for VFCWs. Interestingly, when used in HSF CWs, the efficiency was not as high as for the gravel media.

Oyster shell has been proposed as a possible substrate for use in VFCWs to enhance P removal. This material is produced in large amounts in Asia, where the oyster market is rather big. The idea of using oyster shells in CWs could eliminate the cost for its management and disposal, and provides a possible P-sorbing substrate, since it has a high content of Ca (Zhen et al., 2013). Park and Polprasert (2008) compared its adsorption capacity with that of dewatered alum sludge and found that it is higher for the oyster shell. When used in VFCWs, P removal efficiency reached a mean of 96% and its adsorption capacity was measured equal to 24 g/kg, almost double than that of alum sludge. The useful operational time of this substrate was also found to be more prolonged. Oyster shell was also used as substrate in VFCW for swine wastewater treatment with similar promising results, even compared with other effective materials like volcanics, zeolite, and broken bricks (Zhen et al., 2013; Wang et al., 2013). In all cases, Ca-P form was the most abundant P form in the system.

Blast furnace slag is an industrial porous, nonmetallic by-product of the steel and iron industry with a potential high P-sorbing capacity (Sakadevan and Bavor, 1998; Rustige et al., 2003). Korkusuz et al. (2005) compared this media with common gravel material as substrates in VFCW systems treating domestic wastewater. Results showed that the slag-bed achieve on the average 45% removal of P, significantly higher than that of the gravel-bed. Ca-P and loosely bound P were found to be the predominant P-forms in the slag (Korkusuz et al., 2007a). Same conclusions are also drawn by Li et al. (2013). These authors tested steel slag in a planted VFCW system comprising a downflow and an upflow compartment. Overall P removal reached 80% (inlet loadings of 12.2-36.8 $g/m^2/d$), while the authors suggested an HRT of 1 day, indicating—as previously mentioned—that the contact time between wastewater and media affects the P removal efficiency. The effectiveness of *steel slag* in P removal (more than 90%) in VFCWs has also been reported by Zhang et al. (2007) and Wu et al. (2011b), compared to other tested media (e.g., anthracite).

Hydrated oil shale ash is another potential CW substrate material; it is a Ca-rich solid waste produced after the burning of organic-rich sediment oil-shale at large power plants. Particularly, in Estonia, thermal power plants are the only power plants in the world that burn pulverized kerogenous highly calcareous oil shale (calcite and dolomite content 40-60%; Liira et al., 2009). VFCWs containing this media treated the effluent from an activated sludge treatment plant and water from the aerobic-anoxic pond with reported P removal rates of 77% and 94%, respectively, higher that similar beds filled with *peat* (Kõiv et al., 2009). In another experiment, shale was mixed with *polypropylene pellet* in a VFCW system to treat eutrophic river water (Tang et al., 2008), which enhanced the removal rate of all pollutants (organic matter, nitrogen, and phosphorus) compared to the shale-bed. Physicochemical characteristics of *broken bricks* have also been tested in order to use this material in VFCWs (Wang et al., 2012a). A VFCW bed filled with this media treated domestic wastewater; results showed more than 90% removal of P, indicating that such a media could be used for adsorption and precipitation of P.

It is obvious that the issue of finding an appropriate filter media is still of top priority in the research field. The main goal is to identify media with high P-sorbing capacity and an acceptable useful lifetime until saturation. As described above, there is a large variety of different media to choose, and most of them seem to be effective. However, as mentioned, the final choice will always be based not only on the sorbing capacity but also on the predicted useful life time of the material and also, on the local availability of the material (which, respectively, correlates with the expected costs). After all, this is the reason why so different and sometimes unusual materials are tested and investigated, depending on the media availability and abundance in each region.

6.3.3.1 Filter Units with Porous Media

As already mentioned in Section 6.3.2, another alternative to enhance P removal in CWs is the use of these specific materials as filter media in separate filter units treating the CW effluent. Arias et al. (2003b) investigated a separate filter filled with calcite under real operating conditions (residence time applied up to 99 min), and reported a decrease of P concentration of up to 75%. They concluded that the presence of a separate filter unit with a material with a high P-sorbing capacity could improve the P removal rate in the VFCW system. It was clear that the choice of a filter media with a high P-sorbing capacity which could prolong the useful lifetime of the filter (before the media becomes saturated) was the crucial point in this configuration.

Stefanakis and Tsihrintzis (2012b) used additional filter units to further treat the effluent of a VFCW bed, in order to provide an integrated treatment system for wastewater treatment. These filters contained carbonate material, zeolite, bauxite, and a mixture (50-50%) of zeolite-bauxite, while two different residence times (1 and 2 days) were tested. Results showed that the addition of these filters containing specific media could significantly improve the combined VFCW and filter unit system performance. It was found that zeolite and bauxite enhanced the removal of ammonium and P, respectively, while the filter with the mixture of these two materials was found as the preferred case, providing simultaneous optimum removal of all compounds (98%, 95%, and 71% removals for organic matter, nitrogen, and phosphorus, respectively) when treating a strong municipal wastewater. Moreover, no significant differences were observed between the two residence times tested (1 and 2 days), thus 1 day appeared adequate. This combination (pretreatment in settling tanks, single VFCW bed, and a zeolite filter) was also tested in an onsite VFCW system for domestic wastewater treatment (8 pe) and a similar performance was observed (Gikas and Tsihrintzis, 2012). The concept of this setup could represent an ideal alternative for onsite treatment, taking into account that these filter units are simple in construction and the replacement of the media after its saturation could also easily take place.

6.3.4 Removal of Pathogenic Microorganisms

Compared to the other pollutants, research on the removal of pathogenic microorganisms from wastewater in CWs is more limited. Integrated, systematic studies concerning the investigation of the effects of the various CW parameters on the system performance (pathogen removal) are limited. Most of the published research usually focuses only on the potential for public health risks and simply report some average values of indicator organisms (*Escherichia coli*, coliform bacteria) from the system effluent. Common indicators used are total and/or fecal coliforms (TC and/or FC) measured as colony-forming units (CFU) or most probable number. These techniques provide information on the density of the bacteria population and not any insight to community structure. For the latter, it is only during the last decade that new molecular fingerprinting techniques have been developed, e.g., FISH detection and PCR-based techniques.

Thus, it is still difficult to formulate a general conclusion over the optimum design and setup and its effectiveness. This means that it is still difficult to evaluate the role of parameters like flow orientation, the loading rate, the applied residence time, the porous media, the vegetation, etc., but only some general trends from available studies could carefully be formulated. Moreover, as already mentioned in Section 5.7 of Chapter 5, there are many gaps in fundamental knowledge on the pathogen removal processes and the fate of these microorganisms in CWs. Especially for VFCWs, their effectiveness on this topic is far less studied compared to horizontal flow systems (Vymazal, 2005b), and it is only during the last decade that some results are published. Performance of various VFCWs in many countries concerning the removal of pathogenic microorganisms shows that a decrease of the pathogenic microbial population takes place in these systems. Table 6.6 presents a summary of the performance of various VFCW systems concerning the elimination of pathogenic bacteria. Many studies report an almost complete removal of common

TABLE 6.6 Performance of Various VFCW Systems in Different Countries in the Removal of Bacteria Indicators (TC = Total Coliform, FC = Fecal Coliform, FS = Fecal Streptococci)

References	System Setup	Influent Indicators and Concentrations	Effluent Concentrations	Removal Rate
Vandaele et al. (2000)	Two-stage VFCW	FC, FS, clostridia		95.3%, 99%, 99%
Meuleman et al. (2003)	VFCW	*E. coli*:1.4-7.9 × 10^8 CFU/L	0-42 CFU/L	>99% >99% Decrease > 2 log
		F-specific RNA bacteriophages: 1.4-10.7 × 10^6 CFU/L	<20 CFU/L	
Weedon (2003, 2010)	single VFCW with sand (80 cm depth)	FC: 1.6 × 10^7 CFU/100 mL	2.8 × 10^3 CFU/100 mL	4-log reduction Low, high loading rates: >3 log, 2 log
Ausland et al. (2002)	Unplanted filters with sands/LWA	Total bacteria: 6.8 log/mL	4.5-5.1 log/mL	1.6-2.3 log/mL
		FC: 6.3 log/mL	0.0-3.5 log/mL	2.9-6.3 log/mL
		FS: 2.9 log/mL	0.0-0.4 log/mL	2.5-2.9 log/mL
Vacca et al. (2005)	Four VFCWs with/without plants	TC: 8.2 log (CFU/100 mL)	5.9-6.8 log	>99%
		Heterotrophic bacteria: 8.3 log	6.5-7.4 log	
Arias et al. (2003a)	Two-stage VFCW with pre-treatment and calcite-filters	TC: 4-84 × 10^6 CFU/100 mL	0.001-0.01	99.5-99.9%
		FC: 4.5-56 × 10^6 CFU/100 mL	0.001-0.04	
		FS: 190-712 × 10^3 CFU/100 mL	0.2-6	

Ulrich et al. (2005)	1. Septic tank-VF-HSF	*E. coli*, TC: 10^6-10^7/100 mL Enterococci, coliphages 10^5/100 mL Campylobacters/acrobacters: 4.2×10^6/100 mL *Clostridium perfringens*	10^4/100 mL 1.2×10^5/100 mL	VF: 1.5 log 2 log 1 log
	2. Lagoon-VF	*E. coli*, TC: 10^6-10^7/100 mL Enterococci, coliphages: 8.7×10^4 and 2.6×10^4/100 mL Campylobacters/arcobacters: 4.3×10^5/100 mL *Clostridium perfringens*	10^3-10^4/100 mL 10^3/100 mL First VF 5×10^3/100 mL 2nd VF 1×10^3/100 mL First VF 3.5×10^2/100 mL Second VF 1.5×10^2/100 mL	VF: 2 log 1.5 log 1.5-2 log
Keffala and Ghrabi (2005)	Two VF-HSF systems Planted and unplanted	TC: 1.04-1.03 × 10^7/100 mL	1.53-1.94 × 10^5/100 mL	3.87 log
		FC: 8.35×10^5- 4.43×10^6/100 mL	4.06-4.24 × 10^4/100 mL	3.26-3.73 log
		E. coli: 9.24×10^6/100 mL	9.25-2.16 × 10^4/100 mL	4.7-3.92 log
Alexandrino et al. (2007)	Primary treatment-VF-HSF			VF: 1.7-2.1 log/10 mL
Sleytr et al. (2007)	Ten VFCWs (eight planted, two unplanted)	TC: 6.99 log (CFU/mL)	2.69	4.31
		E. coli: 6.59	2.24	4.35
		Enterococci: 6.06	1.26	4.80
		Heterotrophic plate count (HPC): 6.22	3.37	2.85

Continued

TABLE 6.6 Performance of Various VFCW Systems in Different Countries in the Removal of Bacteria Indicators (TC = Total Coliform, FC = Fecal Coliform, FS = Fecal Streptococci)—cont'd

References	System Setup	Influent Indicators and Concentrations	Effluent Concentrations	Removal Rate
Wand et al. (2007)	Lab-scale VFCWs	*E. coli*: 10^8 cells/mL	n.d.—0.3×10^2	3-4 log (>99%)
Gross et al. (2007) and Sklarz et al. (2009)	Recycled VFCW with UV	FC: 5×10^7 CFU/100 mL	10,100 CFU/100 mL 2×10^5 CFU/100 mL	99%
Puigagut et al. (2007)	VFCWs			1.5 log
Masi et al. (2007)	VFCWs	FC: $2.9\text{-}4.2 \times 10^6$ CFU/100 mL	$1.3 \times 10^3\text{-}3.1 \times 10^2$	3-5 log (99%)
		E. coli: $3.4\text{-}6.6 \times 10^6$ CFU/100 mL	$1.1\text{-}1.8 \times 10^2$	
		TC: $4.9\text{-}8.0 \times 10^6$ CFU/100 mL	$3.8 \times 10^3\text{-}3.2 \times 10^4$	
Tsihrintzis et al. (2007) and Tsihtintzis and Gikas (2010)	VFCWs	TC: 4.8×10^7 CFU/100 mL	5×10^4	99.9%
Torrens et al. (2009b)	Six VFCWs	FC: 4.7 (log CFU or PFU/100 mL)	3.1-4.0	0.5 log
		E. coli: 4.4-4.5	2.9-3.9	2 log
		Somatic coliphages: 4.1	2.9-3.5	0.4-1.5 log
		F-specific bacteriophages: 3.6	2.9-3.5	0.2-1 log
Zurita et al. (2009)	VFCWs	TC: 4700×10^3 MPN/100 mL	68-88	97-96.8%

Redder et al. (2010)	Two systems VF-HSF	107 cfu/100 mL (*E. coli*) 150 *Giardia* cysts/100 L 14 *Cryptosporidium* oocysts/100 L		Reduction factors up to 5 with a range of 0.5-5.9
	One system VF-VF	523 *Giardia* cysts/100 L 147 *Cryptosporidium* oocysts/100 L	First stage: <1 *Cryptosporidium* oocyst	*Cryptosporidium* spp. 1.80 log *Giardia* spp. 1.92 log protozoan parasites: 3 log
Olsson (2011)	VFCWs	*E. coli*: 7.6×10^6 MPN/100 mL	1.1×10^5-2.5×10^6	0.7-2.2 log
Zurita et al. (2011)	VFCWs	TC: 4700-7800 $\times 10^3$ MPN/100 mL		96-99.3%
Gikas et al. (2011)	VFCW	TC: 63.8×10^6 CFU/100 mL	0.011×10^6	99.9%
Gikas and Tsihrintzis (2012)	VFCW (onsite treatment)	TC: 198.3×10^5 CFU/100 mL	0.07×10^5 CFU/100 mL	99.97%
García-Pérez et al. (2011)	Recirculating VFCW	FC: 3.0×10^6 CFU/100 mL	34.4×10^3 CFU/100 mL	99.9%
Abou-Elela and Hellal (2012)	VFCW	TC: 2.8×10^7 MPN/100 ml FC: 2.3×10^6 MPN/100 ml *E. coli*: 2.4×10^6 MPN/100 ml	2.6×10^3 MPN/100 mL 1.25×10^3 MPN/100 mL 1.11×10^3 MPN/100 mL	94-99.99%

bacteria indicators, indicating the possibility of further use of the treated effluent; there are many published studies reporting the reuse of the effluent for irrigation. However, this option should always take into account national and international limits defined by legislation, while a more detailed study and monitoring of the system performance with time will provide more safe results.

Meuleman et al. (2003) reported the overall efficiency of one of the oldest VFCW beds in the Netherlands (constructed in 1975), which treated domestic wastewater and wastewater from recreational activities after primary treatment in a sedimentation tank. Although monitoring of the bacterial population was not made on a regular basis, the removal efficiency of the system was always >99% for *E. coli* and F-specific RNA bacteriophages, with respective influent values of $1.4\text{-}7.9 \times 10^8$ and $1.4\text{-}10.7 \times 10^6$ CFU/L.

Arias et al. (2003a) studied the removal of indicator bacteria in a two-stage VFCW system with pre- and post-treatment in a calcite-filter unit in Denmark. They reported that the first VFCW bed had the highest efficiency (1.5 log, 1.7 log, and 0.8 log cycle removals for TC, FC, and fecal streptococci (FS), respectively). Overall, the elimination of bacteria in the system was more than 99%. A system (septic tank, VFCW bed, and zeolite filter) constructed in Greece showed significant reduction of TC (influent—effluent values: 198.3×10^5 CFU/mL and 0.07×10^5 CFU/mL, respectively), with the major part of TC removal taking place in the VFCW bed (Gikas and Tsihrintzis, 2012). Compact single-stage VFCW beds for domestic wastewater treatment showed an efficiency of 3-4 log cycles for the removal of FC, which was reduced to 2 log cycles for higher loading rates (Weedon, 2003, 2010). Full-scale VFCW facilities in Greece also show significant elimination of pathogenic bacteria without use of additional disinfection techniques (Tsihrintzis et al., 2007; Tsihrintzis and Gikas, 2010; Gikas et al., 2011).

Similar removal rates are also reported by Ulrich et al. (2005) for two VFCW systems. The one comprised pre-treatment (septic tank), a VFCW bed, and an HSF CW bed, and the other, a lagoon and two-stage VFCWs. Removal of *E. coli*, coliform bacteria, enterococci, and campylobacters/arcobacters varied between 1.5 and 2.5 log cycles, for influent values of $10^5\text{-}10^6$/100 mL. The same efficiency is also reported for *Giardia* cysts, *Cryptosporidium* oocysts, and *Clostridium perfringens*. The authors stated that this satisfying performance of CW systems overcomes conventional activated sludge systems. Similar performance (1.5 log cycles reduction for FC) was also reported for VFCWs in Spain (Puigagut et al., 2007). Higher removal rates (3-5 log cycles) were reported by Masi et al. (2007) for VFCWs in Italy. Keffala and Ghrabi (2005) also investigated a combined system of VFCW and HSF CW (planted and unplanted), and reported a 3-4 log cycle removal for TC, FC, and *E. coli*, while no significant differences were observed between planted and unplanted systems. A similar system (primary treatment-VFCW-HSF CW) was tested by Alexandrino et al. (2007). These authors evaluated the bacteria removal efficiency by both the cultivation and the microscopic count method. The vertical system was found to possess a higher and more constant performance (1.8-2.1 log cycle/100 mL) compared to the horizontal bed (0.2-0.3 log cycle/100 mL), when the microscopic enumeration was used. On the other hand, the standard cultivation method gave similar performances (2-2.2 log cycle/100 mL). This indicated that there was a decrease in the bacteria culturability in the HSF CW treatment, which resulted in an overestimation of the actual bacteria concentration.

Abidi et al. (2009) also investigated the performance of hybrid systems (one VF-HSF and one HSF-VF) in Tunis. They reported that the major part of bacteria removal takes place in the first stage, while no significant differences were found between the two system removals (TC: 1.98-2.07 log cycles, FC: 1.91-1.78 log cycles, *E. coli*: 2.12-2.13 log cycles). The same is also reported for a three-stage system (VF-HSF-VF) in Turkey, where almost 95% retention of FC and TC took place in the first VF stage (Tunçsiper et al., 2012). Significantly, high elimination of TC (up to 99.3%) was also reported for VFCWs in Mexico, which was found higher compared to respective efficiency of HSF CWs (Zurita et al., 2011). Up to 99.99% removal of bacteria indicators (TC, FC and *E. coli*) were reported for a VFCW system in Egypt (Abou-Elela and Hellal, 2012).

Ausland et al. (2002) investigated the removal of FC and FS in intermittently loaded with septic tank effluent pilot filters containing natural sands and three different types of LWA. Removal rates varied between 2.9 and 6.3 log cycles, while FC removal was 3 log cycles higher in filter with fine-grained sand, as also in the filters which were dosed under pressure and not with gravity. A positive correlation was also found between the FC removal and the residence time applied.

Vacca et al. (2005) tested the removal of bacteria in two planted and two unplanted VFCWs (as well as HSF CWs) filled with expanded clay and sand treating domestic wastewater. The bacteria removal rate reached 1.5-2.5 log cycles, which corresponds to >99% removal. This study showed the effects of different filter media, plant presence, and flow orientation on the microbial community and gave a better insight into the microbial structure using molecular profiling methods (PCR-SSCP). Sleytr et al. (2007) reported a high efficiency of 4.31 log cycles for TC, 4.35 log cycles for *E. coli*, 4.80 log cycles for enterococci, and 2.85 log cycles for heterotrophic plate count (HPC). They also found that the majority of the removal takes place in the first 20 cm of the upper layer, while again no significant differences were reported between planted and unplanted beds. The same observation was also reported by Wand et al. (2007), with similar *E. coli* removal efficiency (>99%). This study gave a better insight into the bacteria removal processes; adsorption was found to be of no significant importance in *E. coli* removal from domestic wastewater, while predation by protozoa and biolytic processes of *Bdellovibrios* play the dominant role in the removal. A study by Redder et al. (2010) reports the removal of pathogenic protozoan parasites which are attracting increasing interest due to their medical and economic effects. Elimination of *Cryptosporidium* and

Giardia spp. reached a 2 log cycle reduction and of *E. coli* up to 5 log cycles in various systems (VF with recirculation, VF-HSF). The two-stage systems were the most efficient for both bacteria and protozoan indicator organisms.

More than 99% removal of FC was also reported for VFCW systems with effluent recirculation (Gross et al., 2007; Sklarz et al., 2009). *E. coli* concentration reduction reached 3 log cycles and, after further treatment with a UV unit, the final concentration measured 10 CFU/100 mL. This effluent is appropriate for urban landscape irrigation. Another VFCW system with effluent recirculation for onsite wastewater treatment, planted with maize, also showed significant (99.9%) FC removal (García-Pérez et al., 2011, 2013). The use of graywater treated in recirculating VFCWs was evaluated for yard irrigation, by monitoring various types of pathogens and FC bacteria (Benami et al., 2013). Results of this study showed that the treated graywater, and soils irrigated with this water and freshwater, contained comparable pathogen types, indicating that irrigation with graywater may not be completely responsible for bacteria detected in yard soils.

Torrens et al. (2009b) investigated the removal of bacterial and viral indicators in VFCWs, as also the effects of plant, filter media characteristics, hydraulic load, and HRT. Reported efficiency from this study was lower compared to other previous studies (up to 2 log cycles). It was found that the bacteria removal is related to the water retention time in the bed (i.e., bed depth) and the hydraulic load, while no significant difference was found between planted and unplanted beds and between the two different sands used. A larger experiment of six pilot VFCWs with different characteristics showed a bacteria elimination of up to 2.2 log cycles (Olsson, 2011). The use of sands instead of gravel as the main filter media was found to enhance the system efficiency, while no significant difference was reported under the presence of plants. The major part of removal took place in the upper layers of the bed.

6.3.5 Effects of Vegetation

The exact role of plants in CWs remains even today a discussion topic. As mentioned in Section 4.1.3 of Chapter 4, the positive effects of plant presence in CWs have been generally recognized and defined for many years now (Brix, 1994b, 1997). The main effects of plant presence are physical (Brix, 1994b, 1997): they stabilize the bed surface and matrix, promote physical filtration, prevent clogging phenomena (especially in VFCWs), and provide high attachment area for microbes. Additionally, plants also contribute to the pollutant removal by the direct uptake of nutrients, the absorbance and accumulation of heavy metal, and other substances from wastewater, as also by the transfer of oxygen to the rhizosphere aiding the growth and reproduction of aerobic microorganisms. Thus, they simulate aerobic decomposition of OM and the growth of nitrifying bacteria. Nevertheless, regarding the microbial processes for OM degradation, the role of plants seems of more limited importance

(Tanner, 2001; Stottmeister et al., 2003; Akratos and Tsihrintzis, 2007). Furthermore, another question that also concerns the scientific community over the last years has to do with the possible different effects in removal efficiency that different plant species may have (Brisson and Chazarenc, 2009). Generally, the contribution of plants to the various pollutant removals in horizontal flow systems has been investigated more and there are published studies reporting the effects of plant presence (e.g., Karathanasis et al., 2003; Akratos and Tsihrintzis, 2007; Caselles-Osorio and García, 2007).

Respective investigation in VFCWs is far more limited. It is expected that in VFCWs, the vertical downward movement of the water should enhance plant-mediated processes (e.g., nutrient uptake and removal) compared to horizontal flow systems, due to increased direct contact between the wastewater and the rootzone, taking into account that in VFCWs, more than 50% of the root biomass is located in the upper 5 cm of the bed (Breen and Chick, 1995). These observations are indications of the possible effects of plant presence in VFCWs. However, reported results from various latest studies give a mixed picture compared to this assumption. Some conclusions can be drawn from studies where planted and unplanted units are compared, as also beds with different plant species.

6.3.5.1 Removal of OM, N, and P and Other Compounds

Many studies report improved performance in planted compared to unplanted VFCWs. Heritage et al. (1995) compared four planted VFCW beds with different plant species (*Typha orientalis*, *Schoenoplectus validus*, *Baumea articulate*, and *Cyperus involucratus*) and one unplanted. Removal of BOD_5, SS, and TKN was higher in the planted beds, while no significant differences were observed for phosphorus. *Baumea* and *Cyperus* were also found to be more tolerant species to various salt levels. Improved TN and TP removal by planted VFCWs with *Cyperus alternifolius* was reported by Cui et al. (2009, 2011). Increased TN removal was also observed by Cui et al. (2010) for a VFCW planted with *Canna indica* compared to unplanted beds. Olsson (2011) reported significantly higher removal of ammonia nitrogen in a planted bed with *P. australis* and sand as substrate, than in an unplanted one. Higher efficiency for COD and nutrient removal was observed for planted upflow VFCWs compared to unplanted beds for the treatment of industrial wastewater (Ong et al., 2010). In this study, the Manchurian wild rice (*Zornia latifolia*) showed higher removal rates for TP, TN, and NH_4^+ -N than the common reed (*P. australis*).

Stefanakis and Tsihrintzis (2012a) found increased removal rates for OM, N, and P in planted VFCW units compared to unplanted ones; however, no significant differences were observed between VFCWs planted with *P. australis* and *Typha latifolia*. VFCWs in Nepal planted with *Phragmites karka* also performed better in OM and N removal compared to unplanted beds, while no difference were found for P, although the plants were not harvested in these systems (Pandley et al., 2013). COD and N removal were higher in VFCWs planted with *P. australis* and operating with intermittent aeration combined with step-feeding strategy (Fan et al., 2013). Ouattara et al. (2011) also observed higher removal efficiencies for all pollutants in planted beds with *Panicum maximum* than in unplanted ones.

A positive effect of plant presence in VFCWs is also presented for pharmaceuticals and PPCPs removal from urban wastewater (Matamoros et al., 2007). The VFCW systems proved to be more resistant to higher loads and more efficient in the removal of these compounds compared to unplanted sand filters and horizontal subsurface flow systems. Planted VFCWs for the removal of molybdenum (Mo) from wastewater were also proved more efficient, while *T. latifolia* showed higher Mo uptake capacity compared to *P. australis* (Lian et al., 2013).

Chang et al. (2012) compared two integrated VFCWs with two different plant species: *T. orientalis* and *Arundo donax*, and *C. indica* and *Pontederia cordata*. These combinations, however, did not result in significantly different performances. Improved efficiency for VFCWs planted with *Amaranthus hybridus* than in unplanted beds was also reported by Coulibaly et al. (2008) for COD, N, and P removal. These authors also reported that increased plant density is correlated with respectively enhanced removal capacity. Besides the comparison of different plant species, the use of more than one species in one system has also been proposed. Zhu et al. (2010) run an experiment in VFCWs for domestic wastewater treatment, where they used from 1 up to 16 different plant species with various combinations in several beds. The authors report that the increased biodiversity in VFCWs resulted in increased produced plant biomass and respectively enhanced N removal.

Kantawanichkul et al. (1999) compared two species (*Vetiveria zizanioides* and *Cyperus flabelliformis*) in VFCWs under tropical climate conditions (Thailand) and they reported that both plants can be used for domestic wastewater treatment with no significant differences, since removal efficiencies were over 90%. Kantawanichkul et al. (2009) also compared planted and unplanted VFCWs for the treatment of high-strength wastewater under the same climate. In this study, planted systems (*C. involucratus* and *Typha angustifolia* were used) showed significantly higher removal rates for N and P, higher oxygen transfer rates, and higher quantities of ammonia-oxidizing bacteria, while COD mass removal did not differ between planted and unplanted beds. Abou-Elela and Hellal (2012) compared three plant species in VFCWs for the treatment of municipal wastewater and concluded that *Canna* performed better that *P. australis*.

Keffala and Ghrabi (2005) investigated two hybrid systems (VF-HSF), one planted and one unplanted. Results of this study showed that the planted system was more efficient in nitrogen removal (however, there was no statistically significant difference between planted and unplanted VFCWs), while oxidized nitrogen removal (nitrate, nitrite) was higher in the unplanted. In another system consisting of one unplanted VFCW, one planted HSF CW and one planted VFCW, planted beds were found more efficient (Tunçsiper, 2009).

On the other hand, there are studies for which the differences with and without plant presence were found not significant. Scholz and Xu (2002a) reported similar performances for planted and unplanted systems concerning the removal of BOD_5, turbidity, SS, Pb, and Co. Lee and Scholz (2007) found a negative impact of the presence of *P. australis* on BOD_5 removal, while only N removal was enhanced. Tietz et al. (2007b) also did not find significant difference in the treatment efficiency of planted and unplanted VFCWs, while they observed similar quantities of microbial biomass in all systems. No differences were also found between tidal flow (TF) unplanted and planted with *C. alternifolius* VFCWs (Chan et al., 2008); plants were not harvested, however. Zhao et al. (2010) found significant difference between planted (with *Lythrum salicaria*) and unplanted VFCWs only for P removal and not for OM and N. The performance of a recirculating VFCW was also not significantly affected by the presence of plants (Sklarz et al., 2011).

6.3.5.2 Bacteria Removal

Although for OM, N, and P removal, the general picture concerning the role of plants is still not totally defined, the effects of plant presence on the elimination of pathogenic bacteria seem to be clearer. Most of the existing studies in the literature report that differences between planted and unplanted VFCWs are not significant and both setups show similar performances. Vacca et al. (2005) did not found any significant difference in the removal efficiency of pathogenic bacteria (TC, FC) between planted and unplanted VFCWs units. The authors reported that effluent from units planted with *P. australis* did not affect the elimination of enteric bacteria. Keffala and Ghrabi (2005) did not observe significant differences between planted and unplanted combined systems (VFCW and HSF CW) in the removal of TC, FC, and *E. coli*. The same conclusions are also reported by Sleytr et al. (2007) for TC, *E. coli*, enterococci, and heterotrophic bacteria, as also by Redder et al. (2010) for the removal of pathogenic protozoan parasites (*Cryptosporidium* and *Giardia* spp.) in various planted and unplanted systems (VF with recirculation, VF-HSF). Torrens et al. (2009a,b) also did not observe significant difference between planted and unplanted VFCW beds concerning the removal of bacterial and viral indicators. The presence of plants did not affect the removal of pathogenic bacteria in an experimental study with various VFCWs of different characteristics (Olsson, 2011).

6.3.5.3 Nutrient Uptake by Plants and Harvesting

It is known that plants uptake nutrients through their roots to cover their growth needs. An amount of nutrients can be removed by plant harvesting but it represents only a small portion of the nutrient load (Brix, 1997; Tanner, 2001; Stottmeister et al., 2003; Langergraber, 2005). Tanner (2001) reports an annual removal by plant harvesting of 2-8% for N and 1.9-5.3% for P, while this amount can be higher in horizontal flow systems than in vertical ones (Stottmeister et al., 2003).

Keffala and Ghrabi (2005) report 1% of N assimilation in a VF-HSF system. Meuleman et al. (2003) reported uptake values for *P. australis* of 15% of total N and 10% of total P input. The authors also state that reed harvesting in October compared to December could increase N and P removal. Langergraber (2005) reported that expected N and P plant uptake is about 1.9% for wastewater treatment, up to 10% of N, and 3% of P load for surface water treatment, while for graywater treatment, the uptake can reach even 46% of N load and 10% of P load, due to low nutrient content. Lee and Scholz (2007) reported that 17.5% of N could be removed by harvesting for a planted (*P. australis*) VFCW system receiving urban runoff. Stefanakis and Tsihrintzis (2012a) reported mean N and P uptake values in nine planted pilot-scale VFCWs of 1.33-4.05% and 0.17-1.05%, respectively. In the same study, *P. australis* was found capable of assimilating higher N and P amounts (3.89% and 1.05%) than *T. latifolia* (2.45% and 0.30%). Morari and Giardini (2009) reported higher values for plant contribution in N and P annual mass removal (>75% and 70%, respectively) for two VFCWs planted with *P. australis* and *T. latifolia*. It seems that during the first growing seasons (initial phase of system operation), these species possess a high storage capacity, as also mentioned by Lantzke et al. (1999). The same observation was also made by Cui et al. (2011), who measured a TP uptake of 17% by *C. alternifolius* in VFCWs which were harvested 4 times per year. Another VFCW planted with *P. australis* showed 7% uptake of N input load, when operated as the last stage in a three-stage CW system (Tunçsiper, 2009). Ouattara et al. (2011) reported 11.2% and 13.5% uptake of N and P, respectively, for *P. maximum* in a VFCW system treating domestic wastewater. Abou-Elela and Hellal (2012) reported higher uptake of N (68.1 g/m^2) and P (32.6 g/m^2) for *Canna* than for *P. australis*. Moreover, the removal of heavy metals by plant harvesting in VFCWs was found to be very small (<1%; Scholz and Xu, 2002a).

As it seems, plants do uptake nutrients from the wastewater but the extent of this process in VFCWs is more or less limited. Many studies indicate that high uptake values may appear during the start-up phase of the system. It seems that during the initial establishment of the plants, nutrient needs are higher. Generally, direct uptake and sequential harvesting should be considered as a secondary removal mechanism of these nutrients. In lightly loaded systems (e.g., tertiary treatment), removal of the aboveground plant biomass could be of more importance (Langergraber, 2005; Vymazal, 2007). It is also reported that a single plant harvesting in autumn, e.g., in October for the Northern Hemisphere, and not in winter months could potentially increase the amount of nutrients removed (Lantzke et al., 1999; Meuleman et al., 2003). An even earlier harvesting, e.g., in the summer, could negatively affect the long-term vitality of the reeds (Tanner, 2001), as also more than one harvesting events in 1 year (Lantzke et al., 1999). On the other hand, Brix and Arias (2005) suggested that reeds should not be harvested, since the dead plant material would act as an insulation layer during cold winter months. These authors also mention that removal of dead plant material in early spring could assist plant growth.

6.3.6 Effects of Different Aeration Configurations

Sufficient aeration of the VFCW bed is a crucial parameter for the overall effectiveness of the system. Availability of sufficient oxygen amounts is necessary for the various aerobic purifications processes (e.g., organic matter oxidation and nitrification) to take place. The presence of plants in CWs enhances the oxygen availability within the bed, since plants transfer oxygen and diffuse it in their rhizomes (Brix, 1997). However, the amount of this transferred oxygen is still a controversial issue, especially in cold climates where plant deactivation is more prolonged. As mentioned in Chapter 3, intermittent loading of VFCWs is usually applied, since, under this feeding regime, higher oxygen amount is entering the system. However, research for additional aeration supply and increased subsurface oxygen availability in VFCWs has come up with various design modifications, such as the use of vertical aeration pipes, artificial aeration with air pumps, recirculation of the treated effluent, or TF design.

6.3.6.1 Aeration Pipes

This is the most common configuration in VFCWs. Usually, the standard VFCW design includes the use of aeration pipes placed within the bottom drainage layer of the bed. These pipes are vertical, perforated, connected with the bottom drainage pipes which collect the effluent and also provide a connection of the deepest layers of the bed with the atmosphere; thus, they operate as a passive aeration supply (Brix and Arias, 2005; Stefanakis and Tsihrintzis, 2012a). As the bed is loaded with wastewater, the rapid downflow drainage and the passage through the bottom effluent pipe create a siphon and fresh air is sucked through the aeration pipes from the atmosphere (Green et al., 1998). This means that the volume discharged from the effluent at the bottom of the bed is replaced by an equal volume of fresh air introduced through the aeration pipes. The effectiveness and necessity of aeration pipes were revealed by the study of Stefanakis and Tsihrintzis (2012a). These authors compared two similar VFCW pilot-scale units which differed only in the presence/absence of aeration pipes. Results showed that the unit without aeration pipes had significantly lower performance concerning the removal of organic matter and nitrogen (decrease of 5% and 10%, respectively, compared to the unit with aeration pipes). Zhu et al. (2013) compared two VFCWs with and without aeration pipes. Aeration pipes in this system were placed in the center of the vertical distance between the surface and the bottom of the bed. The systems operated like batch reactors, i.e., they were intermittently filled and drained and between these actions the wastewater was left in contact with the filter media for a certain residence time.

The unit with the aeration pipes proved to be more efficient in all cases in the removal of COD, SS, TN, and TP (respective rates 85.0%, 81.2%, 89.9%, and 77.9%) and was found to be more robust when receiving high organic loads.

6.3.6.2 *Artificial Aeration (Aerated Wetlands)*

In order to further improve the oxygen availability, the use of artificial means for oxygen supply in CWs has been proposed. The idea is to use an air pump (a blower or injector) to provide compressed air to the bed. This alternative has been tested in HSF CWs, where oxygen availability is more limited (Ouellet-Plamondon et al., 2006; Chazarenc et al., 2009; Zhang et al., 2010). Over the last years, this modification has also been applied to VFCW systems, usually in systems with saturated periods. Although wastewater aeration is a common practice for other treatment methods (e.g., in activated sludge systems), aeration of gravel beds appeared only during the last 10-15 years. In tanks and ponds, aeration creates air bubbles, and thus, a hydrodynamic mixing effect, which results in a more uniform distribution of the dissolved oxygen at a considerable distance from the air diffuser (Nivala et al., 2013). Therefore, aeration in conventional wastewater treatment techniques is provided via high-volume devices. On the other hand, in subsurface flow CWs containing gravel material, the hydrodynamic mixing is limited and the distance an air bubble can reach does not exceed 30 cm. Therefore, effective aeration in SF CWs could be implemented with uniform distribution of small air quantities across the bottom of the bed (Wallace, 2001). The main advantage of bottom additional aeration in VFCWs is that the combined action of the vertical downward drainage of the wastewater with the upflow movement of air bubbles results in a very good water mixture within the bed. Additional aeration of the bed has been promoted as a positive alternative in order to enhance nitrogen removal, especially in cold climate systems (Wallace et al., 2000).

Improved results are reported in intensified saturated VFCWs receiving primarily treated wastewater (Nivala et al., 2012b). In this design, aeration pipes were placed at the bottom of the bed. Aerated VFCWs always showed lower effluent pollutant concentrations (<5 mg/L $CBOD_5$ and <5 mg/L NH_4^+-N) compared to nonaerated. Dong et al. (2012) investigated nonaerated, intermittently and continuously aerated VFCWs to treat heavily polluted river water. Aeration pipes were placed at the height of 30 cm for a total bed height of 80 cm and produced fine air bubbles of 20-30 μm diameter. Results showed increased DO concentration in aerated systems which enhanced the removal of OM and NH_4^+-N. Intermittently aerated VFCW favored TN removal through the denitrification process facilitated by alternating aerobic/anoxic conditions in the bed. Pan et al. (2012) tested a full-scale aerated VFCW bed in combination with an HSF CW for domestic wastewater treatment. Ammonia removal in the aerated VFCW reached 85%, while an HRT of 2-3 days was proposed for the horizontal system in order to facilitate denitrification of the nitrified VFCW effluent. Artificial aeration using an air compressor was applied to an integrated VFCW system for domestic sewage treatment (Tao et al., 2010). Aeration pipes were placed at the bottom of the downflow chamber (60 cm depth) and air flow (0.25 m^3/h) was delivered for 8 h/d. This study also showed that artificial aeration enhances the OM decomposition and the nitrification process, compared to the nonaerated system. Intermittent aeration was also found to promote denitrification.

Due to the increased oxygen availability, these systems have been promoted not only for domestic/municipal wastewater treatment but also for stronger wastewater (e.g., industrial effluents). Wallace et al. (2006) tested a saturated pilot-VFCW with additional aeration for domestic wastewater, while another system was used to treat process water from a gold mine. The efficiency of the systems was investigated at both low and high temperatures. Results showed that these aerated VFCWs are quite effective in ammonia removal via intense nitrification, which was found to be almost 10 times higher in the aerated system compared to the nonaerated. Moreover, this efficiency could also be sustained even at low temperatures. Aerated wetlands possess lower area demands than nonaerated, while the treatment requires only 10% of energy input compared to the activated sludge method. Wallace and Liner (2011) also tested and applied aerated and nonaerated VFCWs for the treatment of airport deicing runoff. This system was found efficient in the treatment of very high organic loads (4500 kg/d BOD_5) with an average removal of 98.3%, since the oxygen consumption rate reached 1027 g/m^2/d. This increased treatment capacity between aerated and nonaerated systems is also indicated by the difference in the calculated volumetric BOD_5 removal rate constant (5.4 per day vs. 0.55 per day).

Generally, artificial aeration appears as an attracting alternative when additional oxygen amount is desirable. These so-called intensified systems are capable of treating strong wastewater at high loading rates, while they have lower area demands. On the other hand, the use of air-producing devices will increase the total cost of the facility as also the energy consumption. For this, the use of renewable energy sources, e.g., wind or solar power, has been suggested in order to reduce the operation costs (Tao et al., 2010; Nivala et al., 2013).

6.3.6.3 *Effluent Recirculation*

Partial or complete recirculation of the effluent of VFCWs has also been proposed as a strategy to increase the system efficiency. This modification increases the wastewater retention time within the bed, reduces the washout of necessary nitrifying microorganisms, and provides additional oxygen input by convection which is directly related to the ratio of recirculated water (Sun et al., 1998; Platzer, 1999). The expected result is improved performance especially concerning the nitrification of the ammonia load. If the recirculation sends the nitrified VFCW effluent back to the primary stage (e.g., sedimentation tank), the carbon content of the raw wastewater coupled with the anoxic conditions in the tank could also favor denitrification (Laber et al., 1997; Platzer, 1999; Arias et al., 2005). For this predenitrification step, Platzer (1999) reported that recirculation rates of up to 200% could be used, but not for very high nitrogen loads (in this case, an HSF CW should be used as a post-denitrification stage, after the VFCW).

Laber et al. (1997) applied recirculation in a single stage VFCW in Austria with a pre-treatment (settling tank). Recirculation rates tested were 50-60% and 80% with simultaneous increase of the water level in the bed from 0 to 20 cm. Removal of TN reached 72% and for OM (BOD_5 and COD) was always >90%. Bahlo (2000) tested a recirculation ratio of 1.1 in a VFCW bed treating domestic wastewater (29 pe), with mean removal rates 98% for BOD_5 and NH_4^+-N 98.7%. The author reports that mean oxygen saturation in the bed increased from 56% to 67% with recirculation, despite the doubling of the loading rate, while he also mentions increased construction costs for this modification. Sun et al. (1998) tested recirculation in the second and third stage of a three-stage VFCW system, which treated a relatively strong wastewater mixture of septic tank effluent, pig slurry, and effluent from a biofilm reactor (mean influent BOD_5 464 mg/L, NH_4^+-N up to 250 mg/L). Mean removal rates for BOD_5, COD, NH_4^+-N, OP, and SS were 99.1%, 73.1%, 95.1%, 48.9%, and 61.7%, respectively. The authors reported significantly enhanced nitrification when the BOD_5 concentration was less than 50 mg/L. Effluent recirculation was found to enhance the oxygen consumption by microorganisms for OM degradation and nitrification.

A Danish two-stage VFCW system with tested effluent recirculation ratios of 100%, 200%, and 300% resulted in a fully nitrified effluent, independently of the ratio applied (Arias et al., 2005). Respective removal of TN increased from 1% (no recirculation) to 52%, 66%, and 68% for the three ratios, indicating that recycling the effluent back to the sedimentation tank improves denitrification in the tank and nitrification in the VFCW bed. The removal of indicator bacteria was not affected by the effluent recirculation (Arias et al., 2003b). The authors suggest an optimum recirculation ratio between 100% and 200% in order to avoid possible negative impacts of the higher loading rates on the performance. Proposed guidelines for single-stage VFCWs in Denmark suggest an effluent recirculation ratio of 50%, since under this configuration lower effluent concentrations were achieved (Brix and Arias, 2005). The use of effluent recirculation has also been proposed for use in the French system. In this case, the system consists of a single bed with recirculation and not of two stages which is the typical layout of the French system (Prost-Boucle and Molle, 2012). The efficiency of the system was similar to that of a classical French system, but with lower area demands (1.5 compared to 2 m^2/pe), for a 100% recirculation ratio. Respective results are also reported for a recirculating VFCW planted with corn for onsite wastewater treatment (García-Pérez et al., 2011). Very high removal efficiencies were achieved (98.4% for BOD_5, 70.5% for TN, 95.3% for NH_4^+-N, 75.5% for TP, and 99.9% for FC).

Gross et al. (2007) developed a combined system of a recirculating VFCW bed with perforated bottom and a container placed beneath the VFCW. Water collected in the container was recirculated to the VFCW, while the treated effluent was used for irrigation. This system was highly efficient in removing BOD_5, TSS, and FC (>98%) and most of the COD, TN, and TP (81%, 69%, and 71%, respectively). The feasibility analysis showed a return over investment after 3 years. This system has been successfully used for the treatment of primary treated domestic wastewater, graywater from 5 pe, and goat cheese factory effluent (Gross et al., 2008). Recirculation is believed to be the main parameter for the effective performance due to prevention of bed clogging, additional aeration, and, respectively, faster biodegradation, even at extreme conditions (e.g., low and high pH values or high pollutant concentrations), while, with the addition of a final UV disinfection unit, effluent standards for irrigation were always met (Gross et al., 2008; Zapater et al., 2011).

Improved performance was also observed in a hybrid three-stage CW system (VF-HSF-VF) where the final effluent was recirculated back to the inflow of the second stage (Tunçsiper, 2009). With 100% of effluent recirculation, both NH_4^+-N and NO_3^--N removal rates increased, implying a positive effect on both nitrification and denitrification, respectively. Another hybrid system (HSF-VF), with 100% effluent recirculation, was proved capable of providing an effluent quality appropriate for safe discharge in the Mediterranean region, with overall efficiency of 98% and 79% for TKN and TN, respectively (Ayaz et al., 2012b).

The increased treatment capacity achieved by recirculation has been exploited not only for domestic and municipal wastewater but also for other types of wastewater with heavier loads and increased pollutant concentrations. Livestock wastewater has been treated in a VFCW with effluent recirculation tested at various ratios (0-150%), with improved removal efficiency for BOD_5, NH_4^+-N and SS compared to operation without recirculation (He et al., 2006b). Recirculation has been combined with TF systems for pig farm wastewater treatment (Zhao et al., 2004b; Sun et al., 2005). A recirculating (at a ratio of 1:3) VFCW has been loaded with landfill leachate with very good results (Lavrova and Koumanova, 2010), while recirculation has also been suggested for olive mill wastewater treatment in order to decrease the number of treatment stages (Herouvim et al., 2011).

Overall, recirculation appears as an attractive alternative for the enhancement of the system performance in VFCWs. Especially for small or single-stage onsite systems, it can be an appropriate configuration. A positive effect of effluent recirculation is also reported in hybrid systems (VF+HSF CWs; Noorvee et al., 2007; Ayaz et al., 2012b), although in HSF CWs alone, the effect of effluent recirculation was found to be of less importance (Stefanakis and Tsihrintzis, 2009a). In any case, it should always be taken into account that—as for other modifications, e.g., artificial aeration—this setup demands higher energy input for its operation. The alternative to achieve a similar high performance, especially concerning nitrogen removal, would be a two-stage VFCW system, as it is pointed out (Langergraber et al., 2008b).

6.3.6.4 TF Configuration

Tidal flow constructed wetlands (TFCWs) are a relatively new configuration of VFCWs. They operate with consecutive fill and drain cycles (upflow and downflow movement of the wastewater) in order to promote passive aeration of the bed. At first, the wastewater is introduced at the bottom of the bed and gradually fills the bed until it reaches and floods the surface. It is left in contact with system components (media, rootzone, biofilm) for a certain period of time and then it is drained vertically. As the bed is emptying, air is introduced within the bed and diffuses in the void space of the substrate media, thus aerobic conditions are gradually restored. The main advantage of this system is that clogging problems of the surface are limited compared to common VFCWs fed on top of the surface. Due to their operational regime, TFCWs possess a higher treatment capacity than common VFCWs, especially concerning nitrification and OM biodegradation. This enabled the use of such system in the treatment of various wastewater types and of higher loading rates and pollutant concentrations.

Enhanced nitrification potential by TFCWs compared to common VFCWs was first mentioned for a pilot system treating a secondary municipal effluent (Green et al., 1997a). Another pilot- and full-scale multistage TFCWs was implemented in the UK for the treatment of agricultural (diluted pig slurry) wastewater (Sun et al., 1999a,b). The tidal flow process was applied once every hour and the operation scheme was 3 days of loadings followed by 3 days of resting. Increased oxygen availability (up to 357 g $O_2/m^2/d$) is reported for this system by the authors, compared to common donwflow VFCWs. Most of the oxygen is consumed for the OM decomposition (Sun et al., 2005). Authors report that the treatment of wastewater with high OM concentration probably limits the nitrification of ammonia nitrogen, for which the main removal process was immobilization by microbial cells and adsorption (Sun et al., 2005). Phosphorus removal was steady independently of the operation parameters (Zhao et al., 2004a). A direct correlation was also found between BOD_5 and COD removal and loading rates (Sun et al., 2006). Moreover, an HRT of 1 h per tide (during the filling phase) followed by a resting time of 3 h per tide appeared to be adequate for the optimum system performance (Zhao et al., 2004a). Similar OTC (350 $g/m^2/d$) in TFCWs is also reported by Wu et al. (2011a), while the authors also mentioned that under high BOD influent loading, nitrification is more limited due to excessive heterotrophic biofilm growth which operates antagonistically to OM degrading bacteria.

The operation of the TFCW system with simultaneous effluent recirculation was found to further increase the system efficiency (Sun et al., 1999b, 2005; Zhao et al., 2004b), while a recirculation ratio of 1:1 was proved to be adequate. Cui et al. (2012) applied effluent recirculation in a hybrid TFW system (practically a two-stage system), with similar results observed concerning TN and TP removal rates improvement.

The use of alternative media has also been tested in TFWs. An electrostatically neutral, HDPE and lightweight expanded shale aggregate with a CEC of 4 meq/100 g have been compared as substrate media (Austin, 2006). Results showed that the material with high CEC significantly improved the system performance, especially in a planted unit, which implied that the CEC and the presence of plants should be considered as design criteria for TFWs. Alum-containing drinking water treatment sludge is another material tested in TFWs, with improved performance in P removal (Zhao et al., 2011a).

6.3.7 Effect of Evapotranspiration

Evapotranspiration (ET) in a wetland system represents water losses through the water and soil (evaporation, E) and from the aboveground parts of the plants (transpiration, T). ET is affected by solar radiation, wind speed, and heat transfer from the surrounding air (Cronk and Fennessy, 2001). In natural and constructed wetlands, ET is considered as the main water removal mechanism and as the most important parameter in the hydrological balance of wetlands (Stefanakis and Tsihrintzis, 2011; Papaevangelou et al., 2012). In subsurface CWs, where no free-water surface exists, direct calculations of transport coefficients are more difficult. In these systems, vapor diffusion through the porous media takes place while the media itself also possesses a certain level of heat storage capacity. Therefore, practically only real data could be used. In CW systems, the *ET* rate depends on several parameters like the type of vegetation, the vegetation species, the density of the plants, and the climatic conditions, among others (Lafleur, 1990; Kadlec and Knight, 1996). Generally, actual quantification of ET rate in CWs still represents a difficult task, while there are many studies with conflicting results (Lafleur, 1990). The complexity of wetland systems is considered as the main reason for this. Various methods have been developed to calculate a reference ET, which is the ET of a specific extensive surface under good agronomic, soil, and water conditions. These methods are usually calibrated to the specific climatic conditions and wetland characteristics of the CW system of interest.

Water losses through ET could account for up to 50% of the inflow volume in central European countries (Stottmeister et al., 2003). High ET rates may become a problem in CWs. For example, during warm months and in arid zones, increased water losses through ET mean that the actual volume of the water to be treated is smaller. This results in increased effluent pollutant concentrations (Lantzke et al., 1999; Randerson et al., 2012), while the water might become more saline (García et al., 2010). Water losses through ET could represent an important obstacle when the treated water is to be reused (Green et al., 2006). ET losses may also result in zero-discharge from the system. In small CWs, continuous water supply and the energy advection to the peripheral area of the plants cause an additional heat input, and thus, higher ET rates than the one driven only from radiation ("clothesline" and "oasis" effects; Vymazal et al., 2006; Kadlec and Wallace, 2009). From this point of view, VFCWs are preferred because they operate with lower HRTs than horizontal flow systems and have an unsaturated porous media layer. However, respective information on VFCWs is limited. Brix (2004) report that annual water losses for single-stage VFCWs in Denmark should be considered as 2.5 times higher the potential ET measured, due to the above mentioned effects.

Morari and Giardini (2009) report that the effluent quality of a VFCW system was strongly influenced by ET. In these systems in Italy, water losses through ET accounted for 45-67% (*T. latifolia*) and 37-71% (*P. australis*) of the inflow volume. The authors suggest that, under the Mediterranean climate, irrigation requirements using the effluent of a small onsite VFCW system could be met only when ET losses are lower than 45% of the inflow volume.

Kantawanichkul et al. (2009) calculated the water loss rate due to *ET* for tropical VFCWs treating strong wastewater equal to 20 mm/d (*T. angustifolia*), 11.9 mm/d (*C. alternifolius*), and 7.9 mm/d for the unplanted systems. High water losses in these planted systems were also attributed to the "oasis" effect (Brix and Arias, 2005). Therefore, ET rate per area unit base can be significantly higher than the theoretically calculated ET in small systems, while in larger full-scale systems, the respective value is expected to be smaller. Higher COD effluent concentrations were observed in the systems planted with *Cyperus* which was attributed to the higher ET rate and the respective concentration effect. The ET effect on effluent concentration is also mentioned by Karathanasis et al. (2003), who report that occasionally increased BOD effluent concentrations from subsurface CWs in summer months could be attributed to higher ET rates (see also Liolios et al., 2014). Therefore, the design of VFCWs (and CWs, generally) in dry and hot climates for water reuse should include the selection of plant species that minimize ET losses (Masi, 2013). For this, VFCWs are preferred than HSF CWs. Additionally, in warm and dry climates, water losses through ET could be exploited by increasing the applied HLR, without affecting the overall system performance (Randerson et al., 2012).

Meuleman et al. (2003) measured the ET contribution in the water balance for a VFCW system in the Netherlands planted with *P. australis* and after 15 years of operation. *ET* accounted for 13% of water losses, while 73% of the inflow volume corresponded to the wastewater supply and 16% to precipitation. Respective ET value for a combined pilot system in Turkey (unplanted VF-planted HSF-planted VF) was 15% for the third stage VFCW planted with *P. australis* at an HLR of 60 L/m^2/d (Tunçsiper, 2009). Generally, in most published studies, ET values are not measured or even not taken into account, thus, available ET data are still limited.

Chapter 7

Treatment of Special Wastewaters in VFCWs

7.1 TREATMENT OF SPECIAL WASTEWATERS

7.1.1 Tannery Wastewater

Tannery wastewaters are characterized by a variety of chemical substances used in leather processing, which are very alkaline and have high oxygen demands (Mant et al., 2006; Calheiros et al., 2009a). Almost 80% of tanneries use in their productive processes toxic chromium salts and other substances, such as bicarbonates, biocides, sulfate, and sulfide (Aguilar et al., 2008). Tannery wastewaters are characterized by high OM and SS content (mean concentrations of total COD and SS are of the order of 4000 and 2000 mg/L, respectively) and have a dark brown color (UNIDO, 2011). Furthermore, tannery wastewaters contain high concentrations of trivalent (Cr(III)) and hexavalent chromium (Cr(IV)) (150 mg/L), sulfide (160 mg/L), chloride (5000 mg/L), TKN (160 mg/L) and sulfate (1400 mg/L) (UNIDO, 2011). As a result of this, tannery wastewaters are very difficult to treat because of their high organic loads and high concentrations of sulfate, sulfide, chloride, and chromium (Reemtsma and Jekel, 1997; Reemtsma et al., 2002).

Various researchers have used CWs to treat tannery wastewater. The main CW type tested is the horizontal subsurface flow (Kucuk et al., 2003; Calheiros et al., 2007, 2009b, 2010, 2012; Aguilar et al., 2008; Dotro et al., 2011). Cr removal from wastewater has been tested in HSF CWs with rather successful results, as the reported removal efficiencies were very high (Kadlec and Wallace, 2009).

Recently, VFCWs have also been used for tannery wastewater treatment. Mant et al. (2006) used a recirculating VFCW system to treat tannery wastewaters. The system consisted of three planted beds and one unplanted (control bed). Each bed contained gravel and sand substrate. The observed removal efficiency of Cr was 97-99.6%. The beds planted with *Pennisetum purpureum* and *Brachiaria decumbens* removed Cr more efficiently than the bed planted with

☆"To view the full reference list for the book, click here"

Phragmites australis. Reasons for this included the pH decrease, due to the precipitation of chromium hydroxide, and the well-developed rootzones of *P. purpureum* and *B. decumbens*. Regarding root system development, the more developed the plant roots, the greater the surface area available for respective biofilm development, chemical precipitation, and metal binding. Moreover, these tropical grasses showed endurance in the high concentrations of chromium present in tannery wastewaters and reduced OM without showing any signs of phytotoxicity. High tolerance of common reeds in Cr(VI) concentrations was observed by Stefanakis and Tsihrintzis (2012d), who used VFCWs to treat sludge contaminated with Cr. The bed with the gravel substrate had hydroponic characteristics which contained a suitable environment for microorganisms close to the root surface.

VFCWs have also been used for tannery wastewater treatment as part of hybrid CWs systems (Saeed et al., 2012). These authors used a hybrid CW system in Bangladesh consisting of three treatment stages in series: a VFCW, an HSF, and a VFCW. All CWs were planted with common reeds (*P. australis*), but they had different porous media, including organic coco-peat, cupola slag, and pea gravel. Mean removal rates for the whole hybrid system were 86%, 50%, 98%, 98%, and 87% for NH_4^+-N, NO_3^--N, BOD_5, COD, and OP, respectively. According to Saeed et al. (2012), in the VFCWs the higher OM removal was related to higher ammonia effluent concentrations, due to limited nitrification process, as available DO was consumed for organic substance oxidation.

Plant species tested in CWs for tannery wastewater treatment, such as common reeds (*P. australis*; Mant et al., 2006; Saeed et al., 2012), *P. purpureum*, and *B. decumbens* (Mant et al., 2006), appear to be rather tolerant to Cr concentrations as they can grow even under high Cr concentrations reaching 30 mg/L (Mant et al., 2006). Cr (97-98%) was mainly accumulated in the belowground plant parts (Mant et al., 2006; Stefanakis and Tsihrintzis, 2012d). Plant tolerance in Cr is not the only factor which should be taken into consideration for CW applications, as other factors (e.g., climatic conditions, plant adjustment, etc.) also affect plant growth and health (Mant et al., 2006). Plants not only contribute to Cr removal by Cr accumulation but also by providing an excellent environment in their root zone for metal binding in the biofilm (Mant et al., 2006).

HSF and hybrid CW treatment systems removed Cr rather efficiently, mainly through Cr accumulation by plants and, specifically, by their root systems, reaching removal efficiencies of almost 100%, while producing only small amounts of by-products (plant biomass). The selection of plant species should be made after examining different parameters such as the root depth, the plant growth/biomass production, and the plant tolerance to high pollutant loads (Brix, 1994b). The main concern when using CWs for tannery wastewater treatment is that, until now, no viable and cost-effective method for plant biomass management has been developed. Harvested plants derived from CWs treating wastewater with Cr should be handled as hazardous wastes. Although this by-product is a hazardous material, it can be easily handled since the produced biomass is very low, compared to other physicochemical or biological methods that produce large quantities of hazardous sludge.

7.1.2 Landfill Leachate Effluents

Landfill leachate is characterized by high organic and inorganic pollutant concentrations (Bodzek et al., 2006) and is extremely toxic to the environment. The significant impacts of landfill leachate could be eutrophication of aquatic systems and toxic effects on fauna which are caused by a variety of contaminants (Lavrova and Koumanova, 2010). Typical constituent concentrations in landfill leachate are 1000 mg/L BOD_5, 18,000 mg/L COD, 225 mg/L TN, 30 mg/L TP, 1000 mg/L Ca, 250 mg/L Mg, 500 mg/L Na, 300 mg/L K, 60 mg/L Fe, 500 mg/L Cl^-, and 300 mg/L SO_4^{2-} (Tchobanoglous and Kreith, 1998). Due to its high toxicity, landfill leachate is a major threat for aquifer and surface water health status (Bulc, 2006). The exact chemical composition of landfill leachate depends on waste composition, climatic conditions, and the age and degradation rate of the solid waste (Bulc, 2006). The major problem in landfill leachate treatment is the extremely high concentrations of ammonia and organic nitrogen (Bulc, 2006; Lavrova and Koumanova, 2010). Various treatment methods (physicochemical and biological) have been used but either they have high operational cost or they produce dangerous by-products (Bulc, 2006; Kadlec and Zmarthie, 2010). CWs are the most desirable technology providing on-site, passive treatment of landfill wastewaters (Kadlec and Zmarthie, 2010).

A pretreatment stage is usually employed before leachate is applied to CWs. The pretreatment stage usually includes sedimentation basins (Lavrova and Koumanova, 2010). However, in HSF and FWS CW applications for landfill leachate other pretreatment methods have also been reported, including trickling filters (Davison et al., 2006), aeration tanks (Barr and Robinson, 1999; Davison et al., 2006), and septic tanks (Nivala et al., 2007). These pretreatment techniques are necessary not only for SS removal but also to remove a portion of the OM and N in order to reduce toxicity for the plants.

A number of plant species used for the treatment of tannery wastewaters has also been used to treat landfill leachate. The majority of the CW systems in the literature are planted with common reeds (*P. australis*; Feo and Belgiorno, 2005; Bulc, 2006; Sun and Austin, 2007b; Justin et al., 2009; Justin and Zupancic, 2009; Lavrova and Koumanova, 2010). Other plant species used are *Phalaris* sp., *Scirpus* sp., *Typha angustifolia*, *Typha latifolia*, *Solidago rigida*, *Rumex crispus*, *Cirsium vulgare*, *Urtica dioica*, *Salix* spp., *Cannabis sativa*, and *Carex acutiformis* (Bulc, 2006; Justin and Zupancic, 2009; Yalcuk and Ugurlu, 2009). All the above-mentioned plant species appear to be tolerant to toxic leachates. However, in the various CWs tested in the literature, leachate treatment efficiencies showed no significant variation with the plant species used.

CWs appear to be rather effective in treating pretreated landfill leachate with VFCWs being the most common type used. Among the other CW types, only one research group applied an FWS CW for leachate treatment to date (Davison et al., 2006), while several applications of HSF CWs have been recorded (Urbanc-Bercic, 1997; Barr and Robinson, 1999; Nivala et al., 2007; Sawait-tayothin and Polprasert, 2007).

The most popular CW type for landfill leachate treatment is the VF one, due to the fact that VFCWs systems are more efficient bioreactors compared to HSF and FWS systems. VFCWs have been proved to be rather efficient in OM removal as they can achieve removal rates upto 90% (Justin and Zupancic, 2009; Kadlec and Zmarthie, 2010), while applied OM surface loads have been significantly high. The main reason that VF systems achieve these high OM removal rates is the retention of SS on the surface of the bed, which can be proved by the respectively high removal efficiencies of SS (from 59% to 99%; Sun and Austin, 2007b; Justin and Zupancic, 2009). Concerning nitrogen, which is also found at high concentrations in landfill leachate, VFCWs present lower removal rates (62-74%). In order to overcome VF system drawbacks, recent CWs applications on landfill leachate treatment focus on hybrid systems which consist of HSF and VF treatment stages. Differences between hybrid CW systems are found only in the series of CW types, as either a VF-HSF system has been used (Bulc, 2006; Yalcuk and Ugurlu, 2009) or an HSF-VF-HSF system (Justin and Zupancic, 2009; Justin et al., 2009). HSF and VF system combination usually aims at enhancing N and heavy metal removal in the HSF stage, while OM removal rate is still high in the VF stage. Although it appears that pollutant removal efficiencies are lower than those achieved with VF systems alone, it should be underlined that none of the experiments with hybrid systems included a pretreatment stage (Bulc, 2006; Justin and Zupancic, 2009; Justin et al., 2009; Yalcuk and Ugurlu, 2009). Heavy metal removals, which are also key pollutants in landfill leachate, have only been reported by Lavrova and Koumanova (2010) and reached 0%, 30%, 58%, 0%, 88%, and 77% for Zn, Mg, Cr, Mn, Fe, and Al, respectively. High removal efficiencies of some heavy metals should be attributed to plant accumulation. Varying removal rates among HM are indicative of the different levels of mobility and bioavailability of the various HM, which affect their removal rates.

The main issue for most of the researches appears to be nitrogen removal. In order to enhance ammonia removal, Yalcuk and Ugurlu (2009) used zeolite as a substrate. The use of zeolite indeed enhanced ammonia removal, as also found in applications treating domestic wastewater treatment (Stefanakis et al., 2009a; Gikas and Tsihrintzis, 2010, 2012; Stefanakis and Tsihrintzis, 2012b). Although plant species and substrate type do not significantly influence the removal efficiency, CW type does. VFCW systems demonstrate higher removal efficiencies than FWS and HSF systems. It appears that the most preferable CW type for leachate treatment is the VF one, which can be used alone or in combination with the other two types.

7.1.3 Azo-Dye and Textile Industries

Textile industries produce colored wastewater which is heavily polluted with dyes, textile auxiliaries, and chemicals used in the dyeing and finishing processes (Roussy et al., 2005). As only 47% of dyes in textile wastewater are biodegradable, color removal from this wastewater type is the major treatment problem (Pagga and Brown, 1986). Until now, in textile wastewater treatment, both biological and physicochemical methods have extensively been used. Biological treatment usually includes a combination of anaerobic and aerobic processes, which are successively carried out to effectively degrade azo-dye effluents. Anaerobic microorganisms participate in textile wastewater treatment by breaking the azo bonds and decolorizing aromatic amines. As a next step, aerobic processes can effectively mineralize the aromatic amine residues and purify the textile wastewater from the remaining OM (Pagga and Brown, 1986).

All CW applications for textile wastewater treatment concern only VFCWs or hybrid systems (Bulc and Ojstrsek, 2008; Nilratnisakorn et al., 2009; Ong et al., 2009). In two experiments, VFCWs (Nilratnisakorn et al., 2009; Ong et al., 2009) were explicitly used, and in the third a hybrid system comprising a VF and two HSF CWs in series was applied (Bulc and Ojstrsek, 2008). Once again, VF and HSF CW systems are selected when the system receives high pollutant loads or the wastewater contains rather toxic compounds, as these CW types are more effective bioreactors than FWS CWs and can degrade even the most toxic pollutants. None of these studies included wastewater pretreatment prior to entering the CW stage. This lack of pretreatment can be attributed to the fact that all the studies used artificial wastewater; therefore, there was no need to remove either SS or to lower the pollutant load before introducing the wastewater to the CW systems.

Generally, the most common plant species used to treat textile wastewater is *P. australis*. Only Nilratnisakorn et al. (2009) used both *P. australis* and *T. latifolia*. Common reed is preferred, as this species is indigenous to most regions of the world and shows extreme tolerance to most toxic compounds contained in all wastewater types. Although most plant species tested were affected by the toxic impact of the azo-dyes, *P. australis* was the most tolerant species with great contribution to azo-dye removal.

The most interesting result from these three experiments is not the rather high removal rates of OM and N (up to 86% and 67%, respectively), which have been observed in most CW applications for various wastewaters, but the rather high color removal rates, which varied from 60% to 97% (Bulc and Ojstrsek, 2008; Nilratnisakorn et al., 2009; Ong et al., 2009). Color removal from textile wastewater is noteworthy because it is the only pollutant that physicochemical and other biological methods cannot efficiently remove. In these experiments, it was also observed that pH values decreased and sulfates were removed. Although the removal of color is rather promising for future application of CWs in textile wastewater treatment, only a few experiments have been performed to date, and the exact mechanism of color removal by CWs has not yet been fully investigated. Therefore, further research is suggested using control experiments to identify the role of biodegradation, plant accumulation, and porous media adsorption of azo-dyes. If new experiments confirm the ability of CWs to remove color, these systems could represent a very promising and attractive treatment method for all textile industry wastewaters.

7.1.4 Other Industrial Effluents

In the literature, CWs have been used for treating other types of industrial wastewater, as well. These wastewaters include metallurgic wastewaters (Xu et al., 2009), chemical industry wastewaters (Domingos et al., 2009), and synthetic wastewaters which simulate various industrial effluents (Scholz and Xu, 2002b; Chen et al., 2006; Ong et al., 2010; Yadav et al., 2010). The common characteristics of all these wastewaters are the high pollutant loads of organic matter and heavy metals.

Microbial and physicochemical processes significantly affect pollutant removal as the vegetation creates suitable environments for microbial growth. Thus, biodegradation, sorption and sedimentation are considered as the main pollutant removal mechanisms. For industrial effluents containing heavy metals, various removal rates were observed, varying from 51% to 100% for Mn, 56% to 97% for Cd, 50% to 98% for Cr, 50% to 96% for Cu, 10.7% to 97% for Fe, 1% to 92% for Ni, 55% to 99% for Pb, 48% to 78% for Zn, 71% to 85.6% for Al, and 9% to 80% for As (Scholz and Xu, 2002b; Chen et al., 2006; Domingos et al., 2009; Xu et al., 2009; Ong et al., 2010; Yadav et al., 2010). These studies mainly aimed at reducing heavy metal concentrations; therefore, the mechanisms and processes have been already thoroughly discussed in previous sections. Differences in removal rates can mainly be attributed to the different surface loads of each metal applied, which affect plant and bacterial accumulation. High surface loads can create a toxic environment and, therefore, inhibit metal accumulation by plants or bacteria.

7.1.5 Dairy Wastewater

Dairy wastewater usually has high BOD_5 (500-2600 mg/L) and COD concentrations (2000-7000 mg/L) and contains fats (90-500 mg/L fats, oil and grease), nutrients (30-100 mg/L TN and 20-100 mg/L TP), suspended solids (200-1000 mg/L SS), and lactose, as well as detergents and sanitizing agents (Wright and Graves, 1998). In thus wastewater, the cheese whey is the most common effluent. The high pollution load of dairy wastewater makes it a potential threat for environmental ecosystems, especially if it is disposed untreated or partially treated, (Montuelle et al., 1992).

In most experiments of dairy wastewater treatment with CWs, the pretreatment stage mainly aims at the removal of SS, which is essential before the dairy wastewater enters the CW bed, to avoid clogging of the porous media and also reduce the organic load. Common pretreatment methods include simple settling basins (Browne and Jenssen, 2005). The plant species used in various dairy wastewater treatment experiments were *Butomus umbellatus*, *Cucurbita maxima*, *P. australis*, *Sotalia fluviatilis*, *Senecio sylvaticus*, *T. latifolia*, and *U. dioica* (Rousseau et al., 2004a; Browne and Jenssen, 2005; Lee et al., 2010).

The use of VFCWs in dairy wastewater treatment is limited, since in most cases it has been observed that the wetland systems were poor in P removal. In fact, VFCWs are only reported in diary wastewater treatment as a part of hybrid CW systems also comprising HSF CWs. Three research groups (Rousseau et al., 2004a; Browne and Jenssen, 2005; Lee et al., 2010) have used hybrid systems with rather high removal efficiencies (83-96% for COD; 65-92% for nitrogen; 52-99% for TP; 83-99% for TSS). These removal efficiencies were achieved even though pollutant surface loads were higher than those usually applied to CW systems. Tanner et al. (1995) states that vegetation enhanced N and P removal efficiencies. This could be attributed to the aeration zone created in the plant rhizosphere, which promoted aerobic degradation processes, increase nitrification and subsequent gaseous losses of N through denitrification (Hansen and Anderson, 1981; Reddy et al., 1989). Ibekwe et al. (2003) reported the existence of a diverse bacterial community (especially, *Nitrosospira* sp.) responsible for ammonia oxidation and possibly for nitrification processes. Lee et al. (2010) also found that combined CW systems achieved higher removal efficiencies of organic matter than VFCW systems alone. Differences in nutrient and solid concentrations across the wetland system could be attributed to a number of removal mechanisms.

HSF CW systems, however, appear to be more efficient in dairy wastewater treatment. Seventeen experiments using HSF CWs for diary wastewater treatment have been published (Tanner et al., 1995, 2005; Newman and Clausen, 1997; Kern and Idler, 1999; Kern et al., 2000; Newman et al., 2000; Ibekwe et al., 2003; Mantovi et al., 2003; Gasiunas et al., 2005; Moir et al., 2005; Smith et al., 2005; Munoz et al., 2006; Farnet et al., 2009; Mustafa et al., 2009; Ghosh and Gopal, 2010; Moreira et al., 2010). HSF CWs have been found rather successful, since removal efficiencies ranged from 28% to 99% for COD, 21% to 99% for N, 2% to 98% for P, and 45% to 95% for TSS.

7.1.6 Animal Farms

Animal farming is one of the main agricultural components worldwide but also one of the main environmental concerns (Hunt and Poach, 2001). Animal farms, as medium or small enterprises, usually lack manpower, financing, and technical competence in order to efficiently treat their wastewater. The treatment problem is also reinforced by the fact that these wastewaters have high organic matter (COD: 38,000-85,000 mg/L), SS (28,500-61,000 mg/L), nitrogen (TKN 1400-3300 mg/L; NH_4^+-N: 500-1300 mg/L), and P concentrations (Babatunde and Zhao, 2010).

Experiments on animal farm wastewater treatment have tested all CW types. FWS CWs have been used only with low surface pollutant loads (Hunt et al., 2003; Poach et al., 2004; Stone et al., 2004; Gottschall et al., 2007). Mainly HSF CWs have been successfully used with removal efficiencies being between 78% and 90% for COD, 22% and 90% for BOD_5, 71% and 98% for TP, 22% and 90% for TN, and 22% and 51% for TSS (Knight et al., 2000; Lee et al., 2004; Kadlec et al., 2005; Meers et al., 2005, 2008; Yeh et al., 2009; Hathaway et al., 2010). Results from other studies also indicate that HSF CWs can effectively remove the pollutants from various animal wastewaters (Knight et al., 2000; Kadlec et al., 2005; Hathaway et al., 2010). Only one attempt using VFCWs for animal farm wastewater treatment is recorded (Zhao et al., 2011a), in which high removal efficiencies were achieved (84% for OM, 78% for TKN, and 97% for TP). VF and HSF hybrid CWs are attracting interest for animal farm wastewater treatment, as they are for most of industrial and agro-industrial wastewater. Even in the first few attempts, hybrid CW systems achieved removal efficiencies of 92-99%, 50-75% and 49-99% for TKN, TP and SS, respectively, while respective surface loads were between 3 and 185, 8 and 312, and 45 and 670 $g/m^2/d$ (Poach et al., 2003; Trias et al., 2004; Kantawanichkul and Somprasert, 2005).

Babatunde and Zhao (2010), in order to overcome the insufficient oxygen supply by the reeds, used a tidal flow VFCW and found increased DO concentrations. The tidal flow regime allowed the application of higher organic matter loads in the VFCWs. On the other hand, Poach et al. (2003) used intermittent wetland feeding in order to improve COD removal. From their experimental results, COD removal was not affected by the increase of the number of drainage periods number.

Babatunde and Zhao (2010) used alum sludge as a top layer in their VFCW. From their experimental results it appears that CWs with alum sludge achieve high SS removal ($76.4 \pm 2.0\%$ to $91.7 \pm 3.2\%$). Furthermore, alum sludge can also enhance phosphorus removal, while the leaching potential of alum sludge is one of the lowest recorded, thus, it can also enhance heavy metal removal retention (Babatunde and Zhao, 2010). Zhao et al. (2005) used an anti-sized arrangement of bed material (i.e., coarser material on top of the bed and finer at the bottom) in order to reduce clogging, and found positive results, as the available bed depth for SS removal was increased. Moreover, they speculated that the coarse material of the upper layer increased air convection and diffusion, resulting in a more effective biosolids decomposition.

7.1.7 Olive Mill Wastewater

Among the agro-industries, olive agriculture is widespread particularly in the Mediterranean region; it produces a large volume of olive mill wastewater (OMW) over a relatively short time period (usually from October to February). During this production period, more than 11 million tons of olives are processed worldwide, resulting in about 9 million tons of OMW (Aktas et al., 2001). OMW is an aqueous, dark, bad-smelling, and turbid suspension, with very high organic load (COD: 15-390 g/L), an acidic pH value (between 4.6 and 5.8), and a high content of phytotoxic and antibacterial phenolic compounds (total phenols: 1.5-14 g/L; Grafias et al., 2010). OMW is formed by the water content of the olives and the water used to wash and process them. Typically, the weight composition of OMW is 83-96% water, 3.5-15% organics, and 0.5-2% mineral salts.

Although CWs have been used to treat a variety of wastewaters, only few attempts to treat OMW have been published (Bubba et al., 2004; Grafias et al., 2010; Yalcuk et al., 2010; Herouvim et al., 2011; Gikas et al., 2013). The main characteristic of all these experiments is that a pretreatment stage is imperative to reduce the high organic and phenol loads which can be toxic to CW vegetation. Before entering the CW treatment stage, the OMW is usually pretreated with various physicochemical and biological treatment methods, such as coagulation (Bubba et al., 2004), electrochemical oxidation (Grafias et al., 2010), biological trickling filters (Herouvim et al., 2011) and open naturally aerated tanks (Gikas et al., 2013). The only experiment on OMW treatment in CWs without a pretreatment stage (Yalcuk et al., 2010), diluted OMW before introducing it into the CW system. The need for an advanced pretreatment stage for OMW treatment firstly derives from the high SS load, which can lead to clogging, and secondly, from the high phenol load, which could be toxic to the vegetation of the CW bed.

Plant species used in OMW treatment experiments are *Cyperus alternatifolius*, *P. australis*, and *T. latifolia* (Bubba et al., 2004; Grafias et al., 2010; Yalcuk et al., 2010; Herouvim et al., 2011).

The only attempt to treat OMW using FWS by Kapellakis et al. (2012) showed a significant removal efficiency for OM (90%), while the influent organic surface load applied was among the highest recorded in the literature (37 g COD/m^2/d). HSF CWs also appear to be efficient in treating OMW (Bubba et al., 2004), since the reported removal efficiencies were 69% for COD, 12% for nitrogen, 55% for TP, 50% for TSS, and 79% for phenols, for an organic load of 220 g COD/m/d. It should be mentioned that these removal efficiencies were achieved with pollutant surface loads higher than those applied to FWS systems, but lower than those applied to VFCW systems. The most efficient CW system for OMW treatment appears to be the VF system (Grafias et al., 2010; Yalcuk et al., 2010; Herouvim et al., 2011). VFCWs show high removal efficiencies for all pollutants (72-86% for COD; 75% for N; 88-95% for P; 79% for phenols), while the pollutant surface loads applied (up to 6600 g COD/m^2/d) were among the highest recorded for all industrial and agro-industrial wastewaters.

Yalcuk et al. (2010) used intermittent feeding VFCWs, which also contain zeolite as substrate material. Zeolites are effective in pollutant removal (especially for ammonia nitrogen), possess a high cation-exchange capacity, and have been used by other research groups as VFCWs substrate material (Stefanakis et al., 2009a; Gikas and Tsihrintzis, 2010, 2012; Stefanakis and Tsihrintzis, 2012a,b). Experimental results reported by Yalcuk et al. (2010) revealed that intermittent feeding enhanced oxygen diffusion in the VFCW bed, and thus, resulted in increased pollutant removal efficiencies.

7.1.8 Summary, Design, and Future Research Suggestions

Special wastewaters examined in this chapter share common characteristics (i.e., high heavy metal concentrations, high organic matter loads, etc.); therefore, CWs treating these wastewaters face almost the same problems and limitations, so common suggestions can be expressed concerning CW design and operation for the treatment of the above-mentioned special wastewater types. The most important issues discussed below include suggestions about pretreatment stage, vegetation, porous media, and CW operation strategy.

One of the common characteristics for some of these special wastewaters is high heavy metal concentrations. CW ability in heavy metal removal is well known and documented (Brix, 1994b; Kadlec and Wallace, 2009). This ability is mainly attributed to the CW vegetation, whose root systems absorb high quantities of heavy metals. Table 7.1 presents design and operational parameters of VFCWs and hybrid CWs treating wastewaters with high heavy metal concentrations. VFCWs either as one stage system or as part of a hybrid system appear to be rather effective in the removal of heavy metals from wastewaters. Various plant species have been tested either in individual or in combined experiments (Table 7.1); however, experimental results showed no significant variations of heavy metal removal between different plant species.

Although various concentrations of heavy metals have been applied to CWs, until now no study tried to examine which are the upper heavy metal influent limit concentrations, which do not cause toxic effects on CW vegetation. Such information is vital for the design and operation of CWs, in order to avoid introducing high HM concentrations, which could destroy the vegetation and put the system out of operation for a long period. Thus, experiments aiming at the determination of heavy metal toxic concentrations are imperative. Another advantage of using CWs for heavy metal removal is that practically the only by-product produced during treatment is plant biomass, which should be removed once a year by harvesting. Harvested biomass from CWs treating wastewater with high heavy metal content should be handled as a hazardous waste. However, this biomass can be more easily handled, since the annualy produced amounts are lower compared to other physicochemical or biological methods that produce large quantities of hazardous sludge.

Another common characteristic of all wastewater types presented in this chapter is the high organic and nitrogen loads. Tables 7.2 and 7.3 present OM and N removal along with other operational parameters of VFCWs treating special wastewaters. Organic matter and nitrogen removal in one stage VFCW

TABLE 7.1 Heavy Metal Removal in VFCWs Treating Special Wastewater

References	CW Surface (m^2)	Plant Species	Wastewater Type	Heavy Metals		HRT (d)
				Influent Concentration (μg/L)	*Removal (%)*	
VF						
Feo and Belgiorno (2005)	3500	–	–		–	
Sun and Austin (2007a,b)	0.07	*P. australis*	Landfill leachate	–	–	
Mant et al. (2006)		*P. purpureum, B. decumbens, P. australis*	Artificial WW with Cr(VI)	Cr: 10,000-20,000	97-99	
Lavrova and Koumanova (2010)	311	*P. australis*	Landfill leachate	Zn: 0.25	Zn: 100	
				Cr: 0.27	Cr: 70	
				Fe: 3.63	Fe: 87	
				Al: 0.325	Al: 77	
Yalcuk and Ugurlu (2009)	0.5	*T. latifolia*	Landfill leachate	Cr: 920	Cr: -	10
				Cu: 900	Cu: -	
				Pb: 8490	Pb: 30-90	
Hybrid						
Kadlec and Zmarthie (2010)	9000	*T. latifolia*	Landfill leachate	Pb: 3	Pb: 33	60
				Zn: 19	Zn: 15	
				Cr: 24	Cr: 67	

Continued

TABLE 7.1 Heavy Metal Removal in VFCWs Treating Special Wastewater—Cont'd

References	CW Surface (m^2)	Plant Species	Wastewater Type	Heavy Metals		HRT (d)
				Influent Concentration (μg/L)	*Removal (%)*	
Justin and Zupancic (2009)	1000	*C. acutiformis*, *P. australis*	Landfill leachate	Al: 0.78	Al: 54.9	
				Cr: 0.66	Cr: 33.1	
				Mn: 0.5	Mn: 12.1	
				Fe: 6.13	Fe: 17.6	
				Ni: 0.13	Ni	
				Cu: 0.15	Cu: 17.6	
				Zn: 0.34	Zn: 32.8	
				Pb: 0.05	Pb	
				As: 0.04	As: 33.6	
				B: 3.78	B: 34.6	

TABLE 7.2 Organic Matter Removal in VFCWs Treating Special Wastewater

References	CW Surface (m^2)	HRT	Plant Species	WW	Pretreatment	COD	
						Surface Load ($g/m^2/d$)	*Removal (%)*
VF							
Lavrova and Koumanova (2010)	0.012	*8*	*P. australis*	LL	Sedimentation, dilution	2800	96
Feo and Belgiorno (2005)	0.07		*P. australis*	LL	Dilution	200	79-84
Sun and Austin (2007a,b)	0.007		*P. australis*	LL	–	690	20-32
Ong et al. (2009)	0.0254		*P. australis*	T	–	450	86
Nilratnisakorn et al. (2009)	0.48		*P. australis, T. latifolia*	T	–	1123	59.6
Grafias et al. (2010)	0.24		–	OMW	Electrochemical oxidation	9740	86
Yalcuk et al. (2010)	0.1256		*T. latifolia, C. alternatifolius*	OMW	Diluted with tap water	86,500	72
Herouvim et al. (2011)			*P. australis*	OMW	Biological filter	14,120	73
Zhao et al. (2004a–c)	0.09		*P. australis*	AF	Diluted feed tank	890-3150	60-80
Babatunde and Zhao (2010)	575		*P. australis*	AF	–	1150	50-86

Continued

TABLE 7.2 Organic Matter Removal in VFCWs Treating Special Wastewater—Cont'd

References	CW Surface (m^2)	HRT	Plant Species	WW	Pretreatment	COD	
						Surface Load ($g/m^2/d$)	*Removal (%)*
Hybrid							
Justin and Zupancic (2009)	1000		*C. acutiformis, P. australis*	LL	Equilibration lagoon	1500	40.9
Bulc and Ojstrsek (2008)	80		*P. australis*		–	276-1379	70-84
Rousseau et al. (2004a,b)	80		*P. australis*	D	–	–	91
Lee et al. (2010)	1.87		*S. fluviatilis*	D	–	–	83-96
Browne and Jenssen (2005)	1990		*P. australis, S. sylvaticus, U. dioica, T. latifolia, Sparganium erectum, B. umbellatus, C. maxima*	D	Settling tanks	84.6 TOC	94
Kantawanichkul and Somprasert (2005)	0.5		*Cacomantis flabelliformis*	AF	Storage tank	1034	45-79

LL, landfill leachate; T, textile wastewater; D, dairy wastewater; AF, animal farm wastewater; OMW, olive mill wastewater.

TABLE 7.3 Nitrogen Removal in VFCWs Treating Special Wastewater

References	CW Surface (m^2)	Plant Species	WW	Pretreatment	TKN	
					Surface Load ($g/m^2/d$)	*Removal (%)*
VF						
Feo and Belgiorno (2005)	0.07	*P. australis*	LL	Dilution	50	62-74
Sun and Austin (2007a,b)	0.007	*P. australis*	LL	–	485	52
Ong et al. (2009)	0.0254	*P. australis*	T	–	62	67
Grafias et al. (2010)	0.24	–	OMW	Electrochemical oxidation	35.2	–
Herouvim et al. (2011)		*P. australis*	OMW	Biological filter	506	75
Zhao et al. (2004a–c)	0.09	*P. australis*		AF	28-159	20-40
Hybrid						
Justin and Zupancic (2009)	1000	*C. acutiformis, P. australis*	LL	Equilibration lagoon	398	41.9
Bulc and Ojstrsek (2008)	80	*P. australis*	T	–	7-82	28-52
Rousseau et al. (2004a,b)	80	*P. australis*	D	–	–	65
Browne and Jenssen (2005)	1990	*P. australis, S. sylvaticus, U. dioica, T. latifolia, S. erectum, B. umbellatus, C. maxima*	D	Settling tanks	49.1	92
Kantawanichkul and Somprasert (2005)	0.5	*C. flabelliformis*	AF	Storage tank	448	55-90
Poach et al. (2003)	120	*Schoenoplectus* spp., *Typha* spp., *Sparganium* spp.	AF	Anaerobic lagoon, holding tank, nitrification unit	330	55-84

LL, landfill leachate; T, textile wastewater; D, dairy wastewater; AF, animal farm wastewater; OMW, olive mill wastewater.

systems appear to be lower than those in hybrid systems, where VFCWs are combined with HSF (Tables 7.2 and 7.3). This was expected as VFCWs, although more efficient than other CW types, need more than one stage to reduce pollutant concentrations to low values. Although these wastewaters are characterized by high organic matter and nitrogen loads, VFCWs seem to be tolerant in these high loads as they are reported to treat wastewaters with COD and N concentrations up to 86,500 (Yalcuk et al., 2010) and 506 mg/L (Herouvim et al., 2011), respectively. Another common characteristic for the majority of VFCWs treating these special wastewaters is the existence of a pretreatment stage, which is used to reduce SS and OM load. In order to maintain the main CW advantage in wastewater treatment, which is the rather low operation cost, a valuable suggestion could be the coupling of VFCWs with other low-cost treatment technologies (e.g., biological trickling filters, coagulation-flocculation; Herouvim et al., 2011), or even better to use a VFCW bed as pretreatment stage based on the French approach, where the porous media in the first stage bed is coarser and can retain higher amounts of SS. Although for OM removal a multistage VFCW system could be efficient, N removal (nitrification and denitrification stages) requires VFCW combination with other types like HSF systems, which are more efficient in denitrifying nitrified effluents.

CW vegetation is a rather controversial subject as the exact contribution of the plants in pollutant removal is not quite clear yet (see Section 6.3.5 in Chapter 6). Many researchers have tested the effects of plant vegetation and plant diversity on CWs performance (Browne and Jenssen, 2005; Nilratnisakorn et al., 2009; Stefanakis and Tsihrintzis, 2012a,c), concluding that although the presence of vegetation increases VFCWs performance, the use of different plant species does not significantly affect CW performance. Thus, plant species used in CWs should be selected based on their tolerance to the local climatic conditions and on the potential to minimize supply costs. Therefore, CWs should be vegetated using indigenous plants.

Porous media selection is another significant issue in VFCWs treating special wastewater, as the need for enhancing ammonia and phosphorus removal and eliminating clogging phenomena is more crucial than in CWs treating municipal wastewaters. Various substrates have been used in VF and HSF CWs (Table 4.3) like zeolite, bauxite, flying ash, river gravel, and quarry gravel (Drizo et al., 1999; Arias et al., 2001; Akratos and Tsihrintzis, 2007; Stefanakis et al., 2009a; Yalcuk et al., 2010; Stefanakis and Tsihrintzis, 2012a,b). These materials, which are characterized either by high cation-exchange capacity or by reactive iron, aluminum, and calcareous material content, have been used and proved to increase nitrogen and phosphorus removal in CWs. Although the main problem regarding porous media is the clogging phenomenon, which is recognized as the major problem for subsurface flow CWs, a lack of studies exists on thoroughly examining clogging or on testing new porous materials. In order to overcome this lack of information, researchers should monitor full-scale CWs for a long time period (more than 5 years) as the majority of the experimental results presented in the literature are based on data, which rarely derive from a more than 2-year monitoring period. Furthermore, it is essential to examine the use of new synthetic materials which could simultaneously increase VFCWs specific surface area and porosity. This is imperative, especially for VFCWs treating special wastewaters, since the use of a material with the above-mentioned characteristics could result in increased biofilm density and clogging could be delayed.

VFCWs treating special wastewaters followed various feeding strategies as intermittent downflow (Kantawanichkul and Somprasert, 2005; Yalcuk and Ugurlu, 2009; Grafias et al., 2010; Lee et al., 2010; Yalcuk et al., 2010; Herouvim et al., 2011), recirculation (Mant et al., 2006; Sun and Austin, 2007a; Lavrova and Koumanova, 2010), saturated upflow (Nilratnisakorn et al., 2009; Ong et al., 2009), and saturated downflow (Bulc and Ojstrsek, 2008; Babatunde and Zhao, 2010; Lavrova and Koumanova, 2010). Although all these different feeding strategies were examined and proposed in the literature, a study comparing them does not exist. As long as information about the effects of these different feeding strategies in VFCW removal efficiencies is lacking, no reliable design and operation guidelines could be given. Furthermore, different recirculation ratios should also be examined.

7.2 GROUNDWATER REMEDIATION

The use of CWs for groundwater remediation appears as another promising application of these systems, besides the treatment of various types of wastewater. Groundwater contamination is a problem in many countries with developed and extensive industrial facilities. Industrial applications and fuel production usually results in groundwater contaminated with petroleum hydrocarbons, BTEX compounds (benzene, toluene, ethylbenzene, *m*-, *o*-, and *p*-xylene), MTBE (methyl tertiary-butyl ether), and chlorinated volatile organic compounds (VOCs). Many of these, for example MTBE, are considered as possible human carcinogens (Johnson et al., 2000). Usually, the approach for groundwater remediation includes high-tech intensive technologies with high energy consumption, e.g., air stripping and filtering with activated carbon, among others (Wilhelm et al., 2002). Recently, CWs attracted interest as an effective and cost-effective option for groundwater remediation. However, successful application and optimization of the system for groundwater remediation is yet to be done. Many studies found in the literature report the efficiency of various CW types, but investigations on the physicochemical and biological transformation processes of the various organic chemicals are limited. HSF CWs have been initially used for the remediation of contaminated groundwater, indicating that rapid degradation of various organic chemicals and chlorinated compounds takes place (Braeckevelt et al., 2011a,b; Seeger et al., 2011a,b; Stefanakis et al., 2013). Recently, research focused on the advantages the vertical flow regime could offer and its potential as remediation technique (Table 7.4). Thus,

TABLE 7.4 Removal of Chlorinated Hydrocarbons, Petroleum Hydrocarbons, and Volatile Organic Compounds in VFCWs for Groundwater Remediation

References	System Setup	Pollutant, Influent Concentration	Efficiency
Ettingh (2002)	Upward VFCW	PCE (perchloroethylene) 0.05 mg/L	
Haberl et al. (2003)	Artificially aerated upward VFCW, HRT = 1d	Benzene	Benzene 87%
		BTEX	BTEX 89-91%
		TPH (total petroleum hydrocarbons)	TPH 95-98%
		MTBE (methyl *tert*-butyl ether)	MTBE 31-28%
Kassenga et al. (2003)	Upward VF columns	*cis*-1,2-DCE (*cis*-1,2-dichloroethene)	More than 90% with time
	Mixtures of sand and peat and of sand, peat, and agricultural waste	1,1,1-TCA (1,1,1-trichloroethane)	
Wallace and Kadlec (2005)	Pilot and full-scale VFCW	BTEX 0.47 mg/L	<0.01 blod[a]
		Benzene 0.17 mg/L	<0.01 blod[a]
		GRO (gasoline-range organics) 2.02 mg/L	<0.05 blod[a]
Waldron (2007)	Upward VFCW	PCE (perchloroethene) 46.5 μg/L	98.9%
Bedessem et al. (2007)	Upward VFCW with artificial aeration	TPH	<0.05 mg/L
		Benzene	>80%
		BTEX	>88%
Amon et al. (2007)	Upward VFCW	PCE 34.0 μg/L	>90%
		TCE 0.6 μg/L	Increase (0.75 μg/L)
		cis-DCE 0.0 μg/L	Blod[a]
		VC 0.0 μg/L	Blod[a]
Eke and Scholz (2008)	Pilot VFCWs	Benzene 1.3 μg/L	85-95%
	Outdoor-indoor	COD, NH_4^+—N,	70-80%, 83-90%,
	(HRT 1d)	NO_3^-—N, OP	88-94%, 58-66%
Tang et al. (2009)	Pilot VFCWs	Benzene 1.0 g/L	88.7-89.8%, 72.7-80.5%
	Indoor-outdoor	Indoor, outdoor systems	

TABLE 7.4 Removal of Chlorinated Hydrocarbons, Petroleum Hydrocarbons, and Volatile Organic Compounds in VFCWs for Groundwater Remediation—Cont'd

References	System Setup	Pollutant, Influent Concentration	Efficiency
Cottin and Merlin (2010)	Filter columns with peat—pozzolana	CBs (chlorinates benzenes: 1,3-DCB, 1,2,3-TCB, 1,2,4-TCB; 1.0, 5.0, 10 μg/L)	>98% 87-95%
Van Afferden et al. (2011)	VF soil filters	MTBE 2970±816 μg/L	1st 70%, total 100% (5 μg/L)
	Roughing/polishing	Benzene 13,966±1998 μg/L	1st 98%, total 100% (0.6 μg/L)
De Biase et al. (2011)	VF filter	Benzene 20 mg/L	>97% (0.4-0.6 mg/L)
		MTBE 3.9 mg/L	>85% (0.3-0.6 mg/L)
		Ammonium 55 mg/L	>73% (15-13 mg/L)
		Nitrate <1 mg/L	135-149 mg/L
		TBA 56×10^{-3} mg/L	>77% ($13\text{-}11 \times 10^{-3}$ mg/L)
		TOC 41 mg/L	>63% (14-15 mg/L)

[a]Below the limit of detection.

VFCWs are gradually becoming popular due to their high oxygen transfer capacity and lower area demands compared to horizontal flow systems.

Chlorinated solvents originating from textile industries and metal degreasing are very common groundwater pollutants, especially in N. America. The majority of the CW treatment systems used for these compounds are upward VFCWs. With this setup, the groundwater inflow is applied at the bottom of the bed and gradually moves upward and fills it. The main benefit is that under this feeding setup, losses by volatilization are minimized, while the bottom anaerobic zone facilitates the removal of recalcitrant chlorinated compounds. The upward movement through the filter media (usually an organic material) favors the sorption of the VOCs, while contact with the biofilm in the rhizosphere enhances biodegradation of these compounds (Kassenga et al., 2003). The proper choice of various parameters like bed depth, soil type, flow rate, etc. could positively affect the system efficiency (Bugg, 2002). The use of mixtures of sand

and peat, and of sand, peat and agricultural waste as filter media showed significant removal of *cis*-1,2-DCE and 1,1,1-TCA compounds, especially in the system with the second mixture, while biodegradation accounted for the major removal (Kassenga et al., 2003). Waldron (2007) investigated the removal in VFCWs of common, very persistent, chlorinated solvents (perchloroethene—PCE and trichloroethene—TCE) found in groundwater. PCE concentration reduced up to 98.9% (effluent value 0.5 μg/L), while it was found that the bottom layers of the bed favored the reductive anaerobic dechlorination of the PCE. The upper layers were better aerated, implying that processes like anaerobic or aerobic oxidation or direct volatilization could take place. Amon et al. (2007) also reports that PCE dechlorination and methane production at the bottom layers of the bed indicate anaerobic or reducing conditions, while the presence of plant roots probably creates some aerobic microzones. The authors also mention the economic benefits of using VFCWs compared to conventional methods (e.g., air strippers), while the overall performance and behavior of the system implies that it can be loaded with even higher loading rates.

Petroleum hydrocarbons, benzene and BTEX are very common contaminants originating from petroleum industries and refineries. BTEX is a highly volatile contaminant, usually found in gasoline, with high solubility and toxicity, thus, it represents a high risk for public health. Especially, benzene is the most toxic compound and known carcinogen. These petroleum wastes are known to be degraded in natural wetland environments. Upward VFCWs with artificial aeration at the bottom have been used to remove these compounds, with very good results, since effluent concentrations were below detection limits for benzene, BTEX, and GRO (Table 7.4; Wallace and Kadlec, 2005; Bedessem et al., 2007). Intermittently loaded VFCWs have also been proved very efficient in benzene removal, no matter which filter media (sand or gravel) was used (Eke and Scholz, 2008; Tang et al., 2009), implying the importance of sufficient oxygen availability. Pilot VFCW columns filled with peat and pozzolana also managed to remove 98% and 87-95% of chlorinated benzenes, respectively (Cottin and Merlin, 2010). Field investigation of MTBE and benzene removal has been implemented in pilot-scale single stage and multistage single-pass VFCWs with different HLRs in a contaminated site in Germany (van Afferden et al., 2011; Figures 7.1 and 7.2). The combination of both filters in a two-stage operation provided a stable and almost complete removal, while about 70% of MTBE and 98% of MTBE were removed in the first VF filter. The design of these filters includes the application of the groundwater about 25 cm below the media surface. This way, toxic emissions to the atmosphere due to compound volatilization are practically eliminated (both MTBE and benzene are highly volatile). Biodegradation was found to be the dominant removal process, implying that the volatile fraction of the compounds is also biodegraded (de Biase et al., 2011). Overall, current results show that VFCWs possess a high treatment capacity for various organic compounds and have the necessary credential for further development in contaminated industrial sites.

Chapter 8

Modeling of Vertical Flow Constructed Wetlands

8.1 INTRODUCTION

Initial CW models included I/O models (Rousseau et al., 2004b) and *k-C* models (Rousseau et al., 2004b; Kadlec and Wallace, 2009). *k-C* models have extensively been used mainly for FWS and HSF CWs. They are based on the following equation:

$$\frac{C_{out} - C^*}{C_{in} - C^*} = e^{-k_y/q} \tag{8.1}$$

where C_{in} is the influent concentration of a constituent (mg/L), C_{out} is the effluent concentration (mg/L), C^* is the background concentration (mg/L), q is the hydraulic loading rate (HLR; $m^3/m^2/d$), and k_y (m/h) is the areal rate constant.

However, these models have limitations in satisfactorily predicting CW performance in different geographic areas and facilities from where they were originally developed or when treating different types of wastewaters. The last few years, several attempts have been made in CW modeling, using different approaches. Among these new approaches is the use of advanced techniques, such as artificial neural networks (ANNs) or fuzzy models. ANNs were first used for HSF CW modeling (Akratos et al., 2008, 2009a,b) to model organic matter, nitrogen and phosphorus removal. Furthermore, ANN results were used in order to develop new modeling equations, which differ from the standard first-order models (e.g., Akratos et al., 2008, 2009a,b). Fuzzy models were used to simulate BOD removal in FWS CWs (Kotti et al., 2013) and were found to be rather successful in predicting the experimental data used for model training and also data from other experimental studies. Although for HSF CWs, many models have been developed and used, including I/O models (Rousseau et al., 2004b), *k-C* models (Rousseau et al., 2004b), models using ANNs (Akratos et al., 2008, 2009a,b), numerical models (Wynn and Liehr, 2001; Mayo and

☆"To view the full reference list for the book, click here"

Bigambo, 2005; Moutsopoulos et al., 2011; Liolios et al., 2012, 2014; Mburu et al., 2013), model development for vertical flow constructed wetlands is more limited (Langergraber, 2008).

VFCW models are divided into two categories: the first includes empirical models, while the second category includes mechanistic models (Langergraber et al., 2009). Empirical models usually include, among others, analytical or regression equations or first-order kinetic equations (Langergraber et al., 2009); they treat the VFCW as a black-box and do not describe adequately the physical, chemical, and biological processes and the effects of dispersion, heterogeneity, and dead zones (Langergraber et al., 2009). On the other hand, mechanistic models are based on the physical, chemical, and microbiological processes governing the operation of VFCWs. These models include, among others, Constructed Wetland 2-Dimensions (CW2D) model (Langergraber, 2001; Langergraber and Šimůnek, 2005), which has already been used and tested by several authors (e.g., Korkusuz et al., 2007b; Mena et al., 2007; Toscano et al., 2009), and FITOVERT (FITO depurazione VERTicale; Giraldi et al., 2009). CW2D was originally developed to describe flow, transport, and reactions of organic matter, nitrogen, and phosphorus in subsurface flow CWs (Langergraber et al., 2009). FITOVERT describes advective and diffusive pollutant transport in VFCWs using the mass conservation equation, while it also can model oxygen transfer in the wastewater and porosity reduction (Langergraber et al., 2009).

8.2 REGRESSION EQUATION MODELS

Modeling of subsurface flow CWs also includes I/O regression equations (Rousseau et al., 2004b). Regression equations are a quite simple and useful tool in predicting CW performance. However, they have two main drawbacks: first, they oversimplify complex processes occurring in CWs by using only a limited number of parameters and do not take into consideration basic operation parameters (Rousseau et al., 2004b). Second, as they are produced using experimental data, their application may be limited only to similar CWs with the ones used for their production.

Regression equations have not only been used for HSF CWs but for VFCWs (Table 8.1) as well. Stefanakis (2011) used a stepwise multiple linear regression (SMLR), in order to produce regression equations for predicting experimental data from 10 pilot-scale intermittent downflow VFCWs. One parameter used was the removed areal BOD_5 load (ALR_{BOD}) as dependent variable, and as independent various operational parameters. SMLR analysis showed that ALR_{BOD} could be predicted by using only the surface BOD_5 load (SLR_{BOD}), the temperature, and the HLR. The prediction was rather good, with R^2 varying from 0.88 to 0.96 (Stefanakis, 2011). The addition of other independent parameters in the regression equation (SLR_{OP}, DO, etc.) did not significantly increase the accuracy. The same independent variables (i.e., HLR and RD-resting days)

TABLE 8.1 Regression Equations for VFCWs

Pollutant	Regression Equation	References
BOD	$ALR_{BOD}=0.947 \cdot SLR_{BOD}-9.285 \cdot SLR_{OP}+0.646 \cdot T+4.96 \cdot DO+86.575 \cdot HLR-36.215$	Stefanakis (2011)
	$ALR_{BOD}=0.807 \cdot SLR_{COD}+0.364 \cdot T-70.220 \cdot HLR-5.684$	
	$ALR_{BOD}=0.758 \cdot SLR_{BOD}+0.647 \cdot T+98.074 \cdot HLR-28.287$	
	$ALR_{BOD}=0.808 \cdot SLR_{BOD}-11.261 \cdot DO-60.321 \cdot HLR-0.298$	
	$ALR_{BOD}=0.807 \cdot SLR_{BOD}+0.364 \cdot T-70.220 \cdot HLR-5.684$	
	$ALR_{BOD}=0.791 \cdot SLR_{BOD}-3.924 \cdot SLR_{OP}+0.209 \cdot SLR_{COD}+0.228 \cdot T-15.499$	
	$ALR_{BOD}=0.874 \cdot SLR_{BOD}+0.252 \cdot T+42.512 \cdot HLR-13.063$	
	$ALR_{BOD}=1.013 \cdot SLR_{BOD}-1.885 \cdot RD-3.583$	
	$C_{BOD,out}=0.430C_{COD}+1.859C_{NH_4-N}-0.973$	Gikas et al. (2011)
COD	$C_{COD,out}=0.599\ C_{BOD}+2.396\ C_{TKN}+0.306\ C_{TSS}-30.329$	Gikas et al. (2011)
	$R_{COD}=26.746\ C_{eff,DO}+55.403$	Chang et al. (2012)
Nitrogen	$C_{TKN,out}=0.042C_{COD,in}+0.823C_{NH_4-N,in}+8.878$	Gikas et al. (2011)
	$C_{NH_4-N,out}=0.034C_{BOD,in}+0.583C_{TKN,in}+3.596$	Gikas et al. (2011)
	$R_{TN}=46.092\times C_{eff,DO}-13.168$	Chang et al. (2012)
Phosphorus	$C_{OP,out}=0.641 \cdot C_{OP,in}-3.395 \cdot HLR+0.690$	Stefanakis (2011)
	$C_{OP,out}=0.656 \cdot C_{OP,in}+0.165$	
	$C_{OP,out}=0.553 \cdot C_{OP,in}-2.351 \cdot HLR+0.522$	
	$C_{OP,out}=0.613 \cdot C_{OP,in}+5.0\times 10^{-6}\ K_t-0.303$	
	$C_{TP,out}=0.023\ C_{TKN,in}+0.257\ C_{OP,in}+0.007\ C_{TSS,in}+2.866$	Gikas et al. (2011)
TSS	$C_{TSS,out}=0.245\ C_{COD,in}-0.043\ C_{EC,in}-8.097\ C_{DO,in}+93.540$	Gikas et al. (2011)

ALR, removed areal load (mg/m^2/d); SLR, influent areal load (mg/m^2/d); HLR, hydraulic load (m^3/m^2/d); *T*, Temperature (°C); DO, dissolved oxygen concentration (mg/L); RD, number of resting days (d); C_{in}, influent concentration (mg/L); C_{out}, effluent concentration (mg/L); R, removal (%); TN, total nitrogen; OP, ortho-phosphate; K_t, hydraulic conductivity.

could be used to also predict the ALR of the various other constituents (i.e., COD, TKN, NH_4^+, TP, OP) (Stefanakis, 2011). Such regression equations can be characterized as a useful tool for the preliminary VFCW sizing, as they use independent variables with known values at the design stage, i.e., HLR, RD, and temperature.

Gikas et al. (2011) used field data from a VFCW in Greece to determine k and C^* values (0.7 and 0.0, respectively) for first-order k-C model for a multistage intermittent downflow full-scale VFCWs (Table 8.1). The k-C model, although it has been mainly used in HSF systems, was found to estimate rather accurately COD, TKN, NH_4^+-N, TP measured data from the full-scale VFCW, with R^2 values of 0.50, 0.71, 0.68, and 0.52, respectively (Gikas et al., 2011). However, it could not predict that accurately TC removal in VFCWs, as the R^2 value was 0.27 (Gikas et al., 2011). Gikas et al. (2011) also used SMLR to produce regression equations for VFCWs, based again on the measured data from a multistage intermittent downflow full-scale VFCW, using as dependent variable BOD_5 effluent concentration and as independent variables various operating parameters (i.e., influent BOD_5 concentration, hydraulic conductivity—HC, EC, DO, HLR, etc.). The effluent concentration of BOD_5 and of other pollutants (i.e., COD, TKN, NH_4^+-N) could be accurately predicted by regression equations which include only pollutant influent concentration and HLR (Gikas et al., 2011). On the contrary, TP and TC were found to depend only on the influent concentration and the CW surface area, respectively (Gikas et al., 2011).

Chang et al. (2012), in order to correlate COD and TN removal with DO effluent concentrations, also developed regression equations. Based on their experimental results, COD removal presented relatively high correlation ($R^2 = 0.523$) only with DO concentration, while temperature seemed not to affect significantly COD removal (Chang et al., 2012). Furthermore, TN removal showed good correlation with DO concentrations ($R^2 = 0.673$).

8.3 MECHANISTIC MODELS FOR VFCWs

According to Langergraber (2008), while for the simulation of HSF CWs, the saturated subsurface flow has been extensively simulated either as a series of continuously stirred tank reactors (CSTRs) or as plug flow reactors (PFR), VFCWs could not be simulated by these models as flow is normally unsaturated. Thus, the majority of the models for VFCWs is based on the model developed by van Genuchten (1982), which describes one-dimensional flow and mass transport under unsaturated-saturated conditions. This model was initially developed to describe water flow and pollutant transport in groundwater and particularly in the vadose zone, but it was also later used to describe VFCW processes, due to the similar conditions encountered (Giraldi and Iannelli, 2009; Giraldi et al., 2010). In order to use the van Genuchten (1982) model, a series of parameters should be defined, including, among others, the residual

TABLE 8.2 Hydraulic Parameters of the van Genuchten-Mualem Model in VFCW Applications

References	Residual Water Content (θ_r)	Saturated Water Content (θ_s)	α (m^{-1})	n	Saturated Hydraulic Conductivity (K_s) (m/h)
Langergraber (2008)	0.045	0.03	14.5	2.68	1.17
Langergraber (2003)	0.056	0.289	12.6	1.92	0.84
Giraldi et al. (2010)	0.035-0.070	0.442-0.480	0.145		3.6-72
Dittmer et al. (2005)	0.05	0.304	0.02	3	0.65

α, n, empirical parameters of van Genuchten equation.

water content (θ_r), the saturated water content (θ_S), the saturated hydraulic conductivity (K_S), and parameters α and n, which are used to model the nonlinear relationship between water content and pressure potential. The two main developed VFCW models, based on van Genuchten (1982) approach, are CW2D and FITOVERT. For these two models, various parameter values of van Genuchten (1982) model have been used, as presented in Table 8.2.

8.3.1 FITOVERT Model

FITOVERT modeling of vertical flow in unsaturated porous media is based on Richards' and van Genuchten equations (Raats, 2001; Giraldi et al., 2010). In addition to the theoretical basis, FITOVERT also includes several practical VFCW aspects, making it a useful design tool. For instance, it operates under a typical nonstationary feeding-emptying schedule, and includes features like evapotranspiration, porosity decrease, substrate clogging, and organic matter and nitrogen removal (Giraldi et al., 2010). Giraldi and Iannelli (2009a) report that the hydrodynamic module in FITOVERT was not affected by the hydraulic load, and the model performance was better when the saturation level was low. According to Giraldi and Iannelli (2009a), either saturated conditions or high hydraulic loads lead to the appearance of dead zones in VFCWs. The FITOVERT hydrodynamic module was also evaluated by Giraldi et al. (2009a), where different saturation conditions were examined, while all model parameters were experimentally determined, except of the dispersivity coefficient.

For describing organic matter and nitrogen biodegradation, Giraldi et al. (2010) used the Activated Sludge Model 1 (ASM1) introduced by Henze et al. (1987), which uses Monod kinetics based on the following equation:

$$r = \mu_{\max} \frac{C}{K_s + C} X \tag{8.2}$$

where r is the reaction rate (mg/L/d), $\mu_{\max}$ is the maximum growth rate (d^{-1}), C is the substrate concentration (mg/L), K_s is the half-saturation constant (mg/L), and X is the biomass concentration (mg/L).

The limitation of ASM1 is that it was developed for modeling activated sludge (i.e., suspended growth biomass), while VFCWs should be simulated as attached growth bioreactor. In order to overcome the difference in the diffusion of dissolved pollutants in the biofilm due to this fact, FITOVERT calibrates the kinetic parameters using half-saturation coefficients and describes the advection-dispersion phenomenon using Bresler (1973) and not van Genuchten (1982) equation (Giraldi et al., 2010). FITOVERT uses an original scheme to describe transport and filtration of pollutants based on Iwasaki (1973) equations.

According to Giraldi et al. (2010), FITOVERT also assumes that oxygen transport is achieved mainly through diffusion (Jury et al., 1991) and neglects the pressure gradient (Massmann and Farrier, 1992); thus, the oxygen diffusion coefficient in free air is corrected using a tortuosity factor (Jury et al., 1991). Furthermore, FITOVERT estimates evapotranspiration using well-established models (Priestley and Taylor, 1972). According to Giraldi and Iannelli (2009a) and Giraldi et al. (2010), transpiration and specifically water uptake from the root zone is described using Richard's equation with the addition of a sink term (Varado et al., 2006), which is correlated with the water potential gradient between the plant and the medium, the hydraulic conductivity, the root density distribution, and the atmospheric demand (Chabot et al., 2002).

8.3.2 CW2D Model

CW2D has also been extensively used in various simulations of VFCWs (e.g., Langergraber, 2001, 2003; Dittmer et al., 2005; Henrichs et al., 2007; Toscano et al., 2009). It mainly focuses on the three major pollutants of municipal wastewater (i.e., organic matter, nitrogen, and phosphorus). CW2D is divided into two submodels, one for flow (quantity model) and one for pollutant transport (quality model) (Langergraber, 2001). These two submodels are interconnected, as the outputs of the flow model (e.g., flow rate, alterations in storage, and internal sources and sinks) are used as inputs to the transport model (Langergraber, 2003). Moreover, outputs of the transport model influence parameters of the flow model. For example, the concentration and temperature (outputs of transport model) influence the kinematic viscosity and water density. The hydrodynamic submodel of CW2D was able to simulate very well water flow and pollutant transport in saturated substrates, while water flow in unsaturated substrates was found very sensitive to various operating parameters (Dittmer et al., 2005).

The transport model of CW2D distinguishes biological oxidation and respective reduction in heterotrophic and autotrophic and attributes processes such as hydrolysis and aerobic oxidations (degradation of organic matter and nitrification) to heterotrophic bacteria (XH), while the two-step nitrification process is attributed to autotrophic bacteria (two species, XANs and XANb) (Langergraber, 2001). CW2D uses kinetic parameters for pollutant reactions using Monod kinetic equations (Langergraber, 2001). The three main assumptions for pollutant reactions in CW2D are that: (a) the temperature influences kinetic parameters and diffusion coefficients, (b) the organic matter mineralization and the denitrification process are results of heterotrophic oxidation, and (c) nitrification is a result of autotrophic oxidation (Langergraber, 2001). According to Langergraber (2001), CW2D includes phosphorus as part of the organic matter, while for the inorganic part of nitrogen and phosphorus, the model uses the adsorption equations based on the Langmuir and other isotherms. Almost all applications of CW2D model used the kinetic parameters of ASM1 model. Langergraber et al. (2009) has also developed Constructed Wetland Model No1 (CWM1), which is based on ASM1, and describes all biochemical and degradation processes occurring in subsurface flow CWs. The main innovation of CWM1 is the consideration of anaerobic degradation occurring simultaneously with aerobic processes (Langergraber et al., 2009). The main kinetic parameters used in CW2D and CWM1 are presented in Table 8.3.

According to Henrichs et al. (2007), simulation of COD degradation using CW2D had good results only for single events and not for a long-term operation. This inadequacy of the model was attributed to the simplifications made for COD decrease (Henrichs et al., 2007). In order to overcome these inadequacies, CW2D has to combine the different organic matter removal processes, such as microbial degradation, filtration, and adsorption (Henrichs et al., 2007). Furthermore, other operating parameters should also be considered, such as the role of the top soil layer in absorption, substrate clogging, and plant root effects (Henrichs et al., 2007). Toscano et al. (2009) also used CW2D to simulate VFCWs and found satisfying results concerning water flow and COD decrease, while simulation results for ammonia were less satisfactory.

Mechanistic model application in VFCWs by various researchers is presented in Table 8.4. These models were mainly tested with data from laboratory-scale VFCWs and they managed to successfully simulate both water flow and pollutant removal. Their main drawback is the overestimation of nitrogen removal at low temperatures. Although CW2D and FITOVERT are the most popular models for VFCWs, as they have a series of advantages compared to other models, they also have certain disadvantages. Table 8.5 includes the main

TABLE 8.3 Kinetic Parameters for Pollutant Removal in VFCWs

References	First order Kinetics								Monod Kinetics		
	COD		BOD		N		P		COD	BOD	N
	k (d^{-1})	C* (mg/L)	k (d^{-1})	C* (mg/L)	k (d^{-1})	C* (mg/L)	k (d^{-1})	C* (mg/L)			
Langergraber et al. (2009)	4.21	55									
Langergraber (2007)											0.1-0.35 for K_X 0.5-25 for K_{s,NH_4}
Sklarz et al. (2010)			39.3-94.1								4.6-12.3[b]
Chang et al. (2012)	99.9[a]				15.3[a]		93.4[a]				
Kantawanichkul et al. (2009)	49.8[a]				30.1[a]		13.5[a]				
Ouyang et al. (2011)					0.1 for nitrification 0.0001 for adsorption 0.0004 for denitrification						
Saeed and Sun (2011b)									34.6[c]	80.7[c]	14.8[c]
Morvannou et al. (2011)											$\mu_{max}=0.0037\ h^{-1}$

K_X, half-saturation constant for hydrolysis.
[a]In m/year.
[b]Ratio of the Monod coefficients X/K_s (h^{-1}).
[c]Kinetic coefficients ($g/m^2/d$).

TABLE 8.4 Model Applications on VFCWs

References	Model	Wastewater Type	Unit	Comments		
				Water Flow	*Tracer Studies*	*Pollutant Removal*
Giraldi et al. (2010)	FITOVERT		Pilot scale	• Increased saturation leads to dispersivity decrease • Porosity and hydraulic conductivity reduction, due to microbial and plant biomass growth, were successfully simulated	Tracer studies proved that both saturated and unsaturated conditions are satisfactorily simulated	ASM1 kinetic parameters could not be successfully used for VFCW simulation, as they refer to suspended growth bioreactors and not to attached growth
Dittmer et al. (2005)	HYDRUS/ CW2D	Combined sewer overflows	Pilot scale	• Satisfying results in single-solute transport and controlled flow filtration rates • Experimental data simulation showed that the unsaturated flow is strongly influenced by the experimental conditions		• Organic matter adsorption and degradation takes place at the upper drainage layer • Organic matter pollutant also occurs in the sediment layer
Langergraber (2007)	CW2D	Municipal	Full scale			COD and NH_4^+-N removal at low temperatures could be easily predicted when half-saturation constants are considered as temperature dependent parameters

Continued

TABLE 8.4 Model Applications on VFCWs—cont'd

References	Model	Wastewater Type	Unit	Comments		
				Water Flow	*Tracer Studies*	*Pollutant Removal*
Henrichs et al. (2007)	CW2D	Municipal	Lab and full scale			• COD simulation provided the same results for both lab and full-scale units • When VFCW operates under steady-state conditions, COD effluent concentrations could be successfully simulated • Simulation of COD degradation should take into consideration biochemical processes, adsorption, and filtration
Langergraber (2008)	CW2D		Lab scale	Effluent flow rates were successfully predicted even during summer		COD and NH_4^+-N removal could not be successfully predicted at low temperatures, due to over-prediction of hydrolysis and nitrification rates
Langergraber et al. (2009)	CW2D	Municipal	Lab scale		Tracer experiment proved that the model could successfully predict flow in VFCWs	
Toscano et al. (2009)	CW2D	Combined sewer overflows	Pilot scale	Effluent flow rates were successfully predicted		COD removal was successfully predicted, while ammonia removal was overestimated

TABLE 8.5 Advantages and Disadvantages of VFCWs Models

Model		CW2D	FITOVERT	HYDRUS
Water flow	+	• Successful simulation of water flow for HSF and VFCWs • Necessary parameters are porosity and hydraulic conductivity	Satisfying hydrodynamic simulation of VFCWs	Tracer experiments showed that HYDRUS could better predict experimental results
	–	• It provides better simulation under saturated conditions (HSF CWs) • Water flow is modeled in one dimension and considered as homogenous	• It cannot successfully model the optimal saturation condition; thus, oxygen transfer and clogging rates are also not successfully modeled. • Water flow is modeled in one dimension and considered as homogenous	
Pollutant removal	+	Successful simulation of single-solute transport and reactive transport for HSF and VF CWs	Simulation of the fate of organic matter, nitrogen, particulates, and oxygen transfer	
	–	It cannot successfully predict pollutant degradation during dry periods		

advantages and disadvantages of these models as they are described in various applications (Langergraber, 2008; Giraldi et al., 2009, 2010; Langergraber et al., 2009), showing the need for further advancing science in the area of VFCW modeling.

8.3.3 Other Mechanistic Models

Other mechanistic models for VFCWs include (Langergraber, 2008):

- HYDRUS (Langergraber and Šimůnek, 2005) can simulate systems receiving high pollutant loads and oxygen transport through the plant root zone (Toscano et al., 2009). HYDRUS is the evolution of CW2D. It can solve the Richards equation for both saturated and unsaturated conditions, while it also includes equations for convection and dispersion of heat and mass transport (Langergraber and Šimůnek, 2005). Most recently, HYDRUS v2 was developed (Langergraber and Šimůnek, 2012), which uses the biokinetic models from CW2D and CWM1.
- McGechan et al. (2005) used a soil model in order to simulate nitrogen loss in a hybrid CW system consisting of one HSF and three VF stages. The model by McGechan et al. (2005) was based on several assumptions, which included: (a) microorganism processes were expressed with first-order kinetic equations; (b) HSF and VFCWs were simulated as plug flow reactors; (c) oxygen transfer through the plant root zone was mainly achieved by diffusion and convection. The simulation results by McGechan et al. (2005) showed an underestimation of ammonia effluent concentration, which was attributed to the underestimation of oxygen transport rate.
- Wanko et al. (2006) also developed a mechanistic model for VFCWs aiming to simulate oxygen transport. Wanko et al. (2006) used Monod equations in order to describe biomass growth, while the hydrodynamic model was based again on Richards' equation.

Ouyang et al. (2010) used the STELLA software environment in VFCW simulation; the code was initially developed for HSF CW simulation by Wynn and Liehr (2001) and predicts organic matter and nitrogen removal, and water and oxygen balances, simulating HSF CWs as either a single CSTR or a series of CSTRs. Wynn and Liehr (2001) model uses Monod kinetics and identifies several variables for organic matter degradation (e.g., plant biomass, plant litter, particulate organic carbon and dissolved organic carbon), while for nitrogen removal, it includes five removal mechanisms (i.e., ammonification, immobilization, nitrification, denitrification, and peat accumulation). A major drawback of this model is that the below ground portion of vegetation is not taken into consideration. The rate of leaf water transpiration was also modeled using empirical equations (Ouyang et al., 2010). Ouyang et al. (2011) also developed a nitrogen removal model using again STELLA software. The major nitrogen removal mechanisms in a VFCW, which this program identifies, include: deposition and hydrolysis of organic nitrogen, nitrification and denitrification processes, leaching through substrate zones, and plant uptake.

Sklarz et al. (2010) developed a model to simulate recirculating VFCWs for TSS, BOD_5, and nitrogen removal. This model simulates recirculating VFCWs as a single CSTR bioreactor. It considers that there is always enough oxygen, and therefore, only nitrification occurs and does not include biofilm processes (Sklarz et al., 2010). According to Sklarz et al. (2010), this model presented satisfactory results on predicting pollutant removal in a VFCW and could give information about the optimal recirculation flow rate.

Kotti (2012) used RZWQM (RZWQM, 2010) model developed by USDA-ARS (United States Department of Agriculture—Agriculture Research Service), which was initially designed for simulating water flow and organic matter, nitrogen and pesticides fate in the vadose zone (Ahuja et al., 1996). RZWQM's (2010) main characteristics include: (a) its ability to simulate different agricultural practices and the effect of them on plant growth; (b) groundwater level fluctuations; (c) groundwater flow through the macro-pores; and (d) nutrient concentrations in and below the root zone (Kotti, 2012). RZWQM is a one-dimensional model (vertical in the soil column). The main RZWQM capabilities, which led Kotti (2012) to use it in VFCWs simulation, were: (a) water movement (Ahuja et al., 1993, 1995), (b) pesticides transport (Ahuja et al., 1993; Ma et al., 1995, 1996; Farahani and Ahuja, 1996; Farahani et al., 1996), (c) evaportranspiration (Farahani and Ahuja, 1996; Farahani et al., 1996), (d) underground drainage (Ma et al., 1998a,b), and (e) plant growth (Nokes et al., 1996; Ma et al., 2000). RZWQM comprises six subsystems, which represent a full agricultural system; the corresponding model inputs include (RZWQM, 2010; Kotti, 2012):

(1) meteorological data (i.e., air temperature, solar radiation, wind speed, relative humidity, rainfall depth),
(2) soil characteristics (i.e., inclination, chemical composition, hydraulic conductivity, porosity, etc.),
(3) plant information (i.e., blooming period).

RZWQM was found by Kotti (2012) to be successful in ammonia prediction ($R^2=0.54$), while for nitrite RZWQM predictions were not as good ($R^2=0.27$).

8.3.4 Pollutant Removal Kinetics

While in initial CW models pollutant removal kinetics were mainly based on first-order equations (Kadlec and Wallace, 2009), recently Monod equations have also been proposed (Saeed and Sun, 2011b). Various parameter values either for first-order or Monod kinetics for pollutant removals in VFCWs are presented in Table 8.3. Saeed and Sun (2011b) developed three models for organic matter and nitrogen removal in a VFCW. These kinetic models were based on the combination of first-order, Monod, and multiple Monod equations, assuming that VFCW is a CSTR. Saeed and Sun (2011b) flow model was based on Wynn and Liehr (2001). In order to overcome the problem of modeling nitrogen removal, which depends on more than one substrates (i.e., food source, oxygen), Saeed and Sun (2011b) used multiple Monod kinetics. Saeed and Sun (2011b) chose $NH_4{}^+$-N and dissolved oxygen concentration to be the two limiting substrates of nitrification, and estimated half-saturation constants of NH_4^+-N and dissolved oxygen (0.05 and 0.25 mg/L, respectively) from previous studies (Hawkes, 1983; Verstraete and Vaerenbergh, 1986; Stenstrom and Song, 1991), while for organic matter removal, the half-saturation constants, which were also retrieved from a previous study by Vaccari et al. (2006), were 60 mg/L for BOD_5, 20 mg/L for COD, and 0.2 mg/L for DO (Saeed and Sun, 2011b). Finally, denitrification, as a more complex phenomenon, had three limiting factors, i.e., nitrate concentration, organic carbon, and anoxic conditions (Saeed and Sun, 2011b). Predicted values by this model and experimental data for both organic matter and nitrogen showed a good correlation (Saeed and Sun, 2011a,b).

Nitrification in VFCWs was also modeled by Morvannou et al. (2011). The model calculated nitrification rates at different depths, based on two previous models (activated sludge models extended to VFCWs) (Henze et al., 1987; Simunek et al., 1999). Literature data were used to test and validate it. The results showed that this model could accurately calculate maximum nitrification rates using influent and effluent pollutant concentrations from the literature (Morvannou et al., 2011).

8.4 CLOGGING MODEL

Austin et al. (2006) developed a model to predict the clogging phenomenon in a tidal flow VFCW. This model uses the Damköhler number (D_a), which expresses the ratio of reaction rate to mass transport for biofilm growth in the CW porous media (Austin et al., 2006). According to Austin et al. (2006), when $D_a \ll 1$, the biofilm growth is limited, while when $D_a \gg 1$, mass transport is limited, thus clogging may occur.

8.5 CONCLUSIONS

Although extensive work has been done in modeling FWS and HSF CWs, the respective literature on VFCW modeling is relatively limited. VFCW hydrodynamic models mostly use Darcy law and van Genucthen equations (van Genuchten, 1982) to simulate wastewater flow in the saturated/unsaturated porous media. The most well-known models for VFCWs are FITOVERT and CW2D. The majority of the available models (Sklarz et al., 2010; Saeed and Sun, 2011b) simulate VFCWs as a CSTR or a series of CSTR and not as PFR, which is mainly used for HSF CWs. Although water flow can be successfully simulated, certain issues exist in pollutant removal simulation. The main issue is the form of kinetics that pollutant removal follows. In most cases of VFCWs (Langergraber, 2003; Giraldi et al., 2010; Sklarz et al., 2010; Saeed and Sun, 2011b), Monod kinetic equations are used for pollutant removal and not the most commonly used first-order equations. The selection of the proper kinetic equation is also a subject of disagreement in HSF CWs as well, as until recently only first-order equations were used (Kadlec and Wallace, 2009), while in the last years other types of equations have been proposed for HSF CWs (Akratos et al., 2008, 2009a,b; Kotti et al., 2013). Another issue concerning pollutant removal modeling is the identification of the parameters influencing nitrification and denitrification, as in some models only DO and ammonia concentration are considered as inhibition factors (Saeed and Sun, 2011b), while other models include organic matter and phosphorus concentrations as well (Saeed and Sun, 2011b).

Chapter 9

General Aspects of Sludge Management

9.1 MUNICIPAL SLUDGE CHARACTERISTICS

Wastewater treatment plants (WWTPs) produce large quantities of various types of sludge and solids in all treatment stages (Metcalf and Eddy, 2003) which include:

- *Screens*: mechanical or manually-cleaned bar screens are used for the removal of large solids.
- *Grit removal*: tanks are used for the removal of grit and scum.
- *Pre-aeration*: wastewater is aerated.
- *Primary sedimentation*: the quantities of produced sludge and scum depend upon the collection system and the possible discharge of industrial waste into the system.
- *Biological treatment*: biological degradation of BOD produces suspended solids, while thickening can be applied to concentrate the sludge stream.
- *Secondary sedimentation*: settling tanks are used for the removal of sludge and scum.
- *Sludge-processing facilities*: sludge, compost, and ashes are produced. The characteristics of the end products depend on the characteristics of the sludge being treated, and the processes used. Regulations for the disposal of sludge residuals are becoming increasingly stringent.

Depending on the wastewater treatment stage, sludge could be categorized in:

- *Primary*: produced during primary wastewater settling.
- *Secondary or biological*: produced during secondary biological treatment.
- *Mixed*: primary and secondary sludge mixtures.
- *Tertiary*: produced during tertiary or advanced wastewater treatment.

☆"To view the full reference list for the book, click here"

TABLE 9.1 Typical Characteristics of Wastewater Sludge (Metcalf and Eddy, 2003)

Parameter	Primary	Secondary
Total solids (TS, %)	2.0-7.0	0.5-2.0
Volatile solids (% of TS)	60-80	50-60
Nitrogen (N, % of TS)	1.5-4.0	2.0-5.0
Phosphorus (P_2O_5, % of TS)	0.8-3.0	0.5-0.7
Potassium oxide (K_2O, % TS)	0-1.0	0.5-0.7
Thermal content (kJ/kg, dry mass)	23,000-30,000	18,500-23,000
Alkalinity (mg/L as $CaCO_3$)	500-1500	580-1100
pH	5.0-8.0	6.0-8.0

Sludge is a glutinous watery material produced during biological aerobic or anaerobic treatment of wastewater. Activated sludge contains high concentrations of nutrients and organic matter with high thermal value; thus, it could be used for a series of applications. This sludge usually has a high moisture content of 98-99% (De Maeseneer, 1997) and contains organic solids (Table 9.1). The moisture content could be even higher in the anaerobic digested sludge. Activated sludge also contains pollutants like heavy metals, synthetic organic compounds, and microorganisms (De Maeseneer, 1997; Kim and Smith, 1997). Therefore, its direct disposal to the environment may cause a series of environmental problems, such as air pollution, public health hazards, and pollution of surface and ground water bodies.

Characteristics of sludge vary depending on the treatment stage, as shown in Table 9.2. Sludge contains a series of valuable compounds (i.e., organic carbon, phosphorus, and nitrogen). The main form of TKN in sludge is ammonia, which could be used as fertilizer. The sludge composition depends on the pollutant load and heavy metal concentration of the treated wastewater (Table 9.3) and is a crucial parameter in sludge management, since it defines the proper treatment method which should be used. Sludge substances could be grouped as follows (Rulkens, 2004):

I. Nontoxic organic substances (~60% of dry mass), TKN, phosphorus.
II. Toxic pollutants: heavy metals (Zn, Pb, Cu, Cr, Ni, Cd, Hg, As), PCBs, PAHs, dioxins, pesticides, etc.
III. Pathogenic microorganisms and microbial pollutants.
IV. Inorganic substances (e.g., silica, aluminum, calcium, and magnesium).
V. Moisture up to 99%.

TABLE 9.2 Typical Characteristics of Wastewater Sludge Per Treatment Stage (EC, 2001)

Substance	Unit	A	B1	B2	C	D
Dry solids (DS)	kg/m^3	12	9	7	10	30
Volatile solids (VS)	%DS	65	67	77	72	50
pH		6	7	7	6.5	7
C	%VS	51.5	52.5	53	51	49
H	%VS	7	6	6.7	7.4	7.7
O	%VS	35.5	33	33	33	35
N	%VS	4.5	7.5	6.3	7.1	6.2
C/N		11.4	7.0	8.7	7.2	7.9
P	%DS	2	2	2	2	2
Cl	%DS	0.8	0.8	0.8	0.8	0.8
K	%DS	0.3	0.3	0.3	0.3	0.3
Al	%DS	0.2	0.2	0.2	0.2	0.2
Ca	%DS	10	10	10	10	10
Fe	%DS	2	2	2	2	2
Mg	%DS	0.6	0.6	0.6	0.6	0.6
Fat	%DS	18	8	10	14	10
Proteins	%DS	24	36	34	30	18
Thermal power	kWh/ton DS	4200	4100	4800	4600	3000

A, primary sludge; B_1, secondary biological sludge (load > 0.20 kg BOD_5/kg MLSS); B_2, secondary biological sludge (load < 0.20 kg BOD_5/kg MLSS; Mixed Liquor Suspended Solids); *C*, mixed sludge; *D*, digested sludge.

A main issue nowadays in wastewater treatment is the treatment of the produced activated sludge. Mean sludge production in European countries is ~0.09 kg dry mass/p.e. with moisture content above 90% (Chen et al., 2002). Today, due to stricter legislation, the use of untreated sludge for agricultural applications is avoided. These stricter legislation rules are imposed mainly due to heavy metal concentration in sludge and respective potential environmental and public health risks.

Although produced sludge usually corresponds to less than 1% of the wastewater treatment stream, its handling and management costs can reach up to 40-50% of the total operational cost of a WWTP (Campbell, 2000). Adding the problems deriving from high pollutant loads, pathogenic microorganisms and

TABLE 9.3 Typical Concentrations of Heavy Metals in Domestic Wastewater Sludge (Chen et al., 2002)

Metal	Range	Average
	mg/kg dry mass	
As	1.0-230	10
Cd	1-3500	10
Cr	10-95,000	500
Cu	84-18,000	800
Pb	10-25,000	500
Hg	0.5-60	7
Ni	2-5500	80
Se	1.2-17.5	5
Zn	100-50,000	1700
Fe	1000-150,000	17,000
Co	11.0-2500	30
Sn	2.5-340	16
Mn	32-9950	250

odors, the sludge treatment becomes a crucial matter to be addressed. The first crucial step in sludge management is the reduction of the water content, in order to achieve an optimal solid concentration (Kim and Smith, 1997). Sludge treatment in the future should combine low-cost, reliable technology and public acceptance (Campbell, 2000). In the USA, the public has accepted sludge treatment facilities after understanding that treated sludge can be used as a source of recovered valuable elements (Campbell, 2000).

9.2 SLUDGE HANDLING AND MANAGEMENT—THE PROBLEM

The main targets of sludge treatment are volume reduction and substance degradation. The most common conventional processes include sludge thickening, dewatering, and drying. Figure 9.1 presents the sludge solid content (% of dry mass) during the sludge dewatering processes. After thickening, the sludge mass is usually reduced up to 70% of the initial value, due to water loss, and the solid content reaches 2-3%. With dewatering, the water content is further reduced and the solids content is increased up to 18-20%. Sludge drying, which

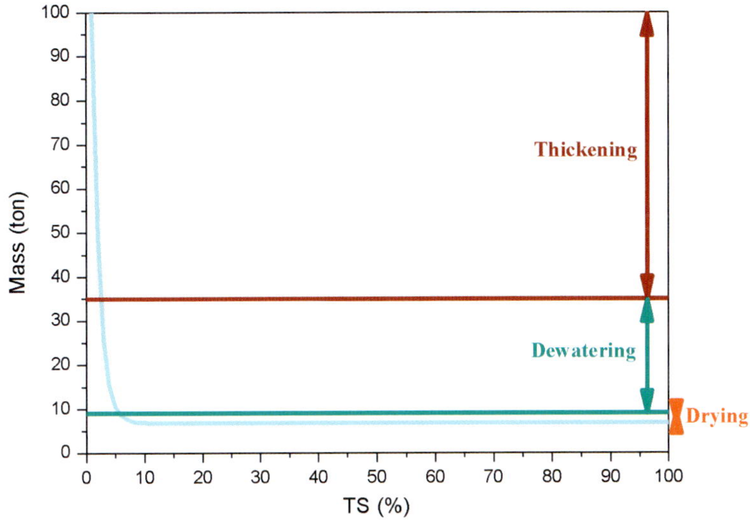

FIGURE 9.1 Sludge treatment processes and solid content (% dry mass) and mass (ton) variations.

is the final step, aims to further reduce the water content (<10%). When thermal drying is used, the sludge is exposed to high temperatures; thus, there is a high energy requirement. Typically, for drying 1 ton of sludge with 10% solids and 90% water content, 2.5-3.0 × 10^6 kJ of thermal energy are needed. Therefore, a sludge drying method, which would reduce the water content without external energy input, is crucial.

Today, the field of sludge dewatering and drying includes various methods and technologies, such as (Dentel, 2001; Chen et al., 2002; Wakeman, 2007):

- *Mechanical dewatering*: vacuum filters, gravity belt thickening, filter belt press, gravity thickening, centrifuge, and membrane press.
- *Direct drying systems*: rotating drums, lamps, belt dryers, spray dryers, solar energy dewatering, etc.
- *Indirect drying systems*: rotaplate indirect dryer, kneading and self-cleaning disc dryer, porcupine processor, K-S Nara paddle and paddle dryer, etc.
- *Fluidized bed dryers.*
- *Combination of drying and incineration.*
- *Aerobic and anaerobic digestion.*
- *Composting.*
- *Calcification.*
- *Drying sand beds.*

According to Dentel (2001), the advantages and disadvantages of the most common methods for sludge dewatering could be summarized as follows:

Gravity thickening: it is the oldest and most simple method for sludge dewatering. Its main advantages include low energy consumption and low operational costs. The main drawback of this method is that the TS content in the final product ranges only between 5% and 8%. Thus, the main applications of gravity thickening include primary solids and solids from attached film processes.

Flotation: Dissolved air flotation is mainly used when the density of the solids is low and, therefore, other mechanical methods could not be used, since the TS content of the final product is only 3.5-4.5%.

Gravity belt thickening: Gravity belt thickening is also a well-established method, since it can provide a final TS content of 5-12%. Furthermore, it requires less area compared to gravity thickening. The main drawback of this method is that it needs the addition of chemicals and has high energy consumption; thus, it possesses increased operational costs.

Thickening by centrifugation: similar to gravity belts, centrifuges can provide a final product with increased TS content, but they have high energy requirements. Furthermore, the capital cost for centrifuges is higher than gravity belt thickeners.

Sand drying bed: it is a very simple method, which still remains in widespread use. The main drawbacks of this method are that it has high area requirements and, usually, odor problems appear, due to the exposure of the sludge in an open area. On the other hand, its capital and operational costs are low, since there is no need for any mechanical equipment or chemical addition.

Belt filter presses: these systems (Figure 9.2) are the most common method used in WWTPs nowadays, as they have lower operational costs, compared to centrifuges, and provide a final product with TS content up to 30-40%. The odor drawback has been solved by placing the belt filter presses in closed areas.

Among the various methods and technologies, aerobic/anaerobic stabilization appears as the most common method in Europe, with applications in 24 out

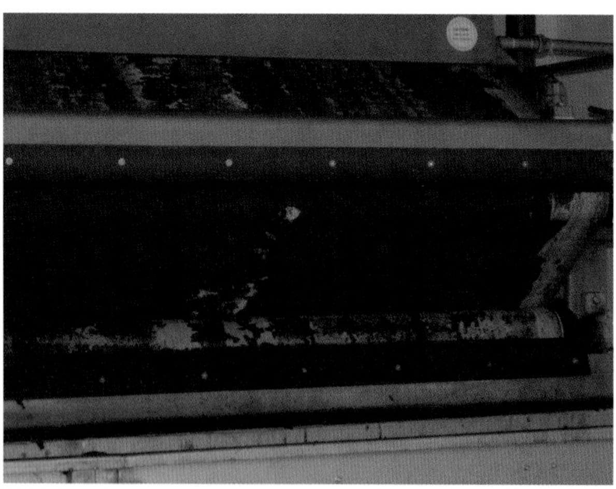

FIGURE 9.2 Photo of belt filter press in a WWTP in North Greece.

of 27 countries (Kelesidis and Stasinakis, 2012). Composting is also widely used, but as a final sludge disposal method. Sludge drying plants also met increasing interest, especially in Germany, Italy, UK, and France.

The selection of the proper disposal method for the dried sludge is based on the following criteria (EEA, 1997):

I. *Environmental*: sludge quality and quantity characteristics, adequacy of treatment, infrastructures for sludge storage, and transportation.
II. *Economical*: capital cost, annual operation cost, income from the exploitation of the commercial value of the final product through reuse.
III. *Technical*: assessment of flexibility, simplicity, and reliability of each alternative.
IV. *Social*: informing the public and stakeholders on the proposed alternatives, presenting the measures taken and the rights and obligations of stakeholders in order to ensure social acceptance.

Table 9.4 presents sludge production and treatment in various countries. Small WWTPs usually face the problem of producing small sludge quantities. Therefore, these plants cannot successfully apply expensive treatment methods, due to high investment and operation costs needed (Kim and Smith, 1997). Based on the directive 91/271/EC (EU, 1991), from 31 December 1998 sludge is not allowed to be disposed in surface water bodies.

9.3 LEGISLATION

The EU legislation focuses on the sludge reuse in agriculture, while other disposal or treatment methods are currently covered by the wastewater legislation. Directive 86/278/EC (EU, 1986) regulates sludge reuse in agriculture, in order to prevent potential negative effects on soil, vegetation, animals, and humans. This directive sets limit values for concentrations of heavy metals in soil and sludge, as well as limits for the annual amount of heavy metals that may be introduced into the soil. Although Directive 86/278/EC (EU, 1986) sets values for heavy metals, these limits concern only original sludge or raw sewage and not the final product after treatment (Andreadakis et al., 2002). Directive's 86/278/EU heavy metal limits have been adopted by all European countries and, in some cases, national legislation (e.g., in Denmark, the Netherlands, Sweden, Greece) imposes even stricter limits (Andreadakis et al., 2002). Concerning pathogenic microorganisms, the European Commission has proposed a limit of 5×10^2 *E. coli* CFU/g as the fecal bacteria indicator (EC, 2000). This proposed limit value has been recently increased to 5×10^5 *E. coli* CFU/g based on the European document on Sludge and Bio-waste (EC, 2010a).

On the other hand, the US EPA has adopted stricter rules for sludge disposal (agricultural and nonagricultural land; USEPA, 1993). USEPA main regulations include (USEPA, 1993):

TABLE 9.4 Production and Treatment of Sludge in Various Countries (Chen et al., 2002; Kelesidis and Stasinakis, 2012; http://epp.eurostat.ec.europa.eu)

Country	Dry Mass (ton/year) (×1000)	Agricultural Use (ton/year) (×1000)	Landfill (ton/year) (×1000)	Compost and Others (ton/ year) (×1000)
USA	5358	1768	1822	1752
Japan	~2000		500	1500
Belgium	176.32	17.27	0	0
Bulgaria	49.76	13.64	25.03	1.08
Czech Republic	196.29	100.57	16.32	72.52
Denmark	141	74	1.41	:
Germany	2170	566.3	0	317.36
Estonia	18.28	0.8	2.09	15.23
Ireland	90	82.67	0.19	7.13
Greece	146.95	5.96	79.81	0
Spain	1205.12	995.06	95.67	–
France	966.38	726.98	42.3	–
Croatia	30.34	–	–	–
Italy	1102.65	315.61	462.2	–
Cyprus	8.12	6.67	0	0
Lithuania	51.31	7.9	0.48	11.2
Luxembourg	9.74	5.16	0	–
Hungary	170.34	56.77	1.36	22.06
Malta	1.24	0	1.24	0
Netherlands	350.95	0	0	0
Austria	262.81	44.36	20.84	83.04
Poland	526.7	109.3	58.9	30.9
Romania	82.07	1.93	40.5	1.33
Slovenia	29.86	0.45	2.96	0.22
Slovakia	54.76	0.92	6.7	35.29
Sweden	203.52	50.46	7.54	65.21
United Kingdom	1419.06	1118.16	8.79	–

- These standards apply only for sludge use and disposal except landfill disposal and co-incineration with other municipal wastes.
- These rules apply for sludge originating only from domestic wastewater treatment and from industrial wastewater treatment.
- Certain limits (specific numerical limits or equations for calculating these limits) for: (a) heavy metals, (b) total hydrocarbons, and (c) pathogenic microorganisms (i.e., *Salmonella* spp., *E. coli*, enteric viruses, and viable helminth ova).
- These limits were set based on previous environmental assessment on human health impact or environmental criteria.
- Pollutant limits are stricter when sludge disposal areas are near surface or drinking water sources.

The main rules for sludge treatment and disposal presented in the Directive 86/278/EU can be summarized as follows (Spinosa, 2001):

- Heavy metal concentrations in sludge used in agriculture should not exceed the limits set by this Directive.
- Sludge used for disposal in agriculture must have been treated (biological, chemical, or heat treatment).
- The total mass of sludge added per area unit in an agricultural field should take into account the nutrient needs for plant growth and should secure soil, surface, and groundwater quality.
- In soils with pH <6, heavy metal mobility should be considered before sludge disposal.

Directive 86/278/EU is expected to be revised to address issues concerning advanced sludge treatment methods, such as (Spinosa, 2001):

- thermal drying,
- thermophilic aerobic stabilization,
- thermophilic anaerobic digestion,
- pasteurization of liquid sludge,
- chemical stabilization with lime at pH equal or above 12 for 3 months.

Although Directive 1991/156 is mainly referring to solid wastes, it contains certain regulations for sludge management, as it sets the basic principles for waste treatment, utilization, and disposal (Spinosa, 2001). Directive 1991/156 imposes certain requirements for wastewater and sludge treatment/disposal (Spinosa, 2001):

- land application, including spreading on agricultural and forest land, disturbed soils for reclamation, dedicated beneficial use sites, etc.
- composting.
- incineration.
- landfilling.

Chapter 10

Sludge Treatment Wetlands—Basic Design Considerations

10.1 INTRODUCTION

Sludge Treatment Wetlands (STWs) are a development of the last 15-20 years and represent a state-of-the-art solution to sludge treatment. The general concept of STW systems is similar to VFCWs for wastewater treatment with some modifications for the introduction and treatment of sludge. The vast majority of STW systems exists in various European countries and the USA (Nielsen, 2011). In Europe, several systems exist in Denmark, France, Germany, Poland, Sweden, Norway and Spain (Figures 10.1 and 10.2). It is reported that in France about 300 STW facilities are in operation (Figure 10.3), while in Denmark more than 120 systems have been implemented (Nielsen, 2011).

Generally, most of the experience comes from systems implemented in Denmark, where STWs are in use since 1988 (Nielsen, 2003). STWs in Denmark are mainly found in small villages with population ranging from 5000 to 10,000 inhabitants. About 30% of the produced sludge in Denmark is treated in STW facilities (Nielsen and Willoughby, 2005). The majority of Danish STWs have a capacity of 200-1000 ton dry mass(DM)/year, while the systems with largest capacity treat 2000-3000 ton DM/year (Nielsen, 2011). Gradually, STWs attract interest from other countries too, especially from countries in the Mediterranean basin (Figures 10.4 and 10.5) (e.g., Greece, Spain, and Italy; Stefanakis and Tsihrintzis, 2012c; Uggetti et al., 2012a; Peruzzi et al., 2013), where the climatic conditions are favorable for these systems, as also from countries in the tropics (Koottatep et al., 2005; Kengne et al., 2009a).

Based on the gained experience until today, the key to the effective operation of STWs is proper design and construction. However, despite the progress and the increasing research performed on STWs over the last years, there are still specific uncertainties on the system design, while there is not a generally accepted setup or design guidelines. STW design is usually based on sludge quality characteristics and on the local climatic conditions (Nielsen and

☆"To view the full reference list for the book, click here"

FIGURE 10.1 A Sludge Treatment Wetland facility in Groß Lafferde, Lahstedt, Niedersachsen, Germany.

FIGURE 10.2 Inlet point of a Sludge Treatment Wetland bed in Groß Lafferde, Lahstedt, Niedersachsen, Germany.

FIGURE 10.3 A Sludge Treatment Wetland bed in a conventional wastewater treatment plant in Alba La Romaine, France.

Willoughby, 2005). It is essential for the designers to know the annual WWTP sludge production (tons of dry mass) and the sludge production duration. In order to maintain a safety factor, it is proposed that STWs should be designed with 10-20% higher capacity than the nominal calculated (Nielsen, 2003). Key factor in STW design is the determination of an appropriate sludge loading rate (SLR; kg DM/m^2/year), usually based on the climatic conditions. Generally, an STW facility should meet the following general parameters (Nielsen, 2003, 2011):

- The first operational phase lasts between 8 and 12 years (depending on the SLR and the design), before the first sludge residue removal.
- The treatment capacity of the whole STW facility should be maintained stable during the sludge residue removal from one or more beds.
- STW vegetation should be regenerated after sludge residue removal without planting new plants.

FIGURE 10.4 A Sludge Treatment Wetland bed in Nea Madytos, Chalkidiki, N. Greece (Gikas et al., 2011).

FIGURE 10.5 A Sludge Treatment Wetland bed in Alcochette, Portugal.

10.2 BASIC DESIGN CONSIDERATIONS

As STWs are a relative new sludge treatment technology, until now there is not a standard methodology for their design. STW dimensioning is mainly based on empirical rules and usually takes into consideration sludge production (tons of dry matter per year), time of sludge production, sludge quality, sludge type, and climatic conditions (Nielsen, 2003). A STW bed operates in loading cycles; each cycle consists of a loading period (usually few days or even few weeks) followed by a resting period (usually longer than the loading period). Thus, in order to accurately design (i.e., size the required surface area and the number of beds) a STW facility, SLR and applied loading and resting period durations are essential (Uggeti et al., 2010). These two characteristics will ultimately determine the life span of each operation phase. However, feeding patterns and resting periods are not standardized.

10.2.1 Operational Lifetime

STWs operational life is usually at least 30 years and is divided into two or three operational phases of 8-12 years (Nielsen and Willoughby, 2005; Nielsen, 2011, 2012; Stefanakis and Tsihrintzis, 2012c). Once the bed is constructed and the reeds are planted, light loadings start (commissioning stage). An initial commissioning period is usually applied, which allows for the reed growth and for their better adjustment to the bed environment. The commissioning period could last up to 2 years (Nielsen, 2003), after which the main operational period with sludge loadings begins. Shorter commissioning periods can also be applied, e.g., a few months in warmer climate like the Mediterranean basin (Stefanakis and Tsihrintzis, 2012c). At the end of each operational phase, the accumulated residual sludge is removed and the STW is emptied. During the sludge residue removal, a sludge layer of 10-15 cm thick is left on the bed surface, in order to allow for vegetation regeneration and the beginning of a new operational phase. When STWs are properly designed, the TS content after 8-12 years of operation could exceed 40% (depending on the SLR and the climate), while sludge mineralization is also achieved, as more than 25% of the organic matter is usually removed (Nielsen, 2003; Nielsen and Willoughby, 2005; Stefanakis and Tsihrintzis, 2012c).

10.2.2 Sludge Loading Rate

The SLR is probably the most crucial design parameter in STW operation and expresses the amount of dry solids (total solids; TS) or dry mass (DM) loaded onto the bed per surface unit in a year, usually expressed as kg TS/m^2/year. The selection of an appropriate SLR is based on the sludge characteristics and the climatic conditions. For example, the SLR proposed for STWs in Denmark is 60 kg DM/m^2/year for activated sludge and 50 kg DM/m^2/year for sludge with high fat content or with low age (<20 days) (Nielsen, 2003). Burgoon et al. (1997) increased the tested SLR from 9.8 to 65 kg DM/m^2/year. Uggeti et al. (2010) reports even higher loading rates up to 125 kg DM/m^2/year tested in Spain, especially in the summer. On the other hand, Nielsen (2003) mentioned that loading rates up to 100 kg DM/m^2/year were proved to be extremely high, and suggested loading rates of 50 kg DM/m^2/year for digested sludge and of

60 kg DM/m^2/year for activated sludge. Obarska-Pempkowiak and Kołecka (2006) reported various SLRs in STW facilities in Europe which range from 15 to 60 kg DM/m^2/year. Higher loading rates, up to 75 kg DM/m^2/year, are suggested by Stefanakis and Tsihrintzis (2012c) in Greece.

Generally, it is suggested that SLRs applied during the commissioning phase should be lower, e.g., half of the designed value (Nielsen, 2011), in order to allow for plant growth and adjustment and avoid potential harmful effects on plants during their first growth period (Nielsen, 2003). Selection of a relatively high SLR could result in system overloading, plant shock, and thus, reduced performance and even operational failure.

10.2.3 Sludge Quality and Origin

Sludge origin and respective sludge quality can also affect the global efficiency of the STW system. For example, lower SLRs (e.g., half than the ones mentioned above) should be applied when the sludge to be treated is not aerobic. High concentration of organic compounds, like fats and oil, in the influent sludge is also indicative of low dewaterability, due to lower density of sludge particles and respective higher volume of capillary water (Nielsen, 2011). It is reported that there is a positive correlation between the reduction of the volatile solids (VS) content and the dewatering efficiency (Kopp and Dichtl, 2000). Therefore, sludge quality and characteristics should be taken into account during the design stage of an STW system.

10.3 FACILITY LAYOUT

STW layout is similar to that of VFCWs and traditional sand drying beds. STWs are usually rectangular concrete basins (Uggetti et al., 2010) or trapezoidal earthen basins (Nielsen, 2003). A typical schematic representation of a STW cross-section is presented in Figure 10.6. The STW bottom is either constructed by concrete or is simply excavated in soil and covered by a low-permeability material (i.e., HDPE geomembrane covered by both sides by geotextile or clay) for sealing off the bed and prevent leaching. The STW bottom has usually a minimum inclination of 1%, in order to collect by gravity the drained water that percolates through the upper porous media to the bottom (Uggeti et al., 2010). Drained water collection is made through a series of perforated plastic pipes placed above the low permeability material. Usually these pipes are placed within a layer of cobbles, in order to avoid clogging of pipe holes from the porous media grains. The drainage system is critical to avoid anaerobic conditions in the drainage layer. This drainage system is practically similar to the one used in VFCWs for wastewater treatment, as shown in Chapter 6.

In order to enhance aeration, another series of vertical perforated pipes are connected to the drainage pipes. Aeration pipes are perforated at the bottom and extend above the media layers (e.g., up to 1 m above the bed surface, so that they

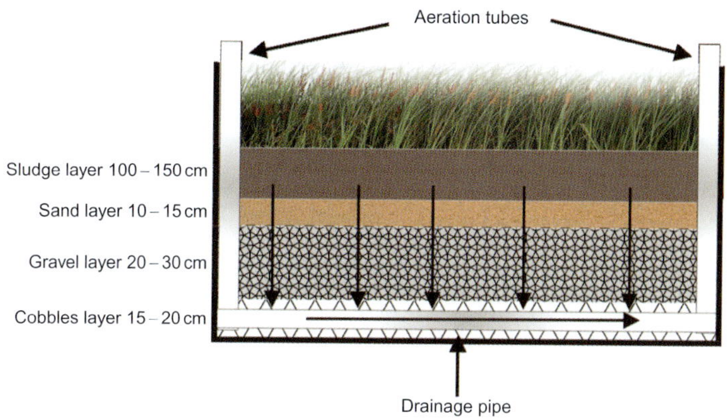

FIGURE 10.6 Cross-section of a typical Sludge Treatment Wetland (schematic representation).

will not get clogged with the gradual formation of the residual sludge layer), thus, enhancing aeration through the gravel filter and sludge layer (Uggeti et al., 2010). The presence of aeration pipes was also found to enhance sludge dewatering (Heinss and Koottatep, 1998; Edwards et al., 2001; Stefanakis and Tsihrintzis, 2012c).

The sludge loading system is also a critical parameter, as it should ensure a uniform distribution of the sludge on the entire STW surface. Thus, the network of feeding pipes should be appropriately designed to cover the entire bed surface. For example, typical dimensioning dictates the placement of a feeding line every 5 m (max) across the bed length. Moreover, pump specifications should be selected after taking into account the following considerations (Nielsen, 2003):

- the operational sludge flow should be at least of 0.15 m/h at the filter surface,
- the sludge viscosity,
- the basin size, and
- the number of sludge distributors per basin (maximum 250 m^2 per sludge distributor).

Various porous media layers are placed above the bottom of the bed. Usually, different gravel layers with increasing grain size from top to bottom are used (Table 10.1). The total thickness of the bed varies from 30-70 cm, as Table 10.1 shows. Usually, in order to avoid clogging and hydraulic failure, the top layer consists of sand (Obarska-Pempkowiak et al., 2003; Nielsen, 2005; Troesch et al., 2009a), although systems without a top sand layer have also been successfully tested (Peruzzi et al., 2009; Gagnon et al., 2012; Stefanakis and Tsihrintzis, 2012c). Typical thicknesses for each gravel layer are 15-20 cm for cobbles or large stones for the bottom drainage layer,

TABLE 10.1 Porous Media Layers and Respective Thicknesses in Various STWs

Number of Layers	Total Bed Thickness (cm)	Porous Media Used, Grain Size (mm), and Respective Layer Thickness (cm)	References
2	30	Sand (15 cm) Drain rock (15 cm)	Burgoon et al. (1997)
4	50	Sand (10 cm) 6 mm washed pea gravel (15 cm) 12 mm round washed gravel (10 cm) 5 cm round washed gravel (15 cm)	Summerfelt et al. (1999)
2	60	Sand (10 cm) Gravel (50 cm)	Zwara and Obarska-Pempkowiak (2000)
3	65	Fine sand (10 cm) Small gravel (15 cm) Large gravel (40 cm)	Koottatep et al. (2001)
2	60	0.5-1 mm coarse sand (30 cm) 3-6 mm pea stone or gravel (20-30 cm)	Begg et al. (2001)
4	50	1-2 mm rounded sea gravel (10 cm) 5-10 mm pea gravel (15 cm) 20-30 mm gravel (10 cm) 30-60 mm stones (15 cm)	Edwards et al. (2001)
3	55-60	Sand (15 cm) Geotextile Pebble gravel (30-45 cm)	Nielsen (2003)
3	60	0.8 mm sand (10 cm) 2/4 mm medium gravel (20 cm) 8/16 mm coarse gravel (30 cm)	Obarska-Pempkowiak et al. (2003)
4	70	Sand (20 cm) 5-10 mm gravel (10 cm) 10-30 mm gravel (20 cm) 30-50 mm gravel (20 cm)	Nassar et al. (2006)
4	60	Coarse sand (10 cm) Fine sand (10 cm) Small gravel (20 cm) Large gravel (20 cm)	Cui et al. (2008)
3	55	50-50% mixture of peat and crushed pine bark (20 cm) 4-16 mm small cobbles (20 cm) 50-150 mm large cobbles (15 cm)	Wang et al. (2009b)

TABLE 10.1 Porous Media Layers and Respective Thicknesses in Various STWs—cont'd

Number of Layers	Total Bed Thickness (cm)	Porous Media Used, Grain Size (mm), and Respective Layer Thickness (cm)	References
2	40	5 mm gravel (15 cm) 40-70 mm gravel (25 cm)	Giraldi et al. (2009b)
1	20	16-32 mm gravel (20 cm)	Chen et al. (2009a)
4	50-55	Sand (5 cm) or green compost (10 cm) 2/6 mm gravel (20 cm) 15/25 mm gravel (10 cm) 30/60 mm peebles (15 cm)	Troesch et al. (2009a)
2	40	5 mm gravel (15 cm) 40/70 mm gravel (25 cm)	Peruzzi et al. (2009)
2	55	0.3-1 mm sand (25 cm) 10-30 mm gravel (30 cm)	Uggetti et al. (2009)
4	45	Sand (10 cm) Fine gravel (10 cm) Coarse gravel (15 cm) Cobbles (10 cm)	Melidis et al. (2010)
4	35	0.4-2.3 mm gravel (5 cm) 1.1-2.3 mm gravel (5 cm) 2-6 mm gravel (20 cm) 20-40 mm gravel (5 cm)	Gagnon et al. (2012)
3	65	Sand (15 cm) Fine gravel (20 cm) Coarse gravel (30 cm)	Matamoros et al. (2012)
4	45	Fine gravel (15, 0 cm) Medium gravel (15 cm) Cobbles (15, 25 cm)	Stefanakis and Tsihrintzis (2012c; 2012d)

20-30 cm for medium gravel, and 10-15 cm for sand (Uggeti et al., 2010; Stefanakis and Tsihrintzis, 2012c). Above the sand layer, pipes for sludge loading are placed either at a corner of the STW bed or along the STW surface (Uggeti et al., 2010). A free board of 1.0-1.5 m above the porous media layer is essential and should be foreseen in the design to allow for the gradual accumulation of the residual sludge.

The upper layers of gravel (diameter from 2 to 10 mm) or sand (diameter from 0.5 to 1 mm), which are used as filtration layers, are also helpful for plant establishment and rooting (Uggeti et al., 2010). Stefanakis and Tsihrintzis (2012c) reported that porous media with higher porosity (i.e., when only coarse-grained materials are used) result in increased drained water volume and decreased evapotranspiration rate.

A STW facility includes several beds, which allows for the alternating operation and application of loading and resting periods, and ensures the continuous operation of the facility (Nielsen, 2003). The number of beds is based on the sludge production, the climatic conditions, and the cycles of loading/resting periods to be applied, and ranges from 3 (Uggetti et al., 2009) to 18 (Nielsen, 2003), which correspond to 400 and 123,000 pe, respectively. According to Nielsen (2003), a minimum of eight beds is usually necessary, while the number of beds should be increased depending on the facility capacity; facilities with capacities of up to 500, 500-1000, and more than 1000 ton DM/year should include 8, 10, and 12-14 basins, respectively. The bed area ranges from 4.5 m^2 (Summerfelt et al., 1999) to 1000 m^2 (Nielsen, 2007). The STW area per capita (Table 10.2) ranges from 1.5 to 4.0 pe/m^2 (De Maeseneer, 1997; Stefanakis and Tsihrintzis, 2012c).

10.4 OPERATION AND FEEDING STRATEGY

The full life cycle of an STW facility includes the following stages, as shown in Figure 10.7 (Nielsen, 2003):

- *Commissioning stage*: this stage includes the time period after reed planting up to the second growing season. During this period, applied SLRs are usually lower, in order to assist plant growth and allow them to adjust to the new environment (Nielsen, 2003). Edwards et al. (2001) recommended a shorter commissioning stage of 1 year. Stefanakis and Tsihrintzis (2012c) applied an even shorter commissioning stage of <6 months, before the start of loadings with the designed SLR. Nielsen (2003) suggests that until the sludge layer reaches a thickness of 30 cm, vegetation should be additionally watered, in order to assist plant growth.
- *Main operational stage* (*loading and resting periods*): this stage includes sludge loading cycles. Each cycle consists of the loading period, which usually lasts a few days, followed by a resting period, which is usually prolonged—compared to the loading period—in order to allow for the dewatering and drying of the loaded sludge, through drainage and evapotranspiration. Each STW bed is loaded with a specific sludge volume during each loading cycle. This amount of sludge is calculated based on the chosen SLR (kg DM/m^2/year), the TS content of the influent sludge, and the chosen loading cycle duration. Loading and resting period durations depend on: (a) the STW facility age, (b) the TS content of the feed sludge, (c) the sludge layer thickness, (d) the number of beds, and (e) climatic conditions and

TABLE 10.2 Number of Beds, Total Surface Area, and Sludge Loading Rate in Different Studies on Sludge Treatment Wetlands

Number of Beds	Surface of Each Bed (m^2)	Sludge Loading Rate (kg TS/m^2/year)	References
25	1000	65	Burgoon et al. (1997)
6	4.5	30	Summerfelt et al. (1999)
–	495	–	Begg et al. (2001)
2	240	–	Obarska-Pempkowiak et al. (2003)
8	500	60	Nielsen and Willoughby (2005)
10	1050	60	Nielsen (2007)
8	468	22-44	Troesch et al. (2009a)
3	66	55	Uggetti et al. (2009)
6	54	51	
7	25	125	
2	140	140	Melidis et al. (2010)
5	75	45	Bianchi et al. (2011)
6	20	25-50	Nielsen and Cooper (2011)
10	818-1308	50	Matamoros et al. (2012)

seasonal variations. A common daily loading schedule includes 1-3 partial loadings per basin (Nielsen, 2003).

– *Final resting stage*: after the main operational stage, a final resting stage is applied without new loadings onto the STW bed. This stage could last from 1 to 2 months up to 1 year (Uggeti et al., 2010; Stefanakis and Tsihrintzis, 2012c). The aim of this stage is to improve the characteristics of the final sludge product. Stefanakis et al. (2011b) reported that this final resting stage significantly improves sludge dewatering, stability, and maturity.
– *Emptying the basin*: during this stage, the residual sludge layer is removed from the bed. A shallow sludge layer of 10-15 cm is left on the bed, in order to allow for the plant regeneration and the beginning of a new operational phase.

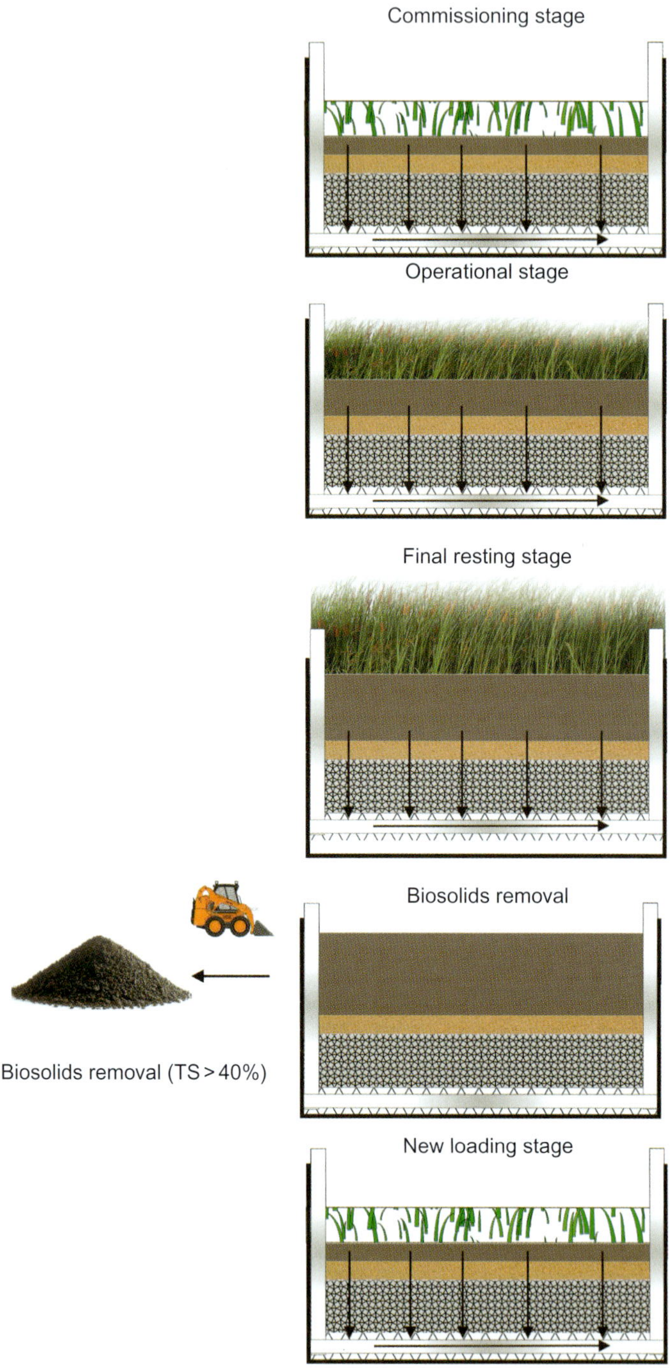

FIGURE 10.7 Sequential stages of operation during the entire lifetime of an STW facility; commissioning, operation phase, final resting phase, sludge residue removal, plant regeneration, and new operation phase.

In Denmark, most of STW systems operate with 7-8 days of loadings followed by 55-65 days of resting. In other cases, loading and resting periods had lower duration of 2-3 and 14-21 days, respectively (Uggeti et al., 2010). Table 10.3 presents various feeding cycles applied in several STWs in different countries. As it is obvious, there is not a generally accepted and applied feeding regime. The loading period usually lasts from 1 to 7 days, while the length of the resting period varies. The level of the applied SLR and the climatic conditions seems to play a dominant role in the selection of an appropriate resting period.

10.5 COMMON OPERATIONAL PROBLEMS

An STW facility can face a series of problems during its lifetime, mainly due to inappropriate design and operation. Common operational problems that have been observed in STWs include (Table 10.4):

- insufficient dewatering,
- poor plant growth,
- odour production,
- insufficient mineralization.

TABLE 10.3 Feeding Regimes (Loading-Resting Periods) Applied in STWs in Different Countries

Country	SLR (kg TS/m^2/year)	Feeding Cycle: Loading-Resting Period (days)	References
France	55	(i) Daily (ii) 1-2/12-13 (iii) 3-7/21-25	Liénard et al. (1995)
NE USA	65	1/14	Burgoon et al. (1997)
Thailand	250	1/7	Koottatep et al. (2001)
UK	67.7	1/2	Edwards et al. (2001)
Poland	60	3-4 times/year	Hardej and Ozimek (2002)
Denmark	60	1-4/10-20 4-5/30-40 5-6/50-55 7-8/55-65	Nielsen (2003)
Italy	10	1/25 1/10 1/13	Barbieri et al. (2003)
Poland	60	1/42	Obarska-Pempkowiak et al. (2003)

Continued

TABLE 10.3 Feeding Regimes (Loading-Resting Periods) Applied in STWs in Different Countries—cont'd

Country	SLR (kg TS/m^2/year)	Feeding Cycle: Loading-Resting Period (days)	References
Palestine	45	1/3 1/7	Nassar et al. (2006)
Cameroon	100-300	1/7	Kengne et al. (2008)
China	252	1/1 1/4	Cui et al. (2008)
France		1/30	Wang et al. (2009b)
Italy	38	1/14 (winter) 1/7 (summer)	Giraldi et al. (2009b)
France	30-36	14/98	Troesch et al. (2009a)
Spain	51 55	2/2 2/10	Uggetti et al. (2009)
Greece	140	1/20	Melidis et al. (2010)
France		1/60 (winter) 1/30 (spring, autumn) 1/15 (summer)	Wang et al. (2010b)
UK	15-50	1-5/35-55	Nielsen and Cooper (2011)
Italy	45	Once in winter Once in summer	Bianchi et al. (2011)
Denmark	55	1/35	Dominiak et al. (2011)
Canada	30	1/6	Gagnon et al. (2012)
Denmark	30-50	5-7/40-50	Matamoros et al. (2012)
Greece	30, 60, 75	7/7 (summer) 7/14 (summer) 7/21 (winter)	Stefanakis and Tsihrintzis (2012c,d)
Germany	20-30	l/7 (activated sludge) 1/14 (aerobic digested) 1/14 (digested sludge)	Obarska-Pempkowiak and Kołecka (2006)
France	15	1/7 (summer) 1/30 (winter)	
Austria	68	1/7 (May to November)	
Denmark	60	1/7	
Poland		8 times/year	

TABLE 10.4 Common Operational Problems in Sludge Treatment Wetlands (Nielsen, 2005)

Insufficient dewatering	• Sludge quality and characteristics • Overloading during the commissioning and/or operational phase • Unequal loading (kg $DM/m^2/year$) • False loading records • Insufficient STW area and beds (improper design) • Prolonged loading periods and short resting periods • Drainage layers composition (e.g., grain size) • High or low vegetation density • Evapotranspiration occurs only from water surface and not from sludge layer
Poor vegetation growth	• Use of nonnative plants • Planting few plants/m^2 (low density) • Plants are nor adjusting to the specific microenvironment • Problems with herbs and pests • Lack of adequate vegetation during the commissioning phase • Intensive loading during the commissioning phase • Intensive loading and anaerobic conditions • Insufficient dewatering ability of specific plant species

Generally, the main reasons for these problems are structural failures and improper design, intensive surface loading during the commissioning phase, and imbalanced duration of loading and resting periods.

As already mentioned, the selection of an appropriate SLR for the specific type of sludge and the climatic conditions is crucial for the effective operation. Higher SLRs can lead to system overloading and, subsequently, to declined efficiency (reduced dewatering and mineralization) and can negatively affect plant growth and development. Limited plant growth, wilting parts, or dryness signs (e.g., yellowing) are indicative of plant stress. An overloaded system also results in an anaerobic sludge layer. Nielsen (2011) reports that limited dewatering efficiency also appears when sludge quality is not taken into account in the selection of the SLR. Additionally, high concentrations of organic compounds in the sludge feed also might affect the system performance. In fact, a VS content higher than 65% TS could result in 5-10% decreased TS content in the residual sludge layer, especially if the fat concentration exceeds 10,000 mg/kg DM.

Moreover, during the facility construction, the composition of the different porous media layers should be taken into consideration. Especially if the connection between the sludge layer and the upper drainage layer gets loose, the "hanging water" phenomenon occurs. This phenomenon appears when resistance from the lower layers is higher than water pressure in the higher saturated layer due to the lack of molecular binding of the two layers.

Such problems in STW facilities also cause management and handling problems, besides the effects on performance. In such cases, the thickness of the residual sludge layer increases faster than the planned increase rate. This means that the general operational schedule of the facility should be changed and the storage capacity of the bed will be reached earlier than the scheduled time; thus, emptying of the bed should also take place earlier.

Over the last years and with the continuous experience gained from operating systems, it becomes clearer that a high number of beds, a low SLR, and long resting periods are of course necessary for the proper and effective operation of the system, but are not the only important parameters in the design of an STW facility. Detailed analysis of the sludge composition, and especially of its organic content (e.g., fats and oil), should also be carried out (Nielsen, 2011).

10.6 VEGETATION

STW vegetation is essential as plants contribute to sludge dewatering through (Brix, 1994b; 1997; Nielsen, 2003):

- increased evapotranspiration,
- stem movement and respective enhancement of sludge porosity and hydraulic conductivity,
- increased dissolved oxygen concentration in drainage and sludge layers,
- absorption of nutrients and metals,
- creation of a dense and extensive root zone, which increases the hydraulic conductivity, and
- promotion of sludge mineralization (Nielsen, 2003).

Generally, as for VFCWs and all CW types, locally available native plants should be preferred. Most common plants used in STWs are the common reed (*Phragmites australis*, Figure 10.8) and the cattails (*Typha latifolia*, Figure 10.9; Stefanakis and Tsihrintzis, 2011). Plants mainly contribute to

FIGURE 10.8 Common reeds (*Phragmites australis*).

FIGURE 10.9 Cattails (*Typha latifolia*).

sludge dewatering through evaportranspiration, as the recorded evapotranspiration rates vary from 4 to 12 mm/day (Stefanakis and Tsihrintzis, 2011). In order to ensure quick adaptation, the initial plant density should range from 4 to 10 rhizomes/m^2 (Stefanakis and Tsihrintzis, 2012c). Generally, plant vegetation in STWs should meet the following requirements (De Maeseneer, 1997):

- rapid growth in various climatic conditions,
- high transpiration rates: typical evapotranspiration values for Western Europe should be ~1500 mm/year,
- tolerance in floods and droughts,
- tolerance in extreme values of salinity and pH,
- extended and deep root system (Liénard et al., 1995),

- ability to create new roots in every knot,
- their stems should stand above sludge during heavy winter, in order to be protected from the wind,
- commercially available, and
- easy planting.

As in CWs for wastewater treatment (see Chapter 4), plants in STWs need and uptake various nutrients and trace elements for their growth, like organic matter, nitrogen, phosphorus, Ca, Mg, Cu, Fe, which are included in sludge.

Common reed (*P. australis*) is probably the best choice, as it has a tall stem and an extended root system, although its growth rates are lower than cattails (*T. latifolia*). Common reed is suitable for sludge dewatering, as it is tolerant in various environmental conditions and has high evapotranspiration rate (USEPA, 1988). It can transfer significant quantities of oxygen within the porous media layers and the sludge layer, creating a diverse bacterial community, which could metabolize organic substances to inorganic. This metabolism is doubled compared to unplanted drying beds. Plants could absorb water from the sludge and evaporate it through their leaves. In general, reeds create aerobic areas around their roots, in an anaerobic environment, thus enhancing sludge stabilization, drying, and mineralization (Summerfelt et al., 1999; Begg et al., 2001). The main drawback of cattails is that its stems are falling during autumn, hindering their function and survival in a sludge layer.

Chapter 11

Processes and Mechanisms in Sludge Treatment Wetlands

11.1 SLUDGE DEWATERING

Sludge, even if it is thickened, still has high moisture content up to 97-98%. Dewatering efficiency of a specific method or technology could be described by the TS content in the sludge layer, which depends on the water type in the sludge. The water in the sludge may appear in four types (Vesilind and Hsu, 1997; Kopp and Dichti, 2001):

1. *Free*: it is not attached to sludge solids and could be separated with settlement,
2. *Interstitial*: it is trapped due to capillary forces inside the flocks and cells. It could be removed through flocks and cells braking,
3. *Vicinal*: it is attached to solids due to water molecular structure and is not mechanically removed, and
4. *Hydration water*: it is chemically bound with particles and it can be removed only through thermo-chemical destruction of particles.

In general, the sludge water could be characterized as bound or bulk (Vesilind and Hsu, 1997; Kopp and Dichti, 2001). As bulk water is characterized the water content that occurs after 30-min settlement of sludge in a cylindrical tube (SVI test) and is the main water form in the sludge. Free and hydration water are the two types left after sludge dewatering. The sludge dewatering rate shows temporal variations. Two stages could be distinguished in sludge water evaporation: (a) high evaporation rates as free water evaporates and (b) lower rates as interstitial water is removed.

In STWs, sludge dewatering occurs under natural processes, which mainly are: (a) water draining and (b) water evapotranspiration (*ET*) (De Maeseneer, 1997; Nielsen, 2003; Giraldi and Iannelli, 2009; Stefanakis and Tsihrintzis, 2011).

The STW vegetation enhances sludge dewatering through one main process. The sludge water content is used by the plants to cover their growth needs.

☆"To view the full reference list for the book, click here"

Plants absorb water through their roots and transpire it through their leaves. Stefanakis and Tsihrintzis (2011) report a higher dewatering efficiency of planted beds compared to unplanted. Generally, it is reported that plant presence increases the sludge volume reduction, as it will be described in Chapter 12 in detail.

11.1.1 Draining

The first dewatering stage in STWs includes water draining through the various media layers, while afterwards water is removed mainly through *ET* (Stefanakis and Tsihrintzis, 2011). Water draining usually lasts 1-2 h after sludge loading, while it takes ~15-24 h for water flow to return to its initial values (Nielsen, 2011). Draining removes a major portion of the sludge water volume, especially in colder climates like North Europe, thus leading to high TS values in the residual sludge layer (up to 30% and higher; Nielsen, 2011). Draining achieves a significant volume reduction, as its maximum water flow values range from 0.008 to 0.02 L/s/m^2 (Nielsen, 2011). It is also influenced by the sludge layer thickness (De Maeseneer, 1997). Stefanakis and Tsihrintzis (2011) report that in moderate climates, like in the Mediterranean basin, draining accounts for up to 20% of water losses, while mean draining rate values vary between 1.33 and 1.66 mm/d for low and high SLRs (up to 75 kg TS/m^2/year).

The evaporation process is slower than the percolation process and is affected by climatic conditions (i.e., air temperature, relative humidity, wind speed, etc.). Water evaporation in sludge dewatering lasts up to 15 days after loading and takes place mainly in the surface layer of the bed, while draining usually lasts only 1-3 days (Nielsen, 2011). Stefanakis and Tsihrintzis (2011) found that in the first 2-3 days, more than 60% of the total drained water volume leaves the bed. The water flow continues even up to 8 days but with smaller daily volumes. Practically, after 6-7 days the water volume is insignificant.

STW vegetation also affects water draining, as the movement of the plant stems creates cracks on the sludge layer, and thus, enhances water flow (Nielsen, 2003; Gagnon et al., 2013). The presence of plants was found to reduce the draining rate; planted STWs showed a lower rate of 1.66 mm/d compared to unplanted beds (2.61 mm/d), as reported by Stefanakis and Tsihrintzis (2011). Also, the water flow in unplanted beds continues for more than 10 days. Moreover, plants enhance sludge dewatering through colloidal organic acid polymerization, through substances produced in their roots (Obarska-Pempkowiak and Kołecka, 2006).

High hydraulic rates could influence water draining, as at high rates water cannot drain and remains trapped between the drainage and the sludge layer. Gustavsson and Engwall (2012) suggest that hydraulic rates should be adjusted when draining rates are low. During the winter period, due to low temperatures, ice may be formed on the sludge surface (Figure 11.1). However, due to plant stem movement, a ring without ice is formed around and along the stems, which

FIGURE 11.1 Formation of ice and snow coverage on the surface of the sludge layer. *Photos taken from pilot-scale STW units in northern Greece.*

permits water percolation (Nielsen, 2003, 2011). Additionally, as Figure 11.1 shows, plant stem movement due to the wind creates cracks on the sludge surface, which also favors the vertical drainage of the water. Through the same mechanism, snow coverage does not eliminate water flow.

11.1.2 Evapotranspiration

STW operation, as mentioned, is based on both the plant *ET* ability and water percolation (Reed et al., 1995; Kim and Smith, 1997; Nielsen, 2003, 2011). *ET* in STWs is achieved through water evaporation from the sludge surface, and mostly through plant transpiration. Plants in STWs absorb water through their root system and transfer it to their stems and leaves, where it is released to the atmosphere (Kengne et al., 2008; Uggeti et al., 2010; Stefanakis and Tsihrintzis, 2012c). *ET* is strongly affected by climatic conditions, like temperature, solar radiation, wind speed, relative humidity, etc. (Stefanakis and Tsihrintzis, 2011; Papaevangelou et al., 2012a,b; Liolios et al., 2014). Temperature values above 15-16 °C could increase respective *ET* values up to 30% (Stefanakis and Tsihrintzis, 2011). Moreover, Giraldi and Iannelli (2009) report that during the summer period, sludge dewatering is increased by 40%.

Since transpiration is a dominant process in STWs, plant function is crucial for sludge dewatering. Plants could significantly increase *ET* values up to 35% in an STW compared to unplanted sludge drying beds (Stefanakis and Tsihrintzis, 2011), since common reed (*Phragmites australis*) possesses a high transpiration capacity (Outwater, 1994; De Maeseneer, 1997). Stefanakis and Tsihrintzis (2011) report that increasing produced plant biomass is accompanied with a respective increase of the *ET* values and decrease of the draining water volume. A positive correlation between water losses through *ET* and produced biomass is also reported by Gagnon et al. (2012) and Korboulewsky et al.

(2012). On the other hand, one of the main problems occurring in STWs is the plant dieback phenomenon (Koottatep et al., 2005; Stefanakis and Tsihrintzis, 2011; 2012c). Dieback could appear either when high sludge loading rates shock vegetation (Stefanakis and Tsihrintzis, 2011) or when *ET* values are very high and sludge moisture cannot provide the needed water. A possible solution to overcome the dieback phenomenon due to high *ET* values is to increase the SLR or to shorten the resting period duration, in order to increase the sludge layer moisture which will be available to the plants. However, this modification should be carefully applied.

In STWs, *ET* represents a key parameter of the water budget and is directly affected by the climatic conditions. De Maeseneer (1997) reports, for early systems in central Europe, that *ET* accounted for 39% of water losses, with respective figures for draining and sludge water content of 46% and 15%. Burgoon et al. (1997) and Summerfelt et al. (1999) report lower *ET* values of 10% and 12-20%, respectively, in arid regions. Begg et al. (2001) found that *ET* accounted for 63.7% of water losses in the water balance (16.7% for drainage) in an STW system after 6 years of operation. Barbieri et al. (2003) mention that during spring and summer months, the increased *ET* values result in a simultaneous limitation of the drained water volume. Nassar et al. (2006) report higher *ET* rate of 6.7 mm/d for an STW system in Palestine, indicating the effect of the climate to this dewatering process. Equal percentage values for *ET* and drainage (50-50%) are reported by Heinss and Koottatep (1998) for a system in central Europe. In moderate climates, e.g., the Mediterranean climate, *ET* accounted for up to 82% of water losses (16.7% for drainage; Stefanakis and Tsihrintzis, 2011). Although these high values might be enhanced due to the "oasis" or "clothesline" effects (Allen et al., 2011), they clearly indicate the importance of this process in the dewatering efficiency of an STW system. Furthermore, porous media size seems also to affect *ET* in STWs. Coarse-grained materials result in higher bed porosities (increased hydraulic conductivity), and thus, in increased draining water volume and decreased *ET* values (Stefanakis and Tsihrintzis, 2011).

11.2 SLUDGE MINERALIZATION

In STW systems, sludge is introduced in growing plant stems under controlled conditions (Figure 11.2). The plants and their root system alter the internal cohesion forces of the sludge, cleaving its colloidal stability and releasing part of bound water. The plants absorb moisture and nutrients from the sludge, while their roots alter sludge characteristics, resulting in a well-dewatered and improved sludge material. The fresh sludge surface is usually black, due to iron sulfide, and of aqueous composition. The lower sludge layers have a brown color and soil texture (Figure 11.3), which indicates the presence of aerobic conditions (Stefanakis and Tsihrintzis, 2012c). If the deepest parts of the sludge layer are still black, this means the mineralization is limited (anaerobic

FIGURE 11.2 Sludge layer surface and growing reed stems in pilot-scale STW units in Northern Greece.

FIGURE 11.3 Core sample from a stabilized sludge in a full-scale STW facility in Groß Lafferde, Lahstedt, Niedersachsen, Germany.

conditions). Generally, due to its longer treatment and stay within the bed, the bottom sludge layer is more mature and stabilized than the top layer (Stefanakis and Tsihrintzis, 2011b).

Sludge stabilization is achieved through a series of biochemical processes in areas near the plant roots. In this area coexist aerobic, anoxic (with NO_3^- without O_2), and anaerobic (without NO_3^- and O_2) zones. Due to these different conditions in the root zone, organic matter and nitrogen are removed through oxidation, ammonification, nitrification, and denitrification (Uggeti et al., 2012a). As it is obvious, the general transformation and removal processes for organic matter and various constituents, like nitrogen and phosphorus, are more or less the same with those occurring in VFCWs for wastewater treatment (see Chapter 5). Oxygen diffusion through perforated tubes, plant roots, and sludge surface enhances the presence of aerobic microorganisms (Nielsen and Willoughby, 2005). The plant role also includes the oxygen transfer through their root system, as they create aerobic micro-areas in the sludge layer, where aerobic degradation is enhanced (Uggeti et al., 2012a). Although the positive effect of plants on sludge stabilization has been proved in several studies (e.g., Stefanakis and Tsihrintzis, 2011b; Peruzzi et al., 2013), vegetation species influence has not been yet thoroughly examined (Gagnon et al., 2013). Furthermore, oxygen forms an oxidative protective film in plant root zone, which protects roots from pollutant toxic effects (Gustavsson and Engwall, 2012). Also, porous media provide the necessary area for microorganism attachment.

Nitrogen is removed in the sludge layer through common processes (ammonification, nitrification and denitrification, plant uptake). STWs have been proved to successfully remove nitrogen from the sludge layer as total nitrogen content decreases (Stefanakis and Tsihrintzis, 2012c; Uggeti et al., 2012a), while nitrite and nitrate content tend to increase (Melidis et al., 2010; Stefanakis and Tsihrintzis, 2012c). Plant species also influence nitrogen removal in STW systems, as it will be described in Chapter 12.

Phosphorus is also removed in STWs through a series of physical, chemical, and biological processes. According to Gagnon et al. (2012), phosphorus is mainly removed in STWs through:

- Direct plant uptake. P removal through plant biomass depends on the species and is usually limited (0.2-5.7%).
- Microbial biomass accumulation.
- Sorption and precipitation in the sludge layer. These processes are favored by the elevated redox potential and the humic substances in the rhizosphere.
- Absorption and precipitation on calcium, aluminium and iron elements present in the sludge layer.

The effects of various parameters, the role of the various processes, as also the efficiency of STWs are described in Chapter 12.

Chapter 12

Performance of Sludge Treatment Wetlands

12.1 DEWATERING EFFICIENCY

The primary target of sludge treatment wetlands (STWs) is the reduction of the sludge volume, i.e., the dewatering of the sludge to be treated. Dewatering transforms sludge from a liquid waste to a final product with characteristics similar to a solid material. The residual sludge layer that is formed in an STW represents the sludge cake, which contains a high amount of total solids (TS). The higher the TS content in the sludge cake, the more efficient the STW is considered (this also counts for all dewatering techniques and methods). The reduction of the sludge volume and the respective increase in the TS content are a critical parameter in the various dewatering methods and techniques, since a good dewatering performance means less expenses (i.e., less final volume) for final transfer and disposal (or further use) of the residual sludge.

At this point, it is useful to clarify some common terms used in the literature. It seems that there is a general confusion among the terms "sludge" and "biosolids." It is true that the use of the word "sludge" possesses a negative concept for the public, which can act as a potential obstacle in the exploitation of this useful material. Therefore, over the last years it is more or less accepted that the term sludge refers only to the solid streams within the wastewater treatment plant and to the material that is simply disposed of without any treatment and/or beneficial exploitation. When this material is further treated, e.g., using one of the various dewatering methods, aiming at the improvement of its characteristics for beneficial use, then the term "biosolids" is used to describe the treated material that leaves the sludge treatment plant, destined for one of the various options for further exploitation, like use as fertilizer, soil amendment, etc. (Spinosa, 2004).

Table 12.1 presents the efficiency of various STW facilities around the world. Dewatering refers to the reduction of the sludge volume treated in each facility, while TS and volatile solids (VS) values are presented for both applied raw sludge and the final biosolids material in the STWs. Moreover, the various

☆"To view the full reference list for the book, click here"

TABLE 12.1 Various Parameters of Full- and Pilot-Scale STWs from Several Countries: Source of Sludge Feed, Sludge Loading Rate (SLR), Dewatering Efficiency (Sludge Volume Reduction, %), Total Solids (TS, %), and Volatile Solids (VS, %) in the Raw and Dewatered Sludge

Source of Sludge	SLR (kg TS/m^2/year)	Dewatering (% Volume Reduction)	TS (%)		VS (%)			Feed/Resting Days	Country	References
			Raw	*STW*	*Raw*	*STW*	*Removal (%)*			
Sludge from extended aeration plant	35.7-44 kg TSS/m^2/year		0.3	10.6-11.4	72.5	58.7-61.6	18.6	1/1 1-2/12-13 3-7/21-25	France	Liénard et al. (1995)
Industrial, aerated lagoons	40 (design 65)		0.7	10-12	79	67	15.2	1/7	USA	Burgoon et al. (1997)
Anaerobically digested sludge	13-60		3	40	49	46	5		USA	Kim and Smith (1997)
Aerobically/anaerobically stabilized	20-30		2	35-40	45-65			1/7, 2/14, 3/21 (weeks)	Belgium	De Maeseneer (1997)
Aquaculture sludge	30		0.75 7860 g TSS/L	156 g TSS/L	82.8	57.2	30.9	Every day	West Virginia, USA	Summerfelt et al. (1999)
Septage and primary pond sludge	80-250		1.6-7	30				1/7	Thailand	Strauss et al. (1999)

Imhoff tank, anaerobically stabilized	60		4-10	38-45	51	42	17.6	Eight times/ year	Poland	Zwara and Obarska-Pempkowiak (2000)
Municipal sludge	8-500		1-2	30-60				1-2/7	Thailand	Koottatep et al. (2001)
Aerobic sludge		92.6%	12,532 mg TSS/L T	14.2	60-70				USA	Begg et al. (2001)
Sludge from settled pig slurry	67.7	84.3-86.3	4	18.7-21.5	74.2	52.5	29.2	1/7	UK	Edwards et al. (2001)
Sewage sludge	21.3		2-4	41.5	51	35.6	30.2		Poland	Pempkowiak and Obarska-Pempkowiak (2002)
Surplus activated sludge	50-60		0.5-1	40				1-8/10-65	Denmark	Nielsen (2003)
Thickened sludge and activated sludge			1.8	25				1/10-25	Italy	Barbieri et al. (2003)
Anaerobically digested sludge, activated sludge	46.7	97.5	0.4	35-40	74.2	52.5	29.2		UK	Cooper et al. (2003)
Activated sludge	59		0.5-1	40				7-14-21/ 49-98-147	Denmark	Nielsen and Willough (2005)
Anaerobically digested sludge	66.7	78	2.9	12.3-20	60	54.7	8.8		USA	NYSERDA (2006)

(Continued)

TABLE 12.1 Various Parameters of Full- and Pilot-Scale STWs from Several Countries: Source of Sludge Feed, Sludge Loading Rate (SLR), Dewatering Efficiency (Sludge Volume Reduction, %), Total Solids (TS, %), and Volatile Solids (VS, %) in the Raw and Dewatered Sludge—cont'd

Source of Sludge	SLR (kg TS/m²/year)	Dewatering (% Volume Reduction)	TS (%)		VS (%)			Feed/Resting Days	Country	References
			Raw	*STW*	*Raw*	*STW*	*Removal (%)*			
Activated sludge	60		0.5-0.8	35				7-8/55-65		Nielsen (2007)
Fermented sewage sludge	53	91.2	4.2	38.8	61.5	44.2	28.1	5-6 times/year Rest 39-176 days	Poland	Kalisz and Salbut (2007)
Sewage sludge	45-135	95.8	1-2	68		39.7		1/14	Palestine	Nassar et al. (2006, 2008)
Thickened activated sludge	252.2		3	35	34.7	30	13.5	1/7	China	Cui et al. (2008)
Aerobic activated sludge	38	93	1	40				1/14 1/7	Italy	Giraldi et al. (2009b), Giraldi and Iannelli (2009)
Mixture of sludge from pit latrines, septic tanks, public toilets	100, 200, 300	90	3.7	51	65.4			1/7	Cameroon	Kengne et al. (2009a, b)
Aerated sludge (thickened and not)	26, 31, 40		2235 9493 (mg/L)	19.4 31	79-80 68	62.9 54-55	21.4 20.6	2/14 (weeks)	France	Troesch et al. (2009a, b)

Activated sludge from extended aeration, Activated sludge from contact-stabilization	51, 55, 125		0.7-1.5 3 0.3-2	22-25 20-28 15-20	52-67 52-42 58-59	39-42 36-40 46-50	33.1 19.1 18.6	2/2 (days) 2/4 (days)	Spain	Uggetti et al. (2009)
Thickened Activated sludge	30-75	90	3.1	33-57	71.6	48-61	25	7/7-21 (days)	Greece	Stefanakis et al. (2009b)
Sludge from Imhoff tank	284	99.6	39.7	55.3		37.9		1/20 days	Greece	Melidis et al. (2010)
Septage Activated sludge	30 50		30 g DM/L 2.4	28-53 15-31	71 59	18.6 14.3	73.8 75.8	5/24 (days)	France	Vincent et al. (2011)
Drinking water sludge	10-50	99	0.134-0.187	28-42		24-28		2-6/30-55 days	England	Nielsen and Cooper (2011)
Aerobic activated sludge	45		1	15, 28, 37				Once in summer and winter	Italy	Bianchi et al. (2011)
Septage sludge	7 cm/ week (100 kg TS/year)		2.6	23				1/7	Canada	Kinsley et al. (2011)
Domestic sludge			30 20	89 86	65.5 73.3	48.3 40.9	26.3 44.2		India	Manderia (2012)
Secondary sludge	0.658 kg TS/m^2/ day			37.2		22.4			China	Zhang et al. (2012c)

(*Continued*)

TABLE 12.1 Various Parameters of Full- and Pilot-Scale STWs from Several Countries: Source of Sludge Feed, Sludge Loading Rate (SLR), Dewatering Efficiency (Sludge Volume Reduction, %), Total Solids (TS, %), and Volatile Solids (VS, %) in the Raw and Dewatered Sludge—cont'd

Source of Sludge	SLR (kg TS/m^2/year)	Dewatering (% Volume Reduction)	TS (%)		VS (%)			Feed/Resting Days	Country	References
			Raw	*STW*	*Raw*	*STW*	*Removal (%)*			
Thickened activated sludge	20, 40	79-81	2.5	16-24	60-72	50	24.2	1/7	Spain	Uggetti et al. (2012a)
Thickened activated sludge	30, 60, 75	89-97	3.1	50-65	73.7	46-51	34.9	7/7-24	Greece	Stefanakis and Tsihrintzis (2012c)
Concentrated fish farm sludge	30	80-89	4	28-33	72	34-42	47.2	1/6 d	Canada	Gagnon et al. (2013)

studies are presented in chronological order, in order to identify possible efficiency improvements as this technology gradually became more understood and research on these systems progressed.

Data reported in Table 12.1 clearly show the overall good effectiveness of STWs in the dewatering of various types of sludge. Dewatering efficiency in most cases exceeds 90% on the initial sludge volume, which is the first and main target of the STW technology. Moreover, TS content in the residual sludge cake can reach high values. Based on Table 12.1, TS content increases from 0.5%–4% in the raw sludge to more than 40% in the residual sludge layer within the STW beds and even higher. Generally, depending on the climatic conditions and the sludge loading rate (SLR), dewatering efficiency and respective TS content can further be improved. Figure 12.1 depicts the range of SLR applied and

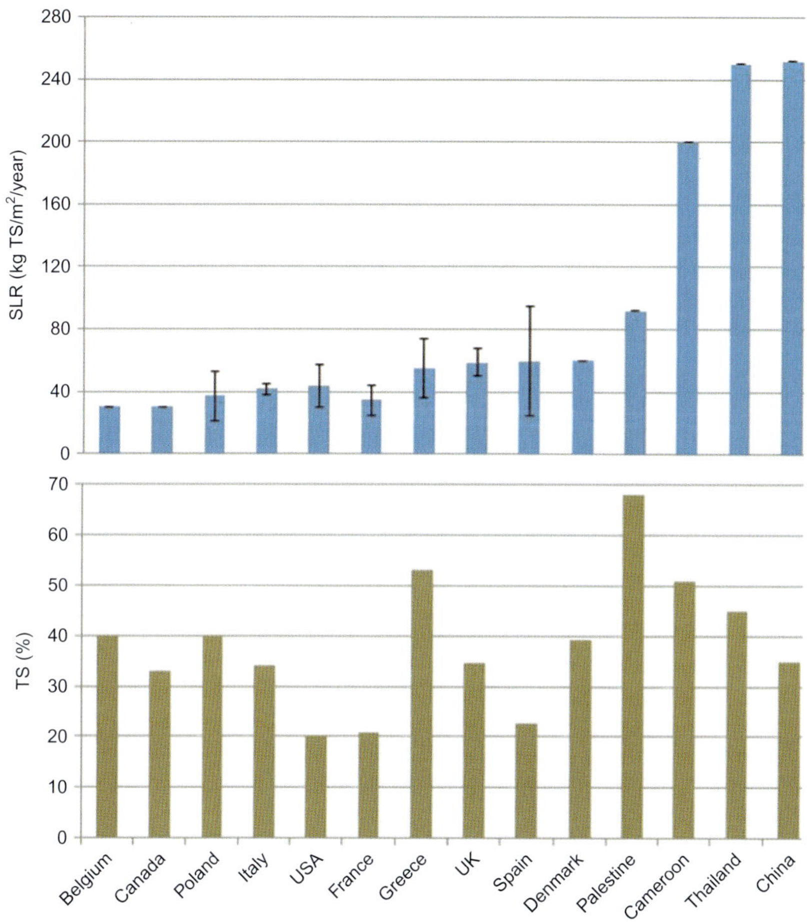

FIGURE 12.1 Typical SLR and respective TS content of the residual sludge in STWs from various countries under different climatic conditions.

the respective TS obtained in STWs in various countries under varying climatic conditions. It is clear that in countries with warmer climates (e.g., Thailand, Cameroon), the SLRs tested are more than double compared to countries with temperate (e.g., Greece, Spain) or colder (e.g., Canada, Denmark) climatic conditions. For example, in countries with tropical climate like Thailand and Cameroon, prolonged seasons with high temperatures and intensive evapotranspiration rates allow for higher SLR applications, e.g., up to 250 kg TS/m^2/year (Koottatep et al., 2001; Kengne et al., 2009a,b). In these regions, higher SLR is also necessary in order to prevent plant death due to insufficient available water amount for plant growth.

However, it is very interesting that effective systems also operate in countries with colder climate, like Canada and Denmark, as Figure 12.1 shows. Dewatering and TS values reported from these countries clearly prove that STW systems can also be effective under colder climatic conditions. On the whole, the key parameter for the proper and effective operation appears to be the applied SLR. In order to ensure a desirable dewatering rate, lower SLR is usually applied under colder climatic conditions and higher in warmer regions. Additionally, the duration of the feeding cycle, i.e., loading and resting periods, seems also to affect the efficiency. Generally, as indicated by Table 12.1, there is not a widely accepted SLR for applications under the various climatic conditions. Usually, the selected SLR is based on the existing knowledge and experience from older facilities. For example, many old full-scale systems exist in Denmark where the proposed SLR is around 60 kg TS/m^2/year (Nielsen, 2003). Thus, taking into account the colder Danish climatic conditions, an estimation of the SLR which could be applied in warmer and colder climates is possible. Investigations in Greece, under the mild Mediterranean climate, showed that higher SLR could be applied up to 75-90 kg TS/m^2/year (Stefanakis and Tsihrintzis, 2012c). However, a rule of thumb does not exist. Although many studies have shown that STWs can indeed be effective in different parts of the world, testing and conducting experiments are the safest road toward the identification of the optimum operational parameters for the various climatic conditions.

Seasonal variations of the TS content are also observed. During the spring and summer months, the climatic conditions (higher temperatures and respectively higher evapotraspiration values) enhance the dewatering of the accumulated sludge, thus, higher TS values are measured. Depending on the duration of the applied resting period, relatively high TS contents are reported during the summer months. For example, Stefanakis and Tsihrintzis (2012c) report TS content up to 65-85% for resting periods of 3-7 weeks in the summer. On the other hand, during the winter months, the decreased temperature and evapotranspiration rates slow down the dewatering process, which results in a temporary increase in the TS values (Barbieri et al., 2003; Nassar et al., 2006; Giraldi and Iannelli, 2009; Uggetti et al., 2009; Stefanakis and Tsihrintzis, 2011, 2012c).

12.1.1 Vertical Profile of the Residual Sludge Layer

Table 12.2 presents TS and VS values from several STW beds for various parts of the residual sludge layer depth. Generally, the bottom parts of the residual sludge layer, which is gradually formed in STWs, contain higher TS content compared to the upper parts. The deeper the sludge layer is, the higher these differences between upper and bottom parts are. Kim and Smith (1997) reported TS contents of 15% and 47% for the upper and bottom part, respectively, of a 71-cm residual sludge layer. A similar trend was reported by several authors (Obarska-Pempkowiak et al., 2003; Kalisz and Sałbut, 2007; Kengne et al., 2009b; Stefanakis et al., 2009; Uggetti et al., 2009; Melián et al., 2010; Stefanakis and Tsihrintzis, 2011, 2012c). Generally, the bottom parts of the sludge layer contain sludge from older applications. This means that the dewatering processes are prolonged in these parts, thus, they are better dewatered and contain a higher TS content.

Table 12.2 also reveals the effects and the importance of the final resting phase, which is applied at the end of an operating cycle of 8-12 years. Stefanakis (2011) and Stefanakis et al. (2011b) reported that during the final resting phase, the TS content of the sludge layer is more than double compared to the loading phase. Under no new sludge loadings, the dewatering processes continue and the dryness of the sludge layer increases significantly. Moreover, the differences between upper and bottom layers are reducing during the application of the final resting phase (without new sludge loadings). Stefanakis and Tsihrintzis (2012c) applied a final resting phase of 6 months in several pilot-scale units under the Mediterranean climate. As Table 12.2 shows, the TS contents during the loading phase were 18-30% for three different levels of SLR, while during the final resting phase these values were significantly increased up to 65%. It was also found that during the final resting phase, the upper layers are better dried and contain similar, if not slightly higher, TS content compared to the bottom layer. The final resting phase is proposed to be applied during the summer months, to take advantage of the high temperatures, when the upper layers are exposed to direct solar radiation and intensive surface evaporation and transpiration by plant roots (which are denser in the upper layers) takes place. Due to these processes, during the final resting phase, the depth of the sludge layer is also further decreased.

Overall, current knowledge and data from existing facilities from various countries clearly prove the positive dewatering performance of these ecological treatment systems. Sludge volume can be significantly reduced more than 90% compared to the initial treated volume, while the solids content of the final residual sludge material increases and reaches more than 40% TS. Once the first operational lifetime of a system is completed (usually 8-12 years), the final resting phase without loadings is applied. During this phase, which can last from few months up to 1 year, the TS content of the residual sludge is further increased and the residual sludge obtains its final characteristics.

TABLE 12.2 TS and VS Content Along the Depth of the Residual Sludge Layer in Various STW Systems

Residual Sludge Layer Depth (cm)		TS (%)	VS (%TS)	References
Total	*Parts*			
71	0-10	15	49	Kim and Smith (1997)
	10-20	21	48	
	20-30	30	44	
	30-40	43	45	
	40-50	40	46	
	50-71	47	46	
12	0-3	36	51	Zwara and Obarska-Pempkowiak (2000)
	3-6	35	49	
	6-9	38	45	
	9-12	41	42	
40	0-10	20.8, 23.5	54.4, 52.1	Edwards et al. (2001)
	10-20	20.8, 19.4	53.4, 53.6	
	20-30	18.2, 21.1	51.2, 51.8	
	30-40	22.1, 22.1	51.0, 51.9	
24	0-2	53.3	40.6	Pempkowiak and Obarska-Pempkowiak (2002)
	6-8	51.4	50.2	
	12-14	59.8	38.6	
	18-20	59.1	36.8	
	23-24	69.9	16.4	
30	0-8	34.6	50.1	Obarska-Pempkowiak et al. (2003)
	8-16	36.1	46.2	
	16-23	39.7	41.6	
	23-30	57	43.3	
100	0-10	26.4, 20.2, 21.6, 14.9		Obarska-Pempkowiak and Kołecka (2006)
	20-30	27.8, 22.5, 25.2, 19.9		
	50-60	29.5, 22.7, 29.4, 22.0		

TABLE 12.2 TS and VS Content Along the Depth of the Residual Sludge Layer in Various STW Systems—cont'd

Residual Sludge Layer Depth (cm)		TS (%)	VS (%TS)	References
Total	*Parts*			
	60-70	31.7, 25.4, 71.5, 21.7		
	70-80	37.0, 29.7, -, 21.6		
	90-100	-, -, -, 25.4		
75	20-35	31.2	47.8	Kalisz and Sałbut (2007)
	40-55	30.5	46.0	
	60-75	35.4	38.7	
35	Upper 5 cm	45.3		Kengne et al. (2009a)
	Center	49.1		
	Bottom 5 cm	58.8		
7.5, 15, 62	Surface	7, 26, 22	55, 35, 47	Uggetti et al. (2009)
	Bottom	23, 31, 17	40, 36, 47	
7.5	Surface	23	72.8	Stefanakis et al. (2009b)
	Bottom	77	40.5	
20	Upper 10 cm	42.0-55.8	40.61	Melidis et al. (2010)
	Bottom 10 cm	55.5-68.0	35.22	
130	0-10	15.2	52.0	Matamoros et al. (2012)
	20-30	19.2	49.6	
	40-50	22.3	45.6	
	70-80	24.0	42.4	
	90-100	24.5	42.8	
	110-120	26.3	41.5	
	120-130	28.5	40.3	

(Continued)

TABLE 12.2 TS and VS Content Along the Depth of the Residual Sludge Layer in Various STW Systems—cont'd

Residual Sludge Layer Depth (cm)		TS (%)	VS (%TS)	References
Total	*Parts*			
8.0, 14.0, 18.8 (SLR: 30, 60, 75 kg TS/ m^2/year)	Upper	18.1, 23.9, 20.5 (LP)[a] 64.2, 60.1, 53.3 (RP)[a]	63.9, 65.7, 64.3 (LP) 52.0, 59.6, 56.9 (RP)	Stefanakis, 2011; Stefanakis et al. (2011b)
	Bottom	24.2, 29.6, 30.0 (LP)	59.5, 60.3, 59.9 (LP)	
		64.0, 57.4, 46.9 (RP)	52.3, 55.4, 56.7 (RP)	
100	0-10	14.9, 26.4, 20.2, 21.6	46.7, 43.5, 46.9, 51.4	Kołecka and Obarska-Pempkowiak (2013)
	20-30	19.9,27.8, 22.5, 25.2	43.1, 43.2, 46.4, 48.8	
	40-50	20.5, 27.7, 23.5, 27.2	40.4, 43.2, 43.2, 43.5	
	60-70	21.7, 31.7, 25.4, 28.5	38.9, 41.6, 43.4, 41.3	
	70-80	21.6, 36.9, 29.7, -	39.8, 36.6, 41.7, -	
	80-90	21.9, -, -, -,	38.8, -, -,	
	90-100	25.4, -, -, -,	38.0, -, -,	

[a]*LP, loading phase; RP, resting phase.*

12.1.2 Comparison with Other Dewatering Methods

Table 12.3 presents the dewatering efficiency of various common dewatering techniques and methods in terms of TS content of the final sludge material. The comparison of the values presented in Tables 12.1 and 12.3 reveals that the STWs possess at least a comparable performance with most of the conventional methods. It is shown that STWs can reach similar dewatering results with mechanical dewatering methods, such as centrifuges, belt filter presses, or vacuum filters. In fact, as these two tables indicate, in many cases, the STWs present higher efficiency compared to these methods.

The results of this dewatering efficiency clearly enhance the advantages of the STWs as a green technology, taking into account that most of the techniques

TABLE 12.3 Typical TS (%) Values of the Final Sludge Material from Conventional and Mechanical Sludge Dewatering Techniques and Methods

Method	Sludge Type	TS (%)	References
Polymer + centrifuge	PRI[a]	26-36	Turonskiy and Mathai (2006)
	SAS[a]	8-20	
	PRI + SAS	18-25	
	Anaerobically digested PRI	25-35	
	Anaerobically digested PRI + SAS	15-27	
	Aerobically digested SAS	8-12	
	Aerobically digested PRI + SAS	11-18	
	Overall	20-35	Al-Malack et al. (2002)
	Digested SAS	28-29	Dimitrova and Carlsson (2011)
	AS, Fresh water sludge	15-16	Pan et al. (2012)
	PRI	25-35	Liu and Lipták (1997)
	Digested PRI	22-35	
	PRI + Biological AS	11-20	
	Digested PRI + Biological AS	20-30	
	SCD[a]	5-15	
	Limed PRI	15-60	
	Pulp and paper sludge	15-40	
	PRI	25-40	USEPA (2000d)
	SAS	16-25	
	PRI + SAS	25-35	
	Aerobically digested PRI + SAS	16-25	
	Anaerobically digested PRI + SAS	22-32	
	Anaerobically digested PRI	25-35	
	Aerobically digested SAS	18-21	
	Hi-temperature aerobic	20-25	
	Hi-temperature anaerobic	22-28	

(*Continued*)

TABLE 12.3 Typical TS (%) Values of the Final Sludge Material from Conventional and Mechanical Sludge Dewatering Techniques and Methods—cont'd

Method	Sludge Type	TS (%)	References
	Lime stabilized	20-28	
	Anaerobically digested AS[a]	15-35	Metcalf and Eddy (2003)
	Aerobically digested AS	8-10	
	AS	14-20	Gonçalves et al. (2007)
Solid bowl conveyor centrifuge	Overall	35	Cheremisinoff (2002)
Hydraulic filter press	Digested SAS (gravity thickening)	30-33	Dimitrova and Carlsson (2011)
	AS, fresh water sludge	13-14	Pan et al. (2003)
	AS	27-33	Gonçalves et al. (2007)
	Anaerobically digested (mixture)	29-35	
Recessed plate filter press	PRI + AS	45-50	Wakeman (2007)
	PRI + AS + $FeCl_3$	40-45	
	Digested PRI + AS + $FeCl_3$	40	
	Tertiary + lime	55	
Electrokinetics-assisted filter press	Biological industrial sludge	35-47	Ouyang et al. (2011)
Electro-osmotic belt filter press	Unthickened SAS	20	Snyman et al. (2000)
Polymer + belt filter press	PRI	26-32	Turonskiy and Mathai (2006)
	SAS	12-20	
	PRI + SAS	18-28	
	Anaerobically digested PRI	24-30	
	Anaerobically digested SAS	12-20	
	Anaerobically digested PRI + SAS	20-25	
	Aerobically digested PRI + SAS	12-25	
	PRI	28-44	Al-Malack et al. (2002)
	SAS, PRI + SAS	20-35	

TABLE 12.3 Typical TS (%) Values of the Final Sludge Material from Conventional and Mechanical Sludge Dewatering Techniques and Methods—cont'd

Method	Sludge Type	TS (%)	References
	PRI+TFi[a]	20-40	
	Anaerobically digested PRI	25-36	
	Anaerobically digested SAS	12-22	
	Anaerobically digested PRI+SAS	18-44	
	Aerobically digested PRI+SAS	12-20	
	Aerobically digested PRI+TFi	12-30	
	Oxygen activated SAS	15-23	
	Thermally conditioned PRI+SAS	25-50	
	AS, Fresh water sludge	12-14	Pan et al. (2003)
	PRI from pig industry	22-25	FSA (2002)
	Digested from pig industry	18-22	
	Slaughterhouse sludge from pig industry	13-15	
	PRI	25-44	Wakeman (2007)
	AS	12-32	
	PRI+AS	20-35	
	Aerobically digested	12-30	
	Anaerobically digested	18-34	
	Thermally conditioned	38-50	
	AS	12-18	Gonçalves et al. (2007)
	Anaerobically digested (mixture)	17-23	
	Anaerobically digested	12-30	Metcalf and Eddy (2003)
	Aerobically digested	12-25	
Rotary drum vacuum filters	PRI	32	Liu and Lipták (1997)
Conditioning chemical+fixed-	PRI	45	Turonskiy and Mathai (2006)
	PRI+SAS	40-45	

(*Continued*)

TABLE 12.3 Typical TS (%) Values of the Final Sludge Material from Conventional and Mechanical Sludge Dewatering Techniques and Methods—cont'd

Method	Sludge Type	TS (%)	References
volume filter press	SAS	45	
	Digested PRI	40	
	Digested PRI+SAS	40-45	
	PRI and SCD	40	Al-Malack et al. (2002)
Sand drying beds	Anaerobically digested PRI and SCD	40	Al-Malack et al. (2002)
	Thermally conditioned PRI and SCD	60	
	Digested sludge	40	Cheremisinoff (2002)
Paved drying beds	Overall	40-50	Al-Malack et al. (2002)
Vacuum-assisted sludge drying beds	Anaerobically digested PRI	12-26	Turonskiy and Mathai (2006)
	Anaerobically digested PRI+SAS	15-20	
	Aerobically digested SAS	10-23	
Vacuum filters	PRI	25-32	Turonskiy and Mathai (2006)
	SAS	12-20	
	PRI+SAS	18-25	
	PRI+TF	20-28	
	Digested PRI	25-32	
	Digested PRI+TFi	20-28	
	Digested PRI+SAS	20-28	
	Anaerobically/aerobically digested municipal sludge	15-30	Al-Malack et al. (2002)
	SAS	16-20	Cheremisinoff (2002)
	Digested PRI	32-40	
	AS	12-18	Gonçalves et al. (2007)
	Anaerobically digested (mixture)	17-23	

TABLE 12.3 Typical TS (%) Values of the Final Sludge Material from Conventional and Mechanical Sludge Dewatering Techniques and Methods—cont'd

Method	Sludge Type	TS (%)	References
Drying lagoons	Activated sludge	20-40	Turonskiy and Mathai (2006)
	Digested sludge	25-30	Al-Malack et al. (2002)
Composting	Septage with sawdust, woodchips, and cow or horse manure, pit latrine sludge, bucket latrine sludge, and MSW[a]	40-60	Strauss et al. (2003)
	Aerobic digestion + centrifuge	83	Bertrán et al. (2004)
	Dewatered raw sludge	44	Ruggeri et al. (2008)
	Dewatered raw sludge with organic amendments	33-51	Komilis et al. (2011a)

Typical inflow TS content: PRI 4-8%, SAS 0.5-4, PRI+SAS 2-4%, aerobically digested PRI+SAS 1.5-3%, anaerobically digested PRI+SAS 2-4%, anaerobically digested PRI 2-4%, aerobically digested SAS 1-4%, Hi-temperature aerobic 4-6%, Hi-temperature anaerobic 3-6%, lime stabilized 4-6% (Al-Malack et al., 2002; Turonskiy and Mathai, 2006; Wakeman, 2007).

[a]Types of sludge: PRI, primary; AS, activated sludge; SAS, surplus activates sludge; SCD, secondary, TF, trickling filter; MSW, municipal solids waste.

and methods reported in Table 12.3 are based on or combined with the use of chemical substances (e.g., polymers, polyelectrolytes for conditioning) and demand high energy input for their operation and maintenance.

12.2 ORGANIC MATTER (VOLATILE SOLIDS)

An important parameter that is monitored and used in sludge treatment and dewatering processes, as an indicator of the biological stabilization, is the VS content. VS represents the portion of TS that has an organic composition and origin and possesses a calorific value. Table 12.1 also contains VS data from various STW systems around the world for both the type of inflow sludge and the final residual sludge material. It is clear that STWs are capable of decreasing the VS content, i.e., the organic portion of the solids. This reduction reaches and often exceeds 30% of the initial VS content in the raw sludge and, depending on the sludge type, this value can be even higher. Generally, the VS content of the final sludge material in STWs varies between 40% and

50% TS, as also reported by Uggetti et al. (2010). Taking into account that during the final resting phase of an STW facility after 8-12 years of operation, the residual sludge material obtains its final characteristics, final VS content of the biosolids to be removed could be even lower (Stefanakis and Tsihrintzis, 2012c). Usually, biosolids removed from STWs (after the final resting phase) comply with the criteria set by the USEPA for a 38% reduction of the VS for proper vector attraction reduction, like flies, birds, etc. (USEPA, 1993, 2003).

12.2.1 Vertical Profile of the Residual Sludge Layer

Table 12.2 presents the VS content in the upper and bottom parts of the residual sludge layer from various STW systems. As it is obvious, differences between the VS content of the upper and bottom parts of the residual sludge layer have been reported by several authors. These differences are smaller compared to the respective differences for TS content. Generally, the bottom parts of the sludge layer, which contain sludge from older applications, are better mineralized, and, thus, the VS content is lower compared to the upper parts. Kim and Smith (1997) report only a 5% decrease in VS between the upper and bottom part of the sludge layer (from 49% to 46%). Other authors report differences of 10-20% (Zwara and Obarska-Pempkowiak, 2000; Obarska-Pempkowiak et al., 2003; Kalisz and Sałbut, 2007; Stefanakis, 2011; Stefanakis et al., 2011; Matamoros et al., 2012), while even higher differences of more than 50% have also been reported (Pempkowiak and Obarska-Pempkowiak, 2002; Stefanakis et al., 2009).

Table 12.2 also reveals, as for TS, the significance of the final resting phase. Based on the work by Stefanakis (2011) and Stefanakis et al. (2011b), the VS content of the total sludge layer is further decreased during the resting phase. This decrease is higher in the upper parts. This means that during the final resting phase, additional mineralization of the organic matter present in the sludge layer takes place. Furthermore, the differences between the upper and bottom parts are also decreased and, as Table 12.2 shows, the VS content along the entire depth of the residual sludge layer has more or less similar values.

12.2.2 Comparison with Other Dewatering Methods

Table 12.4 presents the stabilization efficiency concerning VS removal for other conventional sludge stabilization methods. Generally, reported VS values in the residual sludge between STWs and conventional stabilization methods are comparable, as Tables 12.1 and 12.4 indicate. The range of 40-50% for the final VS content in STWs is relatively close to the respective ranges for anaerobic (35-50%) and aerobic digestion (40-55%; Metcalf and Eddy, 2003). The final VS value also seems to be related to the initial VS content and the origin of the sludge to be treated. Lower initial VS content usually results in respectively lower removal efficiency. Higher values among the various methods are

TABLE 12.4 Typical VS (%TS) Values of the Influent Sludge and the Final Sludge Material from Conventional and Mechanical Sludge Stabilization Methods

Method	Sludge Type	VS (%TS)			References
		Raw	*STWs*	*Reduction (%)*	
Staged anaerobic digestion, thermo-/mesophilic	PRI + TFi/SCD[a]			62-63/47	Schafer et al. (2002)
	10% fermented PRI + 90% WAS[a]			65/43	
	PRI			67-70/57	
Temperature-phased thermo-/mesophilic anaerobic digestion	PRI + WAS			58-68/48-62	Schafer et al. (2002)
	WAS			43-46/34-40	
	PRI + WAS			61.5	Vandenburgh and Ellis (2002)
Three-phase anaerobic digestion	PRI + WAS			56	Schafer et al. (2002)
Single-stage thermophilic anaerobic digestion	1:1 mix of PRI and SCD[a]	(VSS 19-24 g/L)		59.3-64.2	Do et al. (2007)
Anaerobic thermophilic digestion	Prethickened combined PRI and SCD	(35 kg/m^3)		>50	de la Rubia et al. (2006)
Two-phase anaerobic digestion	60% thickened PRI + 40% thickened AS	(VS 27.8 g/L)		57.2	Smith et al. (2005)
Two-phase anaerobic thermo-/mesophilic digestion	56% PRI + 44% WAS	62		31	Rubio-Loza and Noyola (2010)
Anaerobic digestion	WAS pretreated with NaOH			34 (21-52)	Lin et al. (1997)

(*Continued*)

TABLE 12.4 Typical VS (%TS) Values of the Influent Sludge and the Final Sludge Material from Conventional and Mechanical Sludge Stabilization Methods—cont'd

Method	Sludge Type	VS (%TS)			References
		Raw	*STWs*	*Reduction (%)*	
Anaerobic digestion meso-/ thermophilic	PRI/PRI +SCD			47/60-80	Krugel et al. (1998)
Anaerobic mesophilic digestion	75% thickened PRI+25% WAS	68-77		42.3	Ferrer et al. (2012)
Mesophilic anaerobic digestion	Thickened WAS			18 (13-27)	Bolzonella et al. (2005)
	Untreated AS			56-65.5	Appels et al. (2008)
	PRI+SCD	73		56-60	Palatsi et al. (2009)
Stabilization with fly ash and/or lime	WAS	72.1	30-50		Samaras et al. (2008)
Aerobic digestion	AS	73-76	49	50	Farzadkia and Norieh (2005)
Composting	PRI+SCD	74	71		Ruggeri et al. (2008)
	SCD		62		Bertrán et al. (2004)
	Dewatered raw sludge with organic amendments	77-93	66-90	19-45	Komilis et al. (2011a)

[a]*Types of sludge: PRI, primary; AS, activated sludge; SAS, surplus activates sludge; SCD, secondary; TF, trickling filter; MSW, municipal solids waste.*

obtained for composting (up to 70%), which can be attributed to the different nature of the organic materials used for co-composting of sludge.

Based on the above, the advantage of STWs is that they achieve simultaneous stabilization and dewatering of sludge in the same facility. Conventional dewatering methods, e.g., centrifuges, filter presses, etc., are usually combined with a previous stabilization stage; otherwise, their stabilization capacity is practically negligible. This means that conventional sludge treatment demands a two-stage process for stabilization and dewatering, which respectively means increased costs, use of chemical compounds, etc. On the other hand, sludge management and treatment with STWs provide an alternative combined solution, with more than satisfying results for both sludge stabilization and dewatering.

12.3 NUTRIENTS (N, P), pH, AND EC

The sludge to be treated and dewatered contains a certain amount of nutrients like nitrogen and phosphorus. Results from various STW systems indicate that partial removal and transformation of these compounds takes place within the sludge layer. Table 12.5 presents data concerning TN and TP content in the influent sludge and in the residual sludge material obtained from several STW beds. This table shows that the initial nutrient content in the raw sludge is decreased after the treatment in the STW bed. Generally, the nitrogen and phosphorus contents in the residual sludge layer are reduced at a rate which could reach and exceed 46% and 39%, respectively.

As reported by Stefanakis and Tsihrintzis (2012c), this decrease in N and P content proceeds gradually with time with simultaneous increase in $NO_3^{\cdot -}N$, indicating that the nitrification process transforms the ammonia nitrogen to nitrate in aerobic zones that may exist within the sludge layer and along the plant roots. However, lower nitrate production implies that part of the available organic nitrogen is also transformed to ammonium (ammonification). The production of nitrite and its increase compared to the influent sludge also imply that denitrification takes place within the bed. This simultaneous presence of aerobic and anaerobic zones within the sludge layer has been reported by Stefanakis and Tsihrintzis (2012c), who also observed gas bubbles on the sludge layer surface. Peruzzi et al. (2009) and Uggetti et al. (2012c) also reached similar conclusions.

It is also worthy to mention that during the final resting phase, the residual sludge obtains its final characteristics; in the absence of new sludge loadings, TKN concentration is further reduced and nitrification rate is enhanced, as shown by the increased nitrate production (Stefanakis and Tsihrintzis, 2012c). During this phase, the dewatering process still continues (reeds absorb water and nutrients stored in the sludge layer) and direct evaporation from the sludge layer surface takes place (especially when the final resting phase is applied during spring and summer months). These result in rapid drying, as evidenced by the obvious deep cracks of the sludge layer, exposing to the atmosphere large parts of the deposited sludge and promoting this way vertical oxygen diffusion. This is also confirmed by the significant nitrite decrease during the final resting phase of up to 76% (Stefanakis and Tsihrintzis, 2012c).

TP concentration in the residual sludge seems to be less varying compared to nitrogen. Stefanakis and Tsihrintzis (2012c) and Uggetti et al. (2012c) reported a constantly reduced TP concentration in the residual sludge layer for a test period of 3 years, even for higher influent TP content. This behavior is attributed to the phosphate immobilization in microbial cells and the uptake by plants,

TABLE 12.5 Typical TN and TP Concentrations and pH and Electrical Conductivity (EC) Values of the Influent Sludge and the Final Sludge Material in STWs

Source of Sludge	TKN		TP		pH		EC (mS/cm)		
	Raw	*STWs*	*Raw*	*STWs*	*Raw*	*STWs*	*Raw*	*STWs*	References
Soil + anaerobic sludge mixture	0.37 g/kg TS	0.280	0.14 g/kg TS	0.15	7.3	6.6			Guidi et al. (1990)
Sludge from extended aeration plant		4.4-5.6% TS		1.2-1.5% TS	6.5		0.25		Liénard et al. (1995)
Imhoff tank, anaerobically stabilized	3.86 g/kg TS	2.5% TS	0.51 g/kg TS	0.23% TS					Zwara and Obarska-Pempkowiak (2000)
Fermented sewage sludge	3.90% TS[a]	2.5% TS	1.76% TS	2.0% TS	7.0	7.1			Kalisz and Salbut (2007)
Sewage sludge		27 g/kg		3.6 g/kg TS					Nassar et al. (2008)
Thickened activated sludge		2.0%TS		1.25%TS					Cui et al. (2008)
Mixture of sludge from pit latrines, septic tanks, public toilets	1.1 g/L	2.00%TS			7.5	6.01	2.82	0.26	Kengne et al. (2009a)
Activated sludge from extended aeration	6.0%TS	3.25%TS	0.07 g/kg TS	0.1	6.7-7.5	6.9-8.0	140-670	50-230	Uggetti et al. (2009)

Sewage sludge	5.4-6.4% TS[a]	3.5-3.8% TS	1.35-1.66%TS	2.37-2.75%TS					Peruzzi et al. (2009)
Thickened activated sludge	54.9 g/kg TS	36.8-34.3	9.3 g/kg TS	5.7-5.05	6.19				Stefanakis et al. (2009b)
Primary sludge from Imhoff tank	4.45 g/kg TS	4.44	1.81 g/kg TS	1.6					Melidis et al. (2010)
Drinking water sludge		4.8-8.6[a]		8.1-11					Nielsen and Cooper (2011)
Activated sludge from extended aeration	4.96% TS	0.15% TS	0.27-2.68% TS	0.15% TS	6.85	6.1	180	9.04	Uggetti et al. (2012b)
Thickened activated sludge	6.5-6.8% TS	3% TS	1.1-3.9% TS	0.3-0.6% TS					Uggetti et al. (2012c)
Thickened activated sludge	55 g/kg TS	31.5	0.12 g/kg TS	7.3	8.7	6.5	1.97	2.2	Stefanakis et al. (2011b); Stefanakis and Tsihrintzis (2012c)
Fish farm sludge	6.4%TS	2.1-3.7% TS	2.4% TS	1.6-2.2% TS					Gagnon et al. (2013)

[a]*TN.*

which could prevent reintroduction of P in the treatment cycle (Edwards et al., 2001; Peruzzi et al., 2009; Uggetti et al., 2012c).

Table 12.5 also contains data concerning pH and EC values in the sludge layer. Generally, pH values tend to be in the slightly acidic zone in the range of 6.5-7.0 in the residual sludge. Stefanakis et al. (2011b) reported that although pH values remained above 7.5 during the loading phase (influent sludge 8.7), during the final resting phase, the pH values dropped even below 6.5. The authors attributed this behavior to the acid production due to the continuous decomposition of the remaining organic matter, as also to the release of salts by plants. The fact that pH in the unplanted STW units was higher validated this explanation. In the same experiment, an STW bed which was kept unloaded for more than a year always showed a pH value slightly higher than 6.0. This fact confirmed that stabilized sludge may possess a pH value in the range of 6.0-6.5. As these authors stated, this finding is against the common belief that the pH of stable organic waste is usually alkaline (Epstein, 1997). Moreover, further volume reduction of the sludge layer during the final resting phase, coupled with the salt release by plants, could explain the increase in EC values observed by Stefanakis et al. (2011b) during this phase.

12.3.1 Vertical Profile of the Residual Sludge Layer

Table 12.6 presents nitrogen and phosphorus concentration in various depths of the residual sludge layer. Generally, higher concentrations are found in the upper parts, which contain fresh sludge, while the bottom parts contain sludge from older applications. The reduction of nitrogen varies from 12% (Kalisz and Sałbut, 2007) to 44-45% (Pempkowiak and Obarska-Pempkowiak, 2002; Matamoros et al., 2012). The reduction of nitrogen seems to follow the respective availability of organic matter, i.e., VS content, as also implied by Matamoros et al. (2012). The TP profile gives a mixed picture, since different trends are reported. Pempkowiak and Obarska-Pempkowiak (2002) report a decrease of TP concentration with depth, while Matamoros et al. (2012) found an increase. Differences in TP concentration of the influent sludge, as also chemical precipitation at a previous stage in wastewater treatment plants, may explain these differences.

Alterations in residual sludge characteristics during the final resting phase are also observed for N and P (Stefanakis, 2011; Stefanakis et al., 2011). The TN concentration showed a slight decrease during the resting phase for the total depth of the sludge layer, while the differences between upper and bottom parts of the sludge layer remained more or less the same. It is interesting that the TKN concentration in the bottom part was slightly increased during the resting phase, which could be attributed to the dewatering and the respective concentrating effect. It is also noticeable that during the resting phase, the nitrate concentration increased along the total depth of the sludge layer, while the nitrite concentration was decreased (Table 12.6). As mentioned before, this indicates

TABLE 12.6 TN and TP Concentration Variation with the Depth of the Residual Sludge Layer in Various STW Systems

Residual Sludge Layer Depth (cm)		TN		TP		References
Total	*Parts*					
12	0-3	5.0	%VS	0.52	%VS	Zwara and Obarska-Pempkowiak (2000)
	3-6	4.6		0.43		
	6-9	4.0		0.37		
	9-12	3.9		0.33		
24	0-2	3.05	%TS	0.35	%TS	Pempkowiak and Obarska-Pempkowiak (2002)
	6-8	3.86		0.51		
	12-14	2.54		0.26		
	18-20	2.60		0.24		
	23-24	1.72		0.12		
30	0-8	2.48	%VS	0.24	%VS	Obarska-Pempkowiak et al. (2003)
	8-16	2.30		0.22		
	16-23	2.26		0.23		
	23-30	2.30		0.27		
75	20-35	2.5	%TS	2.0	%TS	Kalisz and Sałbut (2007)
	40-55	2.7		2.3		
	60-75	2.2		1.7		
7.5, 15, 62	Surface	3.77, 3.48, 3.37	%TS	0.07, 0.04, 0.18	%TS	Uggetti et al. (2009)
	Bottom	2.72, 2.67, 4.33		0.11, 0.10, 0.10		
20	Upper 10 cm	4.80-4.62[a] 0.08-0.12[b]	mg/g	1.31-1.51	μg/g	Melidis et al. (2010)
	Bottom 10 cm	3.74-4.88[a] 0.29-0.34[b]	mg/g	1.84-1.74	μg/g	
130	0-10	33.5	g/kg	9.4	g/kg	Matamoros et al. (2012)
	20-30	27.6		14.9		
	40-50	24.6		13.8		
	70-80	21.4		10.6		
	90-100	21.8		11.9		
	110-120	21.1		15.0		
	120-130	18.4		16.7		

(*Continued*)

TABLE 12.6 TN and TP Concentration Variation with the Depth of the Residual Sludge Layer in Various STW Systems—cont'd

Residual Sludge Layer Depth (cm)		TN		TP		References
Total	*Parts*					
8.0, 14.0, 18.8 (SLR: 30, 60, 75 kg TS/m[b]/year)	Upper	29.9, 32.5, 29.4 (LP)[a,c] 25.1, 32.5, 31.3 (RP)	mg/g	7.8, 8.2, 8.3 (LP)[c] 7.8, 8.3, 8.0 (RP)	mg/g	Stefanakis (2011); Stefanakis et al. (2011b)
	Bottom	26.0, 28.8, 27.4 (LP)[a,c] 26.9, 30.4, 29.0 (RP)	mg/g	9.3, 10.9, 10.1 (LP)[c] 8.6, 8.9, 8.5 (RP)	mg/g	
	Upper	1.44, 1.73, 1.81 (LP)[c,d] 2.99, 3.00, 3.52 (RP)	mg/g			
	Bottom	1.45, 1.69, 1.92 (LP)[c,d] 3.31, 4.14, 3.78 (RP)	mg/g			
	Upper	3.80, 6.50, 5.95 (LP)[c,e] 4.34, 1.66, 1.58 (RP)	μg/g			
	Bottom	4.06, 6.12, 6.60 (LP)[c,e] 1.45, 1.31, 2.12 (RP)	μg/g			
110	0-40	2.26, 8700[f], 554[d]	%, mg/kg DM[d,f]			Peruzzi et al. (2013)
	40-80	1.97, 1479[f], 741[d]				
	80-110	1.79, 990[f], 542[d]				

[a] *TKN.*
[b] *$NO_3 + NO_2$.*
[c] *LP, loading phase; RP, resting phase.*
[d] *$NO_3^{\cdot-}N$.*
[e] *$NO_2^{\cdot-}N$.*
[f] *NH_4^+-N.*

that the volume reduction of the sludge layer and the deep cracks observed in the vertical profile allowed for the oxygen introduction into deeper parts and, thus, enhanced nitrogen aerobic transformation processes, e.g., nitrification and ammonification.

On the whole, it has to be mentioned that the aim of the sludge treatment in STWs is not to completely eliminate nutrient content in the residual sludge. On the contrary, a certain content of nutrients is desirable since the final biosolids material is usually exploited as a fertilizer. Although there is not a legal limit for nutrient concentration in the biosolids from STWs, however, if this material is further used as a fertilizer or for land application, it should follow the regulations for an organic compost material (Regulation CE No. 2003/2003). For example, Kengne et al. (2009a) mentioned that TN and TP contents in the biosolids from an STW located in the tropics were similar to compost produced from poultry manure. Thus, detailed characterization of the final nutrient concentration after the final resting phase—just before the land application—is necessary, in order to be able to define the appropriate dosage for agricultural applications.

12.4 HEAVY METALS

Heavy metals (HM) represent a critical pollutant present in wastewater sludge. Since biosolids obtained from STWs are usually exploited as fertilizer in agricultural land or for land reclamation, it is always necessary to determine the content of HM (along with other pollutants like pathogens) in the final sludge material, in order to avoid potential environmental and health risks. This prerequisite is reported by the European Sludge Directive 86/278 (EU, 1986) and the proposed third Draft EU Working Document on Sludge (EC, 2000a), which both define thresholds for most common HM.

Table 12.7 presents data concerning the HM content in the influent sludge and in the sludge layer from various STW systems, as also limit values set by the European Commission and the USEPA and some national limits. The major outcome of Table 12.7 is that almost always the HM content in the residual sludge layer is below the respective legal limits. This is a very important parameter of the STW operation and benefits, since these results allow for further use and exploitation of the biosolids material from STWs in agriculture and other land applications.

As it is obvious, depending on the origin and the type of sludge to be treated, there are varying levels of HM contents in the influent sludge. Some metals appear to accumulate at a higher rate (e.g., Cr or Mn) compared to other metals (e.g., Pb, Fe). Additionally, a correlation seems to exist between the HM content in the influent sludge and the respective content in the residual sludge. Generally, HM are gradually accumulated in the sludge layer, but at a slow rate. Such a trend is reported by Melidis et al. (2010) and Stefanakis and Tsihrintzis (2012d). Gradual accumulation of metals could potentially increase the metal

TABLE 12.7 Heavy Metals Content (mg/kg DM) in the Influent Sludge and the Residual Sludge Material in STWs

Legislation/ Source of Sludge		Heavy Metal (mg/kg DM)														References
		As	*Cr*	*Cd*	*Pb*	*Cu*	*Ni*	*Mn*	*Zn*	*Fe*	*Mo*	*Se*	*Co*	*Bo*	*Hg*	
USEPA limit values for land application	High quality		1200		300	1500	420	-	2800							Walsh (1995)
	Ceiling		3000		840	4300	420	-	7500							
European Directive			-	20-40	750-1200	100-1750	300-400	-	2500-4000	-						EU (1986)
EU Working draft			1000	10	750	1000	300		2500						5	EC (2000)
Anaerobically digested sludge	STW	0.57	29	1.4	66.2	408.3	9.5		370		n. d.	n. d.			2.7	Kim and Smith (1997)
Aerobically/ anaerobically stabilized	STW			11.92	341	215			1836						2.79	De Maeseneer (1997)
Imhoff tank, anaerobically stabilized	STW		22.5	1.7	31.4	28.2	16.7		869.3				4.1			Zwara and Obarska-Pempkowiak (2000)
	Polish limits		500	20	500	800	100		2000							
Aerobic sludge	Raw		42	3	154	1906	23	457	684	11,592	4			60		Begg et al. (2001)
	STW		10	2	32	155	9	1443	453	3263	1			10		

Activated sludge	STW (limit value)	39-99 (100)	- (0.8)	- (120)	260-470 (1000)	- (30)		410-1100 (4000)			- (0.8)	Nielsen and Willoughby (2005)
Sewage sludge	STWs	n.d.		0.18	n.d.	0.01			1.72		n.d.	Nassar et al. (2008)
Mixture of sludge from pit latrines, septic tanks, public toilets	STWs	26	2.4	63	575	14	186	703	9579	32		Kengne et al. (2009a)
	EC limits eco label compost	100	1	100	100	50		20				Hogg et al. (2002)
	Limits in Spain	1000	20	750	1000	300		2500				
Sewage sludge	STWs	56	<2	105	432	42		1324				Peruzzi et al. (2009)
Activated sludge from extended aeration, Activated sludge from contact-stabilization	Raw	35.8 36.4 52.1	0.41 0.66 0.76		227 183 232	27.9 50.2 25		348 609 897			4.29 1.99 0.95	Uggetti et al. (2009)
	STWs	55.4 44 58.2	0.63 0.64 0.95		392 147 237.4	29.4 32.5 40.9		551 496 706			5.27 2.39 3.25	
Thickened activated sludge	Raw	20	1.0	40	230	17	183	978	5225			Stefanakis et al. (2009b)
	STWs	140-0	13-14	21-20	61-84	77-0	136-96	726-475	558-920			
Sludge from Imhoff tank	Raw	13.1	0.65	11.8	72.4	36.7	76.5	547.6	8981			Melidis et al. (2010)
	STW	78.3	0.79	31.4	38	17.7	7818	4.7	469			

(*Continued*)

TABLE 12.7 Heavy Metals Content (mg/kg DM) in the Influent Sludge and the Residual Sludge Material in STWs—cont'd

Legislation/ Source of Sludge		Heavy Metal (mg/kg DM)														References
		As	*Cr*	*Cd*	*Pb*	*Cu*	*Ni*	*Mn*	*Zn*	*Fe*	*Mo*	*Se*	*Co*	*Bo*	*Hg*	
Activated sludge from extended aeration	STWs		49- 55	0.6	43- 52	48-55	29- 36		533-551						3.5- 5.3	Uggett et al. (2010)
Aerobic sludge	STW		159	1.42	30.1	563	30		1532						-	Ceccanti et al. (2011)
Activated sludge	STWs		24- 175	1.0- 1.6	29- 68	466- 588	23- 32		1019- 1704							Peruzzi et al. (2011a)
	STWs		25- 70	<3	37- 143	301- 357	23- 32		510-905							Peruzzi et al. (2011b)
Secondary stabilized sludge	Raw		51	1.7	53	252	39		719						<1.5	Uggetti et al. (2011)
	STW		57	0.8	75	265	31		615						<1.5	
SAS and stabilized sludge	Raw			1.7 0.83 -	53 42 -	252 290 -	39 65 -		719 670 -						<1.5 1 -	Uggetti et al. (2012b)
	STWs			0.8, 0.8 1, 1 1, 1	73, 76 57, 72 64, 55	318, 213 305, 440 232, 197	30, 32 76, - -		588, 641 874, 1189 1269,1085						<1.5 1.5, 5 1, 2	

Surplus sludge from secondary sedimentation tank	STW	58.1	1.9	96	259	40	760	1378	5100		Matamoros et al. (2012)
Thickened activated sludge			0.4-0.7	10-18	-	11-12		227-295		3-9	Uggetti et al., 2012a
Thickened activated sludge	Raw	275.8 5000	9.6	117.5	94.7	83.9	140.7	422.9	3354		Stefanakis and Tsihrintzis (2012d)
	STWs	82-206 3469	3.7-4.8	26-35	102-121	29-34	25-115	146-229	4158-4666		
	Greek limits	500	20-40	750-1200	1000-1750	300-400		2500-4000			
Secondary sludge	STWs	11.3 18.1 32.1 17.5	1.07 0.95 0.84 0.74	10.0 10.7 14.6 15.6	165.6 236.6 218.6 80.8	26.3 20.7 20.3 22.4		520.2 416.0 542.2 437.1			Kołecka and Obarska-Pempkowiak (2013)

National and EU limit values are also included.

content in sludge layer, due to dewatering and reduction of the sludge volume, as mentioned by Begg et al. (2001). However, the final content of HM does not usually exceed the legal limits for land application, as also mentioned by other authors (Uggetti et al., 2009). A seasonal variation of heavy metal content is also reported, with lower concentrations occurring during the plant growing season, i.e., spring months (Melidis et al., 2010; Stefanakis and Tsihrintzis, 2012d).

Stefanakis and Tsihrintzis (2012d) added a high Cr concentration in the influent activated sludge (more than 20 times up the typical Cr concentration in municipal sludge) to investigate the accumulation rate under low and high loadings (30, 60 and 75 kg TS/m^2/year) and to observe the system response. These authors reported a faster accumulation of Cr in the residual sludge, which exceeded the legal limits rather fast. This showed that higher HM content in the influent sludge may result in higher accumulation rate in the residual sludge. However, in the activated sludge from municipal wastewater, the influent concentrations are usually low (as shown in Table 12.7), thus, accumulation rates are slow and the HM content remains below the legal limits. Peruzzi et al. (2009) report a stable more or less HM content in the sludge layer with time. Kołecka and Obarska-Pempkowiak (2013) report low the HM content in residual sludge in Danish STW beds operating for 7-15 years. Matamoros et al. (2012) concluded the same for a STW bed after 20 years of operation. Even after many years of operation, the content of HM to be removed in the biosolids was below the legal limits, which shows that the accumulation rate in STWs is indeed very slow and the final sludge material is appropriate for land application and usage.

12.4.1 Vertical Profile of the Residual Sludge Layer

Variations are also reported for HM concentrations between the upper and bottom parts of the residual sludge layer (Table 12.8). Kim and Smith (1997) found higher metal contents at the bottom parts of a 71-cm deep residual sludge layer. The same is also reported by Melidis et al. (2010), for a 20-cm thick sludge layer, by Matamoros et al. (2012) for a 20-year-old STW system with a sludge layer depth of 130 cm, and by Stefanakis and Tsihrintzis (2012d) for residual sludge layer depths of 8-20 cm. Increased HM content with depth can be attributed to the mineralization of the older sludge material in the deeper parts of the sludge layer and the higher dewatering rate of this layer, which concentrate the present HM amount (Stefanakis and Tsihrintzis, 2012d). Zwarka and Obarska-Pempkowiak (2000) and Obarska-Pempkowiak et al. (2003) found fluctuating differences with depth, but for a shallower sludge layer (12 cm). Generally, heavy metal concentrations increase with the depth of the residual sludge layer due to mineralization of organic matter, which concentrates the accumulated metals.

The application of the final resting phase alters the content of HM in the residual sludge (Stefanakis and Tsihrintzis, 2012d). While for Cr, Cd, Pb, and Ni, concentrations were decreased, the opposite was observed for Cu,

TABLE 12.8 Heavy Metal Concentrations Along the Depth of the Residual Sludge Layer in Various STW Systems

Residual Sludge Layer Depth (cm)		As	Cr	Cd	Pb	Cu	Ni	Mn	Zn	Fe	Mo	Se	Co	Hg	References
Total	*Parts*						*mg/kg DM*								
71	0-10	n.d.	12	n.d.	27	200	n.d.		220		n. d.	n. d.			Kim and Smith (1997)
	10-20	0.31	19	n.d.	33	270	8.1		290		n. d.	n. d.			
	20-30	0.46	24	n.d.	60	370	12		570		n. d.	n. d.			
	30-40	0.65	35	n.d.	110	490	20		540		n. d.	n. d.			
	40-50	0.86	36	n.d.	77	490	17		600		nd	nd			
	50-71	n.d.	48	8.2	90	630	n.d.		n.d.		n. d.	n. d.			
30	0-8		22.8	1.64	31.3	28.6	19.0		780				4.5		Obarska-Pempkowiak et al. (2003)
	8-16		21.8	1.80	30.3	28.0	16.6		856				4.2		
	16-23		27.0	1.92	37.2	28.4	18.8		1094				4.5		
	23-30		18.3	1.40	26.7	28.6	12.4		748				3.2		

(Continued)

TABLE 12.8 Heavy Metal Concentrations Along the Depth of the Residual Sludge Layer in Various STW Systems—cont'd

Residual Sludge Layer Depth (cm)		As	Cr	Cd	Pb	Cu	Ni	Mn	Zn	Fe	Mo	Se	Co	Hg	References
Total	*Parts*						*mg/kg DM*								
7.5	Surface	54.8		0.61	49.3	392	29.2		537					5.67	Uggetti et al. (2009)
15		45.6		0.73	46.0	174	36.0		568					1.44	
62		62.0		0.90	80.9	245	44.7		790					3.3	
	Bottom	56.0		0.65	54.3	392	29.6		565					4.87	
		42.4		0.54	33.2	120	29.0		425					3.34	
		54.3		1.0	79.3	230	37.1		622					3.2	
20	Upper 10 cm		69.9	0.50	28.7	55.3	15.5	7420	3.73	240					Melidis et al. (2010)
	Bottom 10 cm		86.8	1.08	34.2	63.4	20.0	8216	5.75	697					
130	0-10		42	1.8	61	184	30	510	1098	5900					Matamoros et al. (2012)
	20-30		56	2.3	78	251	43	550	1443	4600					
	40-50		55	2.2	87	251	41	620	1506	4500					
	70-80		54	1.3	86	239	38	710	1252	6300					
	90-100		65	1.7	104	263	41	700	1299	5200					
	110-120		71	2.1	135	305	42	690	1453	4500					
	120-130		69	2.2	152	330	41	760	1521	5400					
8.0-18.8	Upper		116.5 7632[a]	4.1	30.5	123.6	33.2	195.8	424	4103					Stefanakis and Tsihrintzis (2012d)
	Bottom		144.5 8589[a]	4.2	38.7	104.8	36.5	198.9	506	4576					

[a] Additional Cr added.

Mn, Zn, and Fe. The characteristics as also the mobility and bioavailability levels of each specific metal could explain this difference. However, in all cases the HM content was below the legal limits.

Studies have also been carried out concerning the bioavailability (i.e., potential for plant uptake and leaching) of HM in STWs and their fractionation status. Peruzzi et al. (2011a,b) distinguish four different fractions of HM in sludge: (1) exchangeable fraction, where metals are adsorbed onto Mn and Fe hydroxides; (2) reducible fraction, where metals are bound to Fe and Mn oxides; (3) oxidizable fraction, where metals are bound to organic matter; and (4) residual fraction, where metals are associated with solids. The first two fractions (1 and 2), which are most available, were found to be significantly reduced after the first 3 years of sludge treatment in STWs (Peruzzi et al., 2011a,b). These authors also report low bioavailability of HM due to the strong and gradually enhanced bond between organic matter and HM, which makes them less available for plant uptake. Kołecka and Obarska-Pempkowiak (2013) report that most HM of the sludge layer in STWs after 7-15 years of operation were bound with the stable residual fraction.

12.5 PATHOGENS

The presence of pathogenic microorganisms in the residual sludge material represents another critical pollutant which possesses a potential risk for public health, if the biosolids material is to be used for land application, as fertilizer, etc. Therefore, the European Sludge Directive (EU, 1986) states the necessity to measure fecal microbial indicators before the land application of treated wastewater sludge. The third Draft EU Working Document on Sludge (EC, 2000) defines the legal limit for *Escherichia coli* equal to 5×10^2 CFU/g, while the Working Document on Sludge and Biowaste (EC, 2010a) proposes two indicators: the absence of salmonella in 25-50 g or reduction of *E. coli* to $<5 \times 10^5$ CFU/g (wet weight) of treated sludge. Table 12.9 presents data concerning the content of fecal microbial indicators found in the influent sludge and the residual sludge material in several STW beds. It is clear that the final biosolids material after sludge treatment in STWs complies with the respective legal limits. In all cases, the bacteriological status of the sludge is improved and *E. coli* and *salmonella* bacteria are more or less destroyed.

12.5.1 Vertical Profile of the Residual Sludge Layer

Table 12.10 presents the concentrations of various pathogenic indicators within various depths of the residual sludge layer. The bottom parts of the sludge layer appear to be clearly better hygienized compared to upper parts. Nielsen (2007) reported a reduction in microbial indicators in the residual sludge even after 1-3 days since the last sludge loading, while within 3 months of resting, the concentrations of *Salmonella*, *Enterococci* and *E. coli* were quite low and below the

TABLE 12.9 Microbial Indicator Concentrations in the Influent Sludge and the Residual Sludge Material in STWs

Legislation/Source of Sludge	Pathogenic Microorganisms	References
European Directive	-	EU (1986)
EU Working draft	***E. coli*** 5×10^2 CFU/g	EC (2000)
Working document on sludge and biowaste	Absence of ***Salmonella*** in 25-50 g or Reduction of ***E. Coli*** to $<5 \times 10^5$ CFU/g (wet weight)	EC (2010)
Digested sludge from Imhoff tank	Sludge from Imhoff tank: ***E. coli*** 1×10^7, *Salmonella* detected, ***Ascaris lumbricoides*** 140/kg DM Sludge feed: *E. coli* 8×10^{-7}, ***Salmonella*** detected, ***Ascaris lumbricoides*** 80/kg DM Sludge stored in the bed: ***E. coli*** 2×10^{-4}, ***Salmonella*** not detected, ***Ascaris lumbricoides*** 4050/kg DM Sludge after 8 months: ***E. coli*** 2×10^{-3}, ***Salmonella*** not detected, ***Ascaris lumbricoides*** 4500/kg DM	Obarska-Pempkowiak et al. (2003)
Activated sludge (results 6-9 months after the last loading)	***E. coli*** influent 10^4-10^6/g, residual sludge < 20/g, removal 10^6 units ***Streptococci*** influent 10^4-10^6/g, residual sludge < 100/g, removal 10^6 units ***Salmonella*** influent 10-10^2/g, residual sludge n.d., removal 10^6 units	Nielsen and Willoughby (2005)
Aerobic digested sludge (sampling 3 days after each loading)	***E. coli*** influent 7.9×10^4-1.4×10^6 (CFU/mL) Top layer: 1.2×10^6-3.5×10^7 CFU/g DM, Bottom layer: 1.9×10^4-1.2×10^6 CFU/g DM	Roe and Cardinale (2005)
50% AS + AS from settling tanks, 50% concentrated AS	***E. coli***: influent 7.9×10^5 CFU/g DM, sludge residue: <200 CFU/100 g, reduction: log 6-7 ***Salmonella***: influent 900 CFU/100 g DM, sludge residue: <2 CFU/100 g ***Enterococci***: influent 1.1×10^4 CFU/g DM, sludge residue: <10 CFU/100 g, reduction: log 5	Nielsen (2007)
Sewage sludge	**Helminth eggs**: detected in 3 out of 30 samples	Nassar et al. (2008)

TABLE 12.9 Microbial Indicator Concentrations in the Influent Sludge and the Residual Sludge Material in STWs—cont'd

Legislation/Source of Sludge	Pathogenic Microorganisms	References
Mixture of sludge from pit latrines, septic tanks, public toilets	**Helminth eggs**: influent 10,409/L Residual sludge (1 month after the last sludge application): total eggs 78.9/g TS	Kengne et al. (2009)
Septage from septic tanks + aerated sludge	***E. coli*** (MPN/g) influent 4.2×10^6 MPN/g DM, reduction: 2-4 log units ***Enterococci*** (MPN/g) influent 5.6-5.9×10^6 MPN/g DM, reduction: 2-4 log units	Troesch et al. (2009a; 2009b)
Activated sludge from extended aeration	***Salmonella* spp.** (in 25 g) influent: absence, residual sludge: absence ***E. coli*** (MPN/g) influent: 1100, 1100, 4, residual sludge: <3 to >2400	Uggetti et al. (2009)
Aerobic sludge	***E. coli*** residual sludge: <1000 CFU/g	Ceccanti et al. (2011)
Thickened activated sludge	**Total Coliform** influent: 3.8×10^5 CFU/g DM Residual sludge: 15-102 CFU/g after the final resting phase (~4 months)	Stefanakis et al. (2011b)
Thickened activated sludge	At the beginning and the end of final resting phase ***Salmonella* spp.** (in 25 g) residual sludge: absence ***E. coli*** (MPN/g) residual sludge: beginning (6600, 4500, 3900), end (93, 240, 150)	Uggetti et al. (2012a)

legal limits. He also mentioned that the deeper parts (>25 cm) of the sludge layer are not recontaminated during the successive sludge loadings. Kim and Smith (1997) also report a significant reduction in Total Coliform number with depth. Kengne et al. (2009a) reported a 73% reduction of Helminth eggs between the upper 5 and the bottom 5 cm of a 35 cm residual sludge layer of a STW system located in the tropics. Stefanakis and Tsihrintzis (2011b) ended with the same conclusion, mentioning that the concentration of total coliforms in the bottom parts of the residual sludge layer was always lower compared to the upper parts for both the loading and the resting phases.

These authors also noticed the significance of the final resting phase (without sludge loadings); 3 months after the last sludge application they observed a

TABLE 12.10 Microbial Indicator Concentrations Along the Depth of the Residual Sludge Layer in Various STW Systems

Residual Sludge Layer Depth (cm)		Total Coliform (CFU/g)	*E. coli* (MPN/g)	Helminth Eggs/g TS	*Salmonella* spp. in 25 g	References
Total	*Parts*					
71	0-10	4700				Kim and Smith (1997)
	10-20	270				
	20-30	19				
	30-40	10				
	40-50	8				
	50-71	103				
35	Upper 5 cm			127.4		Kengne et al. (2009)
	Center			75.8		
	Bottom 5 cm			34		
7.5, 15, 62	Surface		460, 1100, <3		Absence	Uggetti et al. (2009)
	Bottom		460, 1100, <3		Absence	
8.0, 14.0, 18.8	Surface	1592, 3437, 3816 (LP)[a] 473, 2142, 1401 (RP)				Stefanakis et al. (2011b)
	Bottom	1823, 2276, 1043 (LP)[a] 184, 1195, 878 (RP)				

[a] *LP, loading phase; RP, resting phase.*

reduction in the total coliform concentration of the residual sludge layer, below the legal limit, which showed that during this phase the residual sludge obtains its final characteristics for further exploitation. A similar behavior is also reported by Uggetti et al. (2012b).

12.6 SPECIFIC MICROCOMPOUNDS

The municipal sludge also contains some toxic organic compounds, like LAS (linear alkylbenzene sulphonates), NPE (nonylphenolethoxylates), DEHP (di (2-ethylhexyl) phthalate), and PAH (polycyclic aromatic hydrocarbons), which represent a potential environmental risk if they are found at high concentrations in biosolids used as fertilizer or for land application. The Working Document on Sludge, third Draft (EC, 2000) suggested limit values of 2600, 50, 100, and 6 mg/kg DM for LAS, NPE, DEHP, and PAH, respectively. LAS and NPE are originated from various cleaning agents (Nielsen, 2005). Table 12.11 presents data of mineralization of these compounds in STW systems.

Nielsen (2005) investigated the mineralization of LAS and NPE in an STW bed, by anaerobic storage in a container and by storage in a sludge pile which was regularly mechanically mixed. He found that aerobic conditions enhance the mineralization of these compounds. LAS and NPE are largely mineralized after 5-6 months of treatment in the STW bed, with respective reduction of 98% and 93% (Nielsen, 2005), resulting in concentrations below the proposed legal limits. Moreover, as Table 12.11 shows, the treatment in the STW bed achieved similar mineralization rates for both LAS and NPE compared to composting and mechanical aeration. A high removal of organic compounds (LAS, NPE, PAH, and DEHP) below the legal limits is also reported by Peruzzi et al. (2011a, 2013) for systems after 4 years of operation. The aerobic conditions within the systems are considered by these authors to enhance the removal of these toxic compounds.

Matamoros et al. (2012) investigated the fate of polycyclic and macrocylic musks in a 30-year-old STW system. These compounds are also used as ingredients in cleaning agents and in personal care products (PCPs) and need to be removed before the land application of biosolids. These authors focused on specific compounds like galaxolide (HHCB), tonalide (AHTN), cashmeran, celestolide, ambrettolide, DEHP (a compound used in polymer products). They found that the removal of these compounds varied with depth of the residual sludge between 40% and 98% (Table 12.11). An increase was found only for tonalide which was attributed to the different composition of sludge in the past.

Chen et al. (2009a,b) also investigated the fate of DEHP along with other xenobiotic compounds like triclosan, OTNE, HHCB, HHCB-lactone, and AHTN. In this pilot study, reported removal rates after 12 months were 20-30% for AHTN and HHCB, 70% for triclosan, and 70% for OTNE. They reported different behavior of these compounds in the sludge layer parts for this first period of the experiment (Table 12.11) and stated that after a full operation period of 10 years, a considerable reduction of these compounds is expected.

TABLE 12.11 Mineralization Rate of Various Organic Micropollutants in STW Systems

Source of Sludge	Organic Micropollutants		References
Mesophilic anaerobic digested sludge	**LAS (mg/kg DM)** Influent: 5000 STW: 100 (98% removal) Sludge pile: 480 (90% removal) Storage in containers: 2900 (42% removal) Depths 0-120 cm: 41% removal Depth 0-20 cm: 75% removal Depth 20-120 cm: 0% removal Composting: 100% removal Mechanical aeration: 95% removal	**NPE (mg/kg DM)** Influent: 54 STW: 4 (92% removal) Sludge pile: 27 (43% removal) Storage in containers: 63 Depths 0-120 cm: 10% removal Depth 0-20 cm: 61% removal Depth 20-120 cm: 0% removal Composting: 78-95% removal Mechanical aeration: 75-95% removal	Nielsen (2005); Nielsen and Willoughby (2005)
Surplus sludge from secondary sedimentation tank (20-year-old STW bed)	**Musk fragrances: Galaxolide (HHCB), Tonalide (AHTN), Cashmeran, Celestolide, Ambrettolide, DEHP** Concentrations in residual sludge layer: 10-9000 ng/g DM in the order DEHP > galaxolide > tonalide > cashmeran > ambrettolide Decrease with depth: ambrettolide (98%), cashmeran (93%), galaxolide (90%) Tonalide (increase with depth by a factor of 2)		Matamoros et al. (2012)

Excess sludge from WWTP (13 months experiment)	Xenobiotics: **triclosan, OTNE, HHCB, HHCB-lactone, AHTN, DEHP** triclosan: initial 1400 (top), 1900 (middle), 2000 (bottom) ng/g DM After 13 months: 800 (top), 900 (middle), 600 (bottom) ng/g DM HHCB: 8000-12,000 ng/g DM, AHTN: 1500-2300 ng/g DM, HHCB-lactone: 1400-2100 ng/g DM OTNE: initial 2500 (top), 2500 (middle), 2400 (bottom) ng/g DM After 13 months: reduction of 42% (top), 53% (middle), 70% (bottom) DEHP: initial 11,500 (top), 10,500 (middle), 7200 (bottom) ng/g DM After 13 months: reduction of 70% (top), 71% (middle), 40% (bottom)	Chen et al. (2009a)
Activated sludge	**LAS** (mg/kg DM): influent 405-2359, residual sludge after 6 (18-405) and 48(3-78) months **NPE** (mg/kg DM): influent 41-132, residual sludge after 6 (29-143) and 48(19-48) months **DEHP** (mg/kg DM): influent 7-38, residual sludge after 6 (2-25) and 48(1-4) months	Peruzzi et al. (2011a)

Biodegradation of these organic compounds seems to be the main removal mechanism, as in VFCWs for wastewater treatment.

Gustavsson et al. (2007) tested the efficiency and capacity of STWs to treat industrial sludge, which contains pharmaceuticals and explosives substances, mainly nitroaromatic compounds. These compounds are considered as environmental hazards, since they can cause genotoxicity and cytotoxicity. During this study, embryos of zebrafish (*Danio rerio*) which are sensitive to the presence of nitroaromatic compounds were used to test the sludge toxicity. Results showed a reduced toxicity compared to the influent sludge, while no genotoxicity was detected in the effluent water. However, the authors report that the actual removal mechanisms still remain largely unknown.

Overall, STW systems appear to provide adequate treatment for various organic micropollutants that may exist in sludge. Results obtained from several studies clearly indicate the treatment capacity of the systems, since in most case the final content of these compounds in the biosolids was below the legal limits for land application.

12.7 DRAINED WATER

Drained water represents an important parameter of the operation of STW systems. It appears a short time after the sludge application at the bottom of the bed, as a result of the filtration process taking place within the bed when the added water moves downward through the sludge layer and the porous media layers. The flow can last up to several days, depending on the sludge characteristics and the applied SLR. Koottatep et al. (2001) reported that for an STW located in the tropics and receiving an SLR of 250 kg TS/m^2/year, the drainage flow lasted up to 6 days. Under temperate climate, the flow could last between 6 and 9 days, depending on the applied SLR, but the majority of the drained water volume appears during the first 3 days (Stefanakis and Tsihrintzis, 2011). The drained water contains most of the pollutants present in the sludge, but with varying concentrations with time. Table 12.12 presents data obtained from various STW systems concerning the drained water quality.

As Table 12.12 clearly shows, the vast majority of the solids added to the system are trapped within the bed through physical processes, like filtration through the porous media layers. The drained water is practically free of solids, since <2% of the solids content in sludge leaves the bed through the drained water. Significant reduction also occurs for organic matter. COD and BOD_5 concentrations in drained water are more than 90% reduced compared to respective concentrations in influent sludge. This also implies that the major portion of the organic matter added to the system remains in the bed. De Maeseneer (1997) reported that effluent BOD_5 concentration is usually below 100 mg/L, while a lower effluent value has also been reported (Begg et al., 2001). According to Stefanakis and Tsihrintzis (2012c), BOD_5 concentration in drained water was always below the threshold of 10-15% of COD concentrations. The fact that

TABLE 12.12 Drained Water Quality in Various STW Systems

References		TSS (mg/L)	VSS (%)	COD (mg/L)	BOD_5 (mg/L)	TKN (mg/L)	NH_4^+-N (mg/L)	$NO_3^{\cdot-}$N (mg/L)	$NO_2^{\cdot-}$N (mg/L)	TP (mg/L)	OP (mg/L)	pH	EC (μS/cm)
Liénard et al. (1995)	Raw sludge	2750	75.2									6.5-7	250
	Effluent	19.4, 34.3, 103.4		43, 85.3, 102.1									
Burgoon et al. (1997)	Raw sludge	14,567		14,131		974	126	1.0				7.0	
	Effluent	75		713		95	73	6.0				7.0	
	%	99		95		90	42	-				-	
Summerfelt et al. (1999)	Raw sludge	7860	6204[a]	6855		234		0.057		238	106		
	Effluent	156	93.3	539		26.9		45.4		30.9	7.07		
	%	98	98.5	92.1		89				90	93		
Koottatep et al. (2001)	Raw sludge	16,300	13,500[a]	16,000	2800	830	340	8				7.5	
	Effluent	3120		660		102.2	77.9	180-260					
	%	76-86		94-99		93	80-92						
Begg et al. (2001)	Raw sludge	12,352			1342								
	Effluent	14.2			6								
	%	98			99	9-94[b]		90		80			

(Continued)

TABLE 12.12 Drained Water Quality in Various STW Systems—cont'd

References		TSS (mg/L)	VSS (%)	COD (mg/L)	BOD_5 (mg/L)	TKN (mg/L)	NH_4^+-N (mg/L)	$NO_3^{\cdot-}$N (mg/L)	$NO_2^{\cdot-}$N (mg/L)	TP (mg/L)	OP (mg/L)	pH	EC (μS/cm)
Edwards et al. (2001)	Raw sludge	31,364		22,600	7000		659				5650	7.81	
	Effluent	832		2001	1175		511				581	7.95	
Barbieri et al. (2003)	Raw sludge	30,400		29,480		510[b]					35		
	Effluent	<60		<200		179					29.7		
Obarska-Pempkowiak et al. (2003)	Raw sludge			1000		100-150[b]				10-20			
	Effluent			250		12				1			
Koottatep et al. (2006)	Raw sludge	11,820		14,485		993	412	9					
	Effluent	219		475		121	83	51					
Nassar et al. (2008)	Raw sludge	19,468		11,746		570					8.1		3200
	Effluent	22		161		51					21.6		4700
Cui et al. (2008)	Raw sludge			83-440			16-79			2.6-7.9			
	Effluent			32-123			21-69			0.6-2.1			
	%			60-70			47-57			61-77			

Wang et al. (2009b)	Raw sludge	7830		8330	3280	200	15	1.3		33	<0.2	7.3	
	Effluent	<70		209		8		<0.1		<3	<0.2		
	%	98		99.1		99.4				98	98		
Panuvatvanich et al. (2009)	Raw sludge	19,500			2225	950[b]	320	4.81	0.82			7.48	
	Effluent	989			281	169	121	7.55	0.17			7.25	
Troesch et al. (2009a)	Raw sludge	26,955	68	46,255		1546	308			606	48	7.3	3400
	Effluent	274		805		43	12	29				7.9	1572
	%	96		92		88							
Troesch et al. (2009b)	Raw sludge	1671 8890	79 80	1908 11,840		117 739	1.5 14.6	0.4 0.3		39 229	9.4 38.0	7.1 6.7	1063 1174
	Effluent	50 41		84.5 136		7.2 13.5	3.3 9.8	11.9 50.4		13.7 28.9	11.8 24.9	7.6 7.5	1188 1448
	%	97.2 99.5		96.1 98.1		93.5 96.4							
Gagnon et al. (2012)	Raw sludge	8889	6556[a]	21,889		556	150	17					
	Effluent	46	19	145		10	3	13					

(Continued)

TABLE 12.12 Drained Water Quality in Various STW Systems—cont'd

References		TSS (mg/L)	VSS (%)	COD (mg/L)	BOD_5 (mg/L)	TKN (mg/L)	NH_4^+-N (mg/L)	$NO_3^{\cdot -}$N (mg/L)	$NO_2^{\cdot -}$N (mg/L)	TP (mg/L)	OP (mg/L)	pH	EC (μS/cm)
Stefanakis and Tsihrintzis (2012c)	Raw sludge	23,800	17,080[a]	2869			367	44	0.22		76	8.77	1973
	Effluent 10 min 1 h 2 h 1d 2d	3920 3770 3640 2540 310	3240 2870 2840 1990 230	315 300 575 241 179			86.5 77.9 150.9 88.0 84.3	195.2 100.8 88.2 62.8 66.4	0.58 1.05 1.94 11.07 0.74		23.3 20.6 14.1 22.3 9.4	7.50 7.28 7.76 7.61 7.92	1850 1830 3560 2410 3660
	%	83.5-98.7	81.1-98.7	89.0-93.8			76.4-77.0				69.4-87.7		
De Castro et al. (2012)	Raw sludge	13,176		12,417			52.3	1.3	0.1		30.6	7.1	
	Effluent	599,6		103.7			6.5	15.6	0.1		1.3	6.3	

[a] *mg/L.*
[b] *Total nitrogen.*

a rapid high decrease of COD concentration is observed, even after the first minutes of last sludge feeding (Stefanakis and Tsihrintzis, 2012c), indicates that physical processes, like filtration through the porous media layers, also contribute to organic matter removal, along with biodegradation (Gagnon et al., 2012). It is reported that the nature of COD and the relative portions of particulate and soluble parts may affect this removal, since soluble COD passes more easily than the particulate through the media layers (Wang et al., 2009b).

Nitrogen removal is also observed in the drained water. Data presented in Table 12.12 imply that the major portion of the TKN in influent sludge is under the form of organic nitrogen, as also reported elsewhere (Summerfelt et al., 1999; Panuvatvanich et al., 2009). On the other hand, in drained water, NH_4^+-N seems to be the main form of TKN, while nitrate concentration is also increased (Summerfelt et al., 1999). This indicates that both aerobic processes of ammonification and nitrification take place within the bed and the sludge layer. Liénard et al. (1995) reported that high nitrate concentration (>80 mg/L) was observed after 5-6 min of the sludge load. Similar findings were also reported by Stefanakis and Tsihrintzis (2012c), who observed a rapid decrease in ammonia nitrogen concentration even during the first 10 min after the last sludge application, with a simultaneous increase in nitrate at the same time point. This indicates that rapid nitrification takes place almost immediately after the sludge feeding. Decreased pH values at the same time confirmed that nitrification takes place, and produced H^+ results in lower pH values. Fluctuating concentrations of ammonia during the first 1-2 days after sludge feeding imply that ammonification also takes place. Liénard et al. (1995) mentioned that ammonium adsorption to the biofilm, the porous media, and the sludge layer also plays a role in ammonia removal. They also mentioned that nitrification takes place during the resting period between the sludge applications, with a respective increase in nitrate production. High reduction of TKN due to ammonia nitrogen transformation to nitrate (nitrification) is also attributed to the operational mode of the system (vertical flow) and the vertical drainage of the water, which provides good aerations and enhances the aeration-related processes (Koottatep et al., 2001; Panuvatvanich et al., 2009).

An interesting observation was made by Stefanakis and Tsihrintzis (2012c); they reported a sudden, sharp increase in nitrite in drained water 1 day after the last sludge feeding, while at the same point of time, a minimum nitrate concentration was measured. The authors attributed these findings to the denitrification process. Burgoon et al. (1997) also reported that the initial appearance of nitrification was followed by a rapid denitrification. This observation was more intense in beds with a thicker sludge layer, which implies that a deeper sludge layer allows for the appearance of anaerobic microsites which favor denitrification. In shallower sludge layers, i.e., in beds receiving lower SLR, ammonia nitrogen removal is higher, indicating that aerobic processes dominate in the bed.

Phosphorus concentration is also decreased in the drained water, often reaching rates exceeding 95% (Table 12.12). These are far higher than typical P removal rates (up to 30-40%; Verhoeven and Meuleman, 1999; Brix and Arias, 2005; Stefanakis and Tsihrintzis, 2012a) for VFCWs treating wastewater. Different composition of P in sludge could explain this performance. When the major portion of P in sludge exists in its particulate form, retention via adsorption process in the porous media layers takes place at a higher rate (Summerfelt et al., 1999; Wang et al., 2009b). Moreover, the presence of the sludge layer above the porous media layers acts as an additional treatment layer which contributes to the slower vertical drainage of water and increases the contact time with the porous media, thus, enhancing the removal of P through the various physical, biological, and chemical processes.

Table 12.13 presents heavy metal concentrations in drained water of STWs. Although there is not so extensive information on this, reported results show that the majority of the HM content in the influent sludge is retained within the bed. Begg et al. (2001) reported a mean removal of 87% (influent sludge vs. drained water) for 10 HM measured. A similar value (84.6%) was also reported by Stefanakis and Tsihrintzis (2012d). It is interesting that Fe had the highest concentration among the HM measured in one study and the lowest in the other. Probably, higher levels of aeration were achieved in the second study (Stefanakis and Tsihrintzis, 2012d), which enhanced the precipitation of Fe oxides. HM concentrations are higher in drained water during the first hours after the last sludge application, but they follow a decreasing trend with time and they practically reach zero values after 1 day (Stefanakis and Tsihrintzis, 2012d).

Generally, the drained water from STWs does not contain heavy pollutant loads. Usually, this load represents only a few percent of the load received by a WWTP (Obarska-Pempkowiak et al., 2003). Studies have shown that the pollutant load in drained water is related to the level of SLR; beds receiving higher SLR were found to produce drained water containing higher pollutant concentrations (Edwards et al., 2001; Dominiak et al., 2011; Stefanakis and Tsihrintzis, 2012c). However, Troesch et al. (2009b) reported that higher SLR and the respective formation of a deeper sludge layer could improve the quality of the drained water. Apparently, other parameters, like the sludge type and origin, and the concentration of pollutants in fed sludge play a role in the quality of the drained water. Nevertheless, the pollutant load in drained water from STWs is far lower compared to typical wastewater inflow in WWTPs, which allows for recirculation of drained water and its treatment in a nearby WWTP, without any significant effect on its operation (Obarska-Pempkowiak et al., 2003; Uggetti et al., 2009; Stefanakis and Tsihrintzis, 2012c).

TABLE 12.13 Heavy Metal Concentration in Drained Water in STW Systems

References	mg/L	Cr	Cd	Pb	Cu	Ni	Mn	Zn	Fe	Mo	Bo
Begg et al. (2001)	IN	0.14	n.d.	0.5	11	0.82	2	3	19	0.09	0.41
	EFF	n.d.	n.d.	0.29	0.17	0.17	0.29	0.14	0.07	n.d.	0.11
	%	97	100	40	98	80	86	95	98	100	73
Stefanakis (2011); Stefanakis and Tsihrintzis (2012d)	IN	2.78 (24.8[a])	0.10	1.18	0.95	0.84	1.41	4.23	33.54		
	EFF 10 min	0.76 (16.3[a])	0.03	0.41	0.38	0.35	0.11	0.0	0.06		
	2 h	0.40 (11.2[a])	0.02	0.25	0.14	0.20	0.30	0.19	0.63		
	1 day	0.09 (7.8[a])	0.01	0.09	0.06	0.05	0.06	0.06	1.27		
	%	87.0 (49.3[a])	81.2	79.0	77.3	68.6	88.1	97.8	98.1		

[a]Additional Cr added.

12.8 STABILITY AND MATURITY OF RESIDUAL SLUDGE

Since the end product of STW beds, i.e., the residual sludge material or biosolids, is usually exploited and used as fertilizer or for land application, it has to fulfill certain criteria. As we discussed in Sections 12.4 and 12.5, there are legal limits for heavy metals and pathogenic indicators proposed by European directives. Besides these pollutants, however, taking into account that the final biosolids material can be considered an organic waste, it has to be both stable and mature. These two terms have a different content and meaning (Iannotti et al., 1993; Barrena et al., 2006; Komilis and Tziouvaras, 2009; Ceustermans et al., 2010; Stefanakis et al., 2011b).

Although sludge stabilization is often implied by a decrease in VS content (Koottatep et al., 2001; Cui et al., 2008), this expression gives only an indication of the rate of organic matter biodegradation, but does not give any information concerning the inactivation of the microbial community. Stability refers to the microbial activity related to the decomposition of organic matter and is, therefore, expressed by the oxygen consumption or the CO_2 production of the material (Iannotti et al., 1993; Stefanakis et al., 2011b). Maturity is related to the potential toxicity effect of an organic material to plant growth or to seed germination (Wu et al., 2000). Until today, no widely accepted indices for stability or maturity evaluation are accepted, although many indices have been proposed for organic compost materials (Barrena et al., 2006). Regarding the stability and maturity of the biosolids material obtained from STWs, limited work has been carried out. It is only during the last 4-5 years that some studies have been implemented to investigate these characteristics of the final sludge material from STWs.

12.8.1 Stability

As mentioned, various indices have been used by different researchers to estimate the stabilization degree of the residual sludge material which is obtained in STW systems. Table 12.14 presents data from existing studies on stability investigation in STWs.

Cui et al. (2008) estimated the microbial respiration using the TTC (triphenyltetrazolium chloride) as an indicator of bacterial respiration. The detection of DHA (dehydrogenase activity) through the reduction of tetrazolium salts, like TTC, to triphenylformazan (TPF) has been used to measure cell activity. Thus, TTC-DHA was used by these authors as an indicator of sludge stabilization process. Higher TTC-DHA values correspond to limited sludge stabilization degree. Results of this study showed that bottom layers of the residual sludge layer, which contain sludge from older applications, had a lower TTC-DHA value (Cui et al., 2008), which confirms that bottom parts of the sludge layer are better stability, implying that sludge stored in STWs obtains an improved stability degree with time.

TABLE 12.14 Indices Used to Evaluate the Stability of Residual Sludge Material Obtained from STWs

References	Stability Indices				
Cui et al. (2008)	**TTC-DHA (μg TPF/g VS/h)** Top layer: 10,603—5 cm: 6216—10 cm: 896—bottom: 457				
Giraldi et al. (2009b) and Peruzzi et al. (2009)	Months	6	12	18	24
	WSC (μg C/g DM):	23,584	13,301	3978	3379
	Dhase (mg INTF/g dwh):	213	13.57	33.28	31.50
	β-glucosidase (mg PNF/g DM/h):	1963	1230	712	619
	Urease (mg NH_4/g DM/h):	1749	1608	438	928
	O/N (Py-GC):	1.09	0.76	0.87	0.88
	B/E_3 (Py-GC):	0.57	0.64	0.62	0.57
	AL/AR (Py-GC):	0.44	0.48	0.48	0.53
Peruzzi et al. (2009)	**WSC** (μg C/g DM):	53,228-54,945	23,356-23,584	19,534-17,572	11,656-13,301
	Dhase (mg INTF/g dwh):	47.19-47.31	174.17-212.54	74.16-40.54	21.19-13.57
	Fulvic acids (μg C/DM)		56,672-74,446		30,829-35,860
	Humic acids (μg C/DM)		12,471-12,586		10,389-10,654
	O/N (Py-GC):		1.11-1.09		0.83-0.76
	B/E_3 (Py-GC):		0.71-0.57		0.68-0.64
Peruzzi et al. (2011b)	Months	6	48		
	WSC (μg C/g DM):	23,584-23,356-12,961-14,588	2884-3826-5981-6956		
	Dhase (mg INTF/g dwh):	212-174-61.8-42.5	7.65-16.2-4.40-3.14		

(*Continued*)

TABLE 12.14 Indices Used to Evaluate the Stability of Residual Sludge Material Obtained from STWs—cont'd

References	Stability Indices			
	O/N (Py-GC):	1.09-1.11-1.33-1.13	0.96-0.94-0.96-0.85	
	B/E_3 (Py-GC):	0.57-0.71-0.69-0.65	0.85-0.75-0.81-0.78	
Bianch et al. (2011)	**g O_2/kg DM/h** (15 days after sludge feed): 4 (winter), 2.5 (summer)			
Stefanakis et al. (2011b)	(for 75, 60, 30 kg TS/m^2/year)	LP[a]	RP[a]	Limit values
	g **C—CO_2**/dry kg	11.51, 13.7, 10.7	2.22, 2.11, 1.18	1.0, 3.0 (most, least stable)
	SRI_{12}: g O_2/dry kg/h	0.37, 0.37, 0.42	0.14, 0.94, 0.70	0.03, 0.13 (most, least stable)
Uggetti et al. (2011)	**DRI_{24h}**: 1.1-1.4 g O_2/kg TS/h			
Uggetti et al. (2012b)	**DRI_{24h}**: 0.5-0.7 g O_2/kg TS/h			

[a]*LP, loading phase; RP, resting phase.*

Giraldi et al. (2009b) and Peruzzi et al. (2009) measured the following biochemical parameters in STW systems, toward the investigation of the mineralization and stabilization progress of the accumulated sludge:

- humic carbon, humic acids, fulvic acids;
- water soluble carbon (WSC), which corresponds to the readily biodegradable organic fraction;
- DHase (dehydrogenase) activity: an indicator of the microbial activity which occurs intracellularly in the microorganism cells;
- β-glucosidase and urease, enzymes related to the C and N cycle, respectively;
- O/N (Py-GC), pyrolytic index of mineralization (pyrrole to furfural) derived by pyrolysis-gas chromatography;

- B/E_3 (Py-GC), pyrolytic index of humification (benzene to toluene), which increases with the maturity of the organic matter;
- AL/AR (Py-GC), pyrolytic index of energy storage (aliphatic/aromatic compounds), which expresses the ratio of aliphatic to aromatic products (stable part of humic substances).

Results of the study by Giraldi et al. (2009b) showed a gradual significant decrease of the mineralization indices (WSC and DHase), as well as of the C- and N-related enzymes (β-glucosidase and urease), indicating a high degree of stabilization of the sludge layer. Decrease of the mineralization index (O/N) implied the gradual mineralization of the material (Table 12.14). Similar results are reported by Peruzzi et al. (2009), who found a simultaneous decrease of both WSC and DHase in two STW beds, indicating a decrease of the biodegradable organic fraction and a decrease of the respective microbial activity. They also measured parameters related to the formation of humic substances, as an indication of the stabilization and humification of the organic matter. Fulvic and humic acids represent the less and more stable parts of the humic matter, respectively. After 1 year of operation, mineralization and humification of the accumulated sludge was evidenced, based on the results of this study (Table 12.14). Bianchi et al. (2011) also reported a decrease of WSC and DHase with time. After 30 and 48 months of operation, the same trend was observed, implying the continuation of the stabilization process (Peruzzi et al., 2011a,b; 2013). Reduction of the mineralization index (O/N) and increase in the humification index (B/E_3) show that humic substances are formed with time, indicating the gradual humification of the sludge material (Table 12.14).

Stefanakis et al. (2011b) measured two microbial respiration indices based on the static respirometric assay developed by Komilis and Tziouvaras (2009): the cumulative $C\text{-}CO_2$ produced (g $C\text{-}CO_2$/dry kg) and the maximum oxygen consumption rate over a 12-h period (SRI_{12}; mg O_2/dry kg/h) during a 7-day incubation at 35 °C (Komilis et al., 2011b). These indices correspond to the biological activity and stability of an organic substrate. The values of 1.0 and 3.0 g $C\text{-}CO_2$/dry kg and 0.03 and 0.13 g O_2/dry kg/h have been proposed as thresholds for most stable and least stable organic compost materials, respectively (Komilis et al., 2011b). As shown in Table 12.14, values of both indices were above the respective limit values during the loading phase. During the final resting phase, these indices decreased below the threshold values for stable compost material, while after 4 months without new sludge application, measured values approached the limit of most stable compost (Stefanakis et al., 2011b). These authors also observed differences between beds with different SLRs; lower SLR seems to result faster in an even more stable biosolids material. Bianchi et al. (2011) reported a respiration activity of 4 and 2.5 g O_2/kg DM/h during winter and summer period, respectively. This finding confirmed that the microbial activity continues during the colder months, although it proceeds at a slower rate. Uggetti et al. (2011) calculated the dynamic respiration index (DRI_{24h}) between 1.1 and 1.4 g O_2/kg TS/h in a Spanish STW system and between 0.5 and 0.7 g O_2/kg TS/h in Danish systems (Uggetti et al., 2012b). All these reported values for respiration indices are lower compared to the respective values for mixture of primary and activated sludge (6.7 g O_2/kg TS/h) or anaerobically digested sludge (3.7 g O_2/kg TS/h) (Pagans et al., 2006).

Overall results from the various stability indices estimated in the residual sludge material from STWs clearly indicate that sludge treatment in STWs possess a high potential of delivering a stable final biosolids material, appropriate for further exploitation. The level of stability achieved in STW systems is comparable to compost materials, which means that the biosolids can be further used without any additional post-treatment stage.

12.8.2 Maturity

Maturity of a compost material is related to the potential phytotoxicity effects that may arise from the application of the material to plants and seeds. It is usually measured with germination tests. Respective analyses in sludge material from STWs are limited until today. Current studies on STWs are presented in Table 12.15.

TABLE 12.15 Indices Used to Evaluate the Maturity of Residual Sludge Material Obtained from STWs

References	Maturation Indices		
Stefanakis et al. (2011b)	**GI** (%) (for 75, 60, 30 kg TS/m^2/year)	LP[a]: 102.9, 126.7, 128.7 RP[a]: 104.9, 140.4, 145.9	Incubation at 22 °C for 7 days Tomato seeds (*Lycopersicon esculentum*)
Ceccanti et al. (2011)	**GI** (%): 77 ± 4.38	Incubation at 20 °C for 3 days Seeds of *Lepidium sativum*	
Uggetti et al. (2012a)	**GI** (%) Incubation at 20 °C for 7 days Cucumber (*Cucumis sativus*) and lettuce (*Lactuca sativa*) seeds	Lettuce Campaign I: 138, 201, 158 Campaign II: 193, 224, 181	Cucumber Campaign I: 102, 133, 144 Campaign II: 94, 100, 106
Peruzzi et al. (2013)	**GI** (%) Incubation at 20 °C for 3 days Seeds of *Lepidium sativum*	Danish STW 0-40 cm: 141 40-80 cm: 153 80-110 cm: 170 mean: 156	Italian STW mean: 143

[a]*LP, loading phase; RP, resting phase.*

Ceccanti et al. (2011) applied a germination test with seeds of *Lepidium sativum*, a plant sensitive to toxic substances. In these experiments, the petri dishes with the extracted water from the sludge material were kept at 20 °C in dark for 72 h. The germination index (GI; %) was calculated according to the following formula:

$$\mathrm{GI}(\%) = P_r(L_T/L_C) \tag{12.1}$$

where P_r is the mean percentage of seed germination and L_T and L_C are the mean root lengths in the sludge samples and in the control (distilled water) samples, respectively. Results of this study for an STW system after 4 years of operation showed a GI of 77% for the residual sludge material (compared to 100% for the control sample). This value was considered by the authors as an indicator of a mature material, based on the study of Sellami et al. (2008), who reported that a GI value of 80% for composts was indicative of the absence of phytotoxins. Peruzzi et al. (2013) used the same formula to determine the GI in the sludge material from two STW systems in Denmark and in Italy. They also found GI values far above 40%, which was considered as the limit value for phytotoxicity.

Stefanakis et al. (2011b) calculated the GI based on the work by Komilis and Tziouvaras (2009). They used tomato seeds (*Lycopersicon esculentum*) for these tests and the incubation of the petri dishes took place in the dark at 22 ± 0.5 °C for 7 days. Results were expressed as a percentage of the control (distilled water) sample, according to the following formula:

$$\mathrm{GI} = \frac{(\text{Number of germinated seeds (sample)}/\text{total number of seeds in Petri dish (sample)})}{(\text{Number of germinated seeds (control)}/\text{total number of seeds in Petri dish (control)})} \times \frac{(\text{total root length of germinated seeds (sample)}/\text{number of germinated seeds (sample)})}{(\text{total root length of germinated seeds (control)}/\text{number of germinated seeds (control)})} \times 100 \tag{12.2}$$

Based on this equation, GI values above 100% (*ad hoc* phytotoxicity limit) were indicative of potential phyto-enhancing effects, while values below 100% indicate a phytotoxic material. Results of this study in Greece showed that during the loading phase, GI values were fluctuating below and close to 100%, probably due to the release of intermediate metabolites, like acids, during the organic matter decomposition, which could negatively affect the seed germination (Epstein, 1997). During the final resting phase, GI increased and reached values of up to 140% after 4 months, indicating a very mature biosolids material. This again highlights the significance of the resting phase in order to obtain a nonphytotoxic sludge material. A prolonged resting phase delivers an even more mature material, as shown by the GI value (300%) of a bed unloaded for almost 2 years (Stefanakis et al., 2011b).

Uggetti et al. (2012a) used for the GI determination cucumber (*Cucumis sativus*) and lettuce (*Lactuca sativa*) seeds and applied incubation of the petri dishes in the dark for 7 days at 20 °C, according to the procedure of Barrena et al. (2009). The formulas they used for the GI calculation were:

$$\text{Relative seed germination} = \frac{\text{Number of seeds germinated in sludge samples}}{\text{Number of seeds germinated in control samples}} \times 100 \quad (12.3)$$

$$\text{Relative root growth germination} = \frac{\text{Mean root length in sludge samples}}{\text{Mean root length in control samples}} \times 100 \quad (12.4)$$

$$\text{GI}\ (\%) = \frac{(\%\ \text{Seed germination}) \times (\%\ \text{root length})}{100\%} \quad (12.5)$$

Results of this study in Spain (Table 12.15) gave GI values above the limit of 100% at the beginning (campaign I) and at the end (campaign II) of the resting period, indicating that the residual sludge did not cause any toxicity effects on the plant seeds.

Overall, various phytotoxicity tests performed in the residual sludge material from STW systems in different countries and under different climatic conditions showed that the final biosolids that are obtained from these facilities do not cause any harmful effects on plant growth. Especially after the final resting phase usually applied, the characteristics of the biosolids imply a sufficiently mature material.

12.9 EFFECTS OF VEGETATION

Plants represent one of the main components of the STW system. Surely, they are the most easily recognized component since they optically dominate a treatment wetland bed, independently of the application. Thus, as for VFCWs treating wastewater, one of the basic questions concerning the STW technology has to do with the actual role and significance of the plants in the overall performance of the system.

As mentioned in Section 4.1 in Chapter 4, it is generally accepted that the presence of plants brings a series of advantages in wetland systems. However, given the different nature of the application in STWs (sludge instead of wastewater), the first question which needs to be answered, has to do with the capability of common wetland plants to withstand the load of sludge material. The so far gained experience from STW systems shows that, generally, wetland plants can adapt and grow in contaminated sludge. As for VFCWs treating wastewater, *Phragmites australis* is the most widely used plant in STWs and, probably, the optimum choice for its known specific characteristics, i.e., fast growth, high transpiration capacity, deep roots, high tolerance, and local and commercial availability across many countries and regions in the world (De Maeseneer, 1997; Nielsen, 2003).

Hardej and Ozimek (2002) investigated the growth of *P. australis* in sewage sludge compared to natural stands. They found that reed density was more than double in the bed flooded with sludge than in natural stands, probably as a result of the available water and nutrients. Shoot height was smaller in reeds grown in sludge, but they had a larger basal diameter, and thus, higher oxygen transfer capacity to the underground parts. Plant biomass and dry weight was also greater in the beds fed with sludge. Overall, this study showed that the adaptation of the reeds to the conditions created by periodical sludge flooding is considered as a positive effect on sludge treatment and dewatering, since more oxygen is transferred to the roots, while the increased plant biomass enhanced the transpiration needs, and thus, the dewatering process. The same conclusions are reported for two other species, *Echinochloa pyramidalis* and *Cyperus papyrus*, grown in STWs under tropical climate (Kengne et al., 2008). The first species showed increased production of the aboveground aerial parts and the second of the subaerial parts. Among the two, *E. pyramidalis* was proved to be more appropriate for sludge dewatering systems.

However, plants should be handled carefully, especially during the first growing season after their establishment. Application of high loadings directly after planting could harm plant growth, since the development of the sludge layer could overcome the plant growth and, thus, inhibit their development (Nielsen, 2003). Usually, after planting, a commissioning phase is applied, during which the beds are fed with weak wastewater or with lower SLR than the designed one. Some species are more sensitive to high loadings, especially during the first growing period. For example, Stefanakis and Tsihrintzis (2011) and Stefanakis and Tsihrintzis (2012c) reported the death of cattails (*Typha latifolia*) during the first summer period of their growth, although a commissioning phase of more than 4 months was applied. This showed that this species is more sensitive compared to *P. australis* and should be handled more carefully.

12.9.1 Presence of Plants and Plant Species

Generally, the presence of plants appears to be beneficial for the system efficiency. The majority of the studies in the literature, where planted and unplanted systems are compared, shows that planted systems perform better in all tested parameters and indicators. Planted beds, usually, achieve a higher dewatering efficiency, i.e., higher reduction of the sludge volume and higher TS content in the residual sludge, which becomes greater during the final resting phase.

Several tables that follow intend to give a general overview of the available data from the literature, concerning the difference between planted and unplanted systems, as also between beds with different plant species. Table 12.16 shows data from different studies concerning the dewatering efficiency, TS and VS content and nitrogen and phosphorus concentration of the sludge residue from planted and unplanted beds. Table 12.17 contains heavy metal concentrations, while Table 12.18 presents data about pathogens, stability and maturity indices of the sludge residue in planted beds with various plant species, as also in unplanted beds. Finally, Table 12.19 shows the efficiency of planted and unplanted STWs concerning the quality of the drained water.

TABLE 12.16 Comparison of Planted and Unplanted STWs Concerning Dewatering Efficiency and Residual Sludge Pollutant Concentration

References		Dewatering (%)	TS (%)	VS (% TS)	TKN (g/kg DM)	TN	TP (g/kg DM)	TOC (g/kg DM)
Liénard et al. (1995)	*Phragmites australis*		10.6, 11.4	58.7, 61.0	4.4, 4.9[a]		1.2, 1.3[a]	
	Unplanted		11.0	61.6	5.6[a]		1.5[a]	
Edwards et al. (2001)	*Phragmites australis*	84.3, 86.3	20.5, 21.5	52.5, 52.4				
	Unplanted	81.1	18.7	52.5				
Nassar et al. (2008)	*Phragmites australis*	95.9	68	39.7	27	2.71[a]	3.6	
	Unplanted	95.6	70	37.4	27	2.66[a]	4.2	
Kengne et al. (2011) (for SLRs: 100, 200, 300 kg TS/m^2/year)	*Cyperus Papyrus*		30.8% 16.1% 15.6%					
	Echinochloa Pyramidalis		33.5% 21.7% 25.6%					

Wang et al. (2012b)	*Phragmites australis*				12.8	0.02	374
	Typha latifolia				10.6	0.01	384
	Iris pseudacorus				12.9	0.09	385
	Unplanted				12.5	0.12	369
Uggetti et al. (2012a)	*Phragmites australis*	79	13-15	50-60	3.85[a]	0.14[a]	
	Typha latifolia	82, 81	13-15	50-60	3.36, 3.5[a]	0.02, 0.03[a]	
Stefanakis and Tsihrintzis (2012c)	*Phragmites australis*	96.4-96.9	22.3-30.4	52.1-57.5	31.5-35.3	7.1-7.4	
	Unplanted	89.8	17.4	59.9	41.5	7.8	

[a]%TS.

TABLE 12.17 Drained Water Quality in Planted and Unplanted (U) STWs

References	Plant Species	COD (mg/L)	BOD_5 (mg/L)	SS (mg/L)	DS (mg/L)	TKN (mg/L)	NH_4^+-N (mg/L)	TP (mg/L)	OP (mg/L)	pH	EC (ms/cm)
Liénard et al. (1995)	*Phragmites australis*	43, 85.3		19.4, 34.3							
	U	102.1		103.4							
Edwards et al. (2001)	*Phragmites australis*	1822, 2220	1123, 1468	809, 917	3674, 4055		471, 589		569, 590	8.02, 7.65	
	U	1962	935	770	3290		583		583	8.17	
Nassar et al. (2008)	*Phragmites australis*	161		22	2441	51			21.6		4.1
	U	167		33	2324	52			20.4		4.7
Wang et al. (2009b)	*Phragmites australis*	130 g/bed (99.1%)		34 g/bed 99.1% rem		1 g/bed 99.7% rem		<3 98%	<2 98%		
	Typha latifolia	221 g/bed (98.5%)		17 g/bed 99.9% rem		2 g/bed 99.4% rem					
	Iris pseudacorus	275 g/bed (98.2%)		31 g/bed 99.8% rem		21 g/bed 94.4% rem					
	U	372 g/bed (97.5%)		192 g/bed 98.7% rem		43 g/bed 87.4% rem					

Kengne et al. (2011) [for SLRs: 100, 200, 300 kg TS/m²/year]	*Cyperus papyrus*	97.8% 98.7% 99.2%	86.0% 93.7% 93.4%	92.6% 95.7% 93.4%	85.4% 90.9% 85.0%		
	Echinochloa pyramidalis	98.7% 97.9% 97.8%	98.9% 97.9% 95.8%	91.5% 90.1% 89.5%	87.2% 77.6% 79.9%		
Gagnon et al. (2012)	*P. australis*	2.7 (g/m^2)	1.0 (g/m^2)	0.2 (g/m^2)	0.03 (g/m^2)	0.1 (g/m^2)	0.04 (g/m^2)
	Typha latifolia	21.2	6.1	1.2	0.3	3.7	1.00
	Scirpus fluviatilis	54.8	16.5	3.7	1.7	6.0	1.70
	U	175.5	38.6	24.0	24.0	7.5	1.83
Korboulewsky et al. (2012)	*Phragmites australis*	177 g/m^2 (96.7%)	25.0 g/m^2 (99.5%)	2.1 g/m^2 (99.2%)		1.02 g/m^2 (98.0%)	
	Typha latifolia	394 g/m^2 (92.0%)	66.0 g/m^2 (99.2%)	3.5 g/m^2 (98.6%)		1.51 g/m^2 (96.8%)	
	Iris pseudacorus	604 g/m^2 (87.4%)	391 g/m^2 (94.7%)	35.9 g/m^2 (89.7%)		2.34 g/m^2 (94.9%)	
	U	804 g/m^2 (82.0%)	728 g/m^2 (90.6%)	58.0 g/m^2 (82.8%)		1.52 g/m^2 (96.9%)	

(Continued)

TABLE 12.17 Drained Water Quality in Planted and Unplanted (U) STWs—cont'd

References	Plant Species	COD (mg/L)	BOD_5 (mg/L)	SS (mg/L)	DS (mg/L)	TKN (mg/L)	NH_4^+-N (mg/L)	TP (mg/L)	OP (mg/L)	pH	EC (ms/cm)
Stefanakis and Tsihrintzis (2012c)	*Phragmites australis*	203-379 (86.8-92.9%)		2.25-2.75 (88.4-90.5%)			40-113 (69-89%)		13.4-18.7 (75.5-82.4%)	7.58-7.72	1.91-3.07
	U	301 (89.5%)		3.43 (85.6%)			138 (62.5%)		24 (68.5%)	7.39	2.56
Gustavsson and Engwall (2012)	*Phragmites australis*	256.4 (80.7%)	209.1 (71%)	1277 (90%)		83.0[a] (69.7%)		8.9 (69.8%)			
	U	274.8 (87.2%)	283.5 (72.2%)			81.8 (73.7%)		5.9 (75.8%)			

[a]TN.

TABLE 12.18 Comparison of Planted and Unplanted (U) STWs Concerning Heavy Metal Concentration in Residual Sludge

References		Cr	Cd	Pb	Cu	Ni	Mn	Zn	Fe	Hg
Nassar et al. (2008)	*Phragmites australis*			0.18		n.d.			1.72	
	U			0.26		n.d.			1.46	
Uggetti et al. (2012a)	*Phragmites australis*		0.7	13	n.d.	11		238		9
	Typha latifolia		0.4; 0.7	10; 18	n.d.	12; 11		227; 295		3; 5
Stefanakis and Tsihrintzis (2012d)	*Phragmites australis*	132-176	5.7-7.6	46.5-59.8	82.8-90.7	43.3-51.3	141-158	361-407	2625-3161	
	U	232	9.3	85.9	94.8	57.6	160	417	3136	

n.d., not detected.

TABLE 12.19 Comparison of Planted and Unplanted (U) STWs Concerning Pathogenic, Stability, and Maturity Indicator Contents in Residual Sludge

References	Plant Species	Salmonella (in 25 g)	*E. coli* (MPN/g)	Total Coliform (CFU/g)	Respiration Index			GI (%)
					mg O_2/g VS/h	*mg O_2/g TS/h*	*g C—CO_2/g TS*	
Uggetti et al. (2012a)	*Phragmites australis*	absence	6600, 93[a]		3.69, 0.26[a]			102-138, 193-94[a]
	Typha latifolia	absence	4500, 240[a] 3900, 150[a]		2.24, 0.62[a] 3.33, 0.70[a]			201-133, 224-100[a] 158-144, 181-106[a]
Korboulewsky et al. (2012)	*Phragmites australis*					0.2-1.1[b]		
	Typha latifolia					0.2-0.7[b]		
	Iris pseudacorus					0.2-0.6[b]		
	U					0.1-0.2[b]		
Stefanakis and Tsihrintzis (2011b)	*Phragmites australis*			LP: 1567-2442[c] RP: 354-1504		LP: 374-418[c] RP: 70-144	LP: 10.7-13.7[c] RP: 1.18-2.22	LP: 103-129[c] RP: 105-146
	U			LP: 3275[c] RP: 26,199		LP: 473[c] RP: 260	LP: 19.6[c] RP: 3.0	LP: 73[c] RP: 82

[a]*At the beginning and the end of the resting phase. GI is calculated based on Equation (12.5).*
[b]*During a 2 weeks resting period after sludge application.*
[c]*LP, loading phase; RP, resting phase. GI is calculated based on Equation (12.2).*

Liénard et al. (1995) reported that the drying process proceeds more slowly in unplanted beds, especially during the winter period when the systems behave like a sludge lagoons (Table 12.16). The presence of plants resulted in improved system performance: higher OM biodegradation and better quality of the drained water (higher redox potential, DO, $NO_2^{\cdot -}$N, $NO_3^{\cdot -}$N, and org-N, and lower BOD_5, COD, TOC, TN, and NH_4^+-N; Table 12.17). The denitrification rate was also higher in the planted than in the unplanted bed (237.1 and 14.1 g N/m^2/year). In the deeper parts of the sludge layer, the bacteria population was found to be more than 10 times higher in planted than in unplanted beds, while the opposite was the case for the drained water. The findings of this study indicated that microbial activity is higher in planted than in unplanted beds, due to the presence of the root system.

Edwards et al. (2001) found an improved performance of planted beds compared to unplanted, in terms of dewatering efficiency and TS content (Table 12.16). They also reported that the oxidized part of the sludge layer was deeper in the planted beds. However, the quality of the drained water was slightly better in the unplanted bed (Stefanakis et al., 2011b; Table 12.17), which—according to the authors—was attributed to the higher hydraulic conductivity of the planted beds and the respective limited retention time of the water within the bed.

Nassar et al. (2008) reported similar dewatering efficiency for planted and unplanted beds in Palestine (Table 12.16); however, the planted bed was loaded with 15% higher sludge volume. It is also interesting that the sludge density was smaller in the planted bed (782 kg/m^3) than in the unplanted one (844 kg/m^3). The quality of drained water was better in the planted bed in terms of COD, TKN, and SS (Table 12.19), while higher NO_3^-N concentration was found in drained water of the planted bed than the unplanted, indicating that higher oxygen availability in the planted bed enhanced aerobic transformation processes of nitrogen (nitrification). Concerning HM content in the residual sludge of the planted and unplanted beds, no significant differences were found (Table 12.18).

Gagnon et al. (2012) investigated the role of different plant species on the effluent quality of drained water. Results showed that all planted beds had a higher efficiency in the removal of all pollutants compared to the unplanted ones, with respective rates exceeding 94% (Table 12.17). Among the tested plant species (*P. australis*, *T. latifolia*, and *Scirpus fluviatilis*), beds planted with *Phragmites* were the most efficient, while they presented the highest evapotranspiration rates and the lowest drained water volume. A design difference of these systems was the fact that water level was kept at about 25 cm above the bottom of the bed, within the media layers. The authors mentioned that this design allowed for prolonged contact time between the water and the root system, thus enhancing removal processes like nitrification and plant uptake.

Uggetti et al. (2012a) also tested two STW pilot systems with different plant species (*P. australis* and *T. latifolia*). They reported similar efficiencies concerning dewatering (volume reduction, TS content; Table 12.16), drained water quality (COD), VS, HM content (Table 12.18), and pathogenic microorganisms

(Table 12.19). These results suggest that the use of *T. latifolia* can also be promoted for use in STWs, besides *P. australis.*

Korboulewsky et al. (2012) compared planted and unplanted STW beds under Mediterranean climate conditions (SE France). They reported that all planted beds had a higher performance compared to the unplanted ones, in terms of increased removal rates for organic matter, nitrogen, and phosphorus constituents in drained water (Table 12.17). Moreover, solid retention in the bed was 22% higher in planted than in unplanted beds. Among the plant species tested (*P. australis, T. latifolia*, and *Iris pseudacorus*), the bed with *Phragmites* showed the highest efficiency.

Results of Gustavsson and Engwall (2012) showed that planted beds (*P. australis*) had a lower efficiency, concerning the drained water quality, compared to unplanted beds in the treatment of sludge containing pharmaceuticals and nitroaromatic compounds in Sweden (Table 12.17). The fact that part of the experiment took place outdoors at low temperatures, as well as the fact that the SLR applied under these conditions was relatively high (76-153 kg DM/m^2/year), possibly affected the planted beds. During the indoor part of this study, planted beds showed an improved performance compared to the unplanted ones. However, the dry weight of the residual sludge layer was higher in the planted (55-70%) than in the unplanted beds (30-45%).

Kengne et al. (2008) compared two plant species, *C. papyrus* and *E. pyramidalis*, in STWs under tropical climate (Cameroon) and for three different SLRs (100, 200, 300 kg TS/m^2/year). Among these two species, the second one was proposed for use in STWs, since it developed almost double biomass and managed to uptake higher nutrient amount from the sludge. The continuation of the study (Kengne et al., 2011) showed that there were no significant differences between the dewatering efficiency (Table 12.16) and the drained water quality of the beds with the two plant species (Table 12.17). However, for higher SLRs, the bed with *C. papyrus* showed some clogging signs, i.e., water retention on the surface of the sludge layer. This was attributed to the higher density and produced biomass, and thus, higher transpiration capacity of *C. papyrus*.

Wang et al. (2009b; 2012b) used three different plant species in STW mesocosms under Mediterranean climate: common reeds (*P. australis*), broadleaf cattail (*T. latifolia*), and yellow flag (*I. pseudacorus*). They found that all planted beds had a better performance compared to the unplanted ones, resulting in drained water of high quality (Table 12.17). The presence of plants was found to stimulate microbial activity, as indicated by increased respiration rates, independently of the plant species. Beds planted with *Phragmites* showed the highest microbial activity stimulation. Among the three species, beds planted with *Phragmites* and *Typha* showed an increased TKN removal and water reduction, i.e., higher ET rates. Plant biomass of *Phragmites* was also 3 times the respective biomass for *Iris*, indicating higher transpiration needs and, thus, water uptake. *Phragmites* appears as the most attractive species with the lowest effluent water volume (which could be even zero in the summer). The authors suggested that for this specific climate, the choice between *Phragmites* and *Typha* should rely on the feeding regime, since lack of water during the summer months could damage plant survival.

Stefanakis and Tsihrintzis (2012c) compared unplanted and planted STW beds (*P. australis* and *T. latifolia*). Cattails were dried during the first operational summer, indicating that for this species, early loadings should be lower to avoid plant shock. Similar dieback incidents of cattails were reported by Koottatep et al. (2005, 2006). In this study, the planted beds demonstrated an increased dewatering efficiency of 7.5% in sludge volume reduction during the loading phase and of 14% during the resting phase. Respectively higher was the TS content in the planted beds (Table 12.16). The improvement of the performance by plant presence was also shown by the reduced VS content, as also by the lower nitrogen and phosphorus concentration in the residual sludge of the planted compared to the unplanted beds (Table 12.16). Heavy metal accumulation in the residual sludge layer was also lower (Stefanakis and Tsihrintzis, 2012d; Table 12.18), while drained water quality was always better in the planted beds (Stefanakis et al., 2011b; Table 12.17). The authors also mention that the bottom part of the sludge layer was brown in planted beds, which implied that aerobic conditions dominated probably as a result of the extensive plant root system, while in the unplanted beds, the deeper parts were black, indicating anaerobic conditions and limited mineralization and humification of the sludge. Similar observations were also made by Uggetti et al. (2009) and Melidis et al. (2010). It is also interesting that these differences between planted and unplanted beds were statistically confirmed, with the exception of phosphorus, which implies that the main removal mechanism for this pollutant (i.e., adsorption to porous media) is not closely connected to the plant activity. This study also reported the efficiency of planted and unplanted beds concerning pathogenic microorganism removal in the residual sludge (Stefanakis et al., 2011b). Results showed that the planted beds had always a lower content of total coliform units during the loading phase, while after 4 months of resting only the planted beds fulfilled the criteria for land application (Table 12.19).

The overall improved performance by the presence of plants in STW systems is also implied not only by the improved removal rates in the various pollutants and indicators but also by the respective microbial activity. Stefanakis et al. (2011b) found that unplanted beds had an increased respiration activity, as expressed by the C—CO_2 consumed and the O_2 produced. This observation, coupled with the fact that unplanted beds always had a higher microbial population, indicates that the sludge layer is less stabilized (Table 12.19). Korboulewsky et al. (2012) reported that right after the sludge application, the respiration index increased in planted beds and then gradually dropped, while in the unplanted bed no significant change occurred. This indicates that the presence of plants enhances the microbial activity, since the root surface acts as a microbial attachment area. The difference between the two studies is that, in the study by Stefanakis et al. (2011b), the respiration activity was measured at the end of each resting period (2-3 weeks), while Korboulewsky et al. (2012) monitored the respiration activity for 2 weeks after one sludge feed.

Additionally, the germination index was lower in the unplanted beds (Stefanakis et al., 2011b), implying a less mature material which could potentially cause phytotoxicity effects if used for land application. After the completion of the final resting phase in this study, both respiration and germination indices were above the respective limits for land application for the sludge material from the unplanted beds.

Moreover, several authors (Gagnon et al., 2012; Korboulewsky et al., 2012; Wang et al., 2012b) found that beds planted with *P. australis* had an improved performance compared to other plant species, which was attributed to the more intense development of this species for both below- and underground parts. On the other hand, Uggetti et al. (2012a) did not find any significant differences between *P. australis* and *T. latifolia*, and suggested that the latter could also be used since it is not invasive like *Phragmites*. Generally, common reeds are more preferred, taking into account that—as mentioned before—cattails have been proved to be more sensitive in their early growth stage. However, the selection of plants should always include the parameter of local availability.

Different plant species have also been tested in the treatment of sludge containing specific compounds. Chen et al. (2009b) investigated pilot STWs with three different plant species (*T. latifolia*, *P. australis*, and *Phalaris arundinacea*) for the removal of personal care compounds (PCCs) from sewage sludge. The comparison of these species did not show any significant differences concerning the removal of organic microcompounds, like AHTN, Triclosan, OTNE, HHCB-lactone, and HHCB (Table 12.20). However, they attributed this to the relatively small size of the pilot systems and the potential influence of boundary effects. The contribution of plants by direct uptake of these compounds was negligible, while $<1\%$ of these substances was detected in drained water of the pilot units. Gustavsson and Engwall (2012) applied industrial sludge containing nitroaromatic and pharmaceutical residues. Results of this study showed that the plant used (*P. australis*) is tolerant to toxic xenobiotic compounds, with simultaneous good treatment performance of the bed, which enhances its general acceptance as the main plant species in STWs.

12.9.2 Plant Uptake

It is known that wetland plants utilize some nutrients and HM present in water or sludge to cover their growth needs. Generally, in STWs plant uptake of these compounds is not significant compared to the total mass loaded to the system. Table 12.21 presents published data from the literature concerning the role of direct plant uptake in the removal of nitrogen and phosphorus in STW systems, as also the amount of assimilated nutrients in the plant biomass. Moreover, Table 12.22 shows that reed species are also capable of assimilating certain amounts of HM to their biomass.

TABLE 12.20 Comparison of Planted and Unplanted (U) STWs in the Removal of Personal Care Compounds (PCCs)

References		AHTN	Triclosan	OTNE	HHCB-lactone	HHCB
Chen et al. (2009b)	Raw	2250 ng/g	800 ng/g	1600 ng/g	1200 ng/g	11,000 ng/g
	Typha latifolia	24%	310 ng/g	73%	1600 ng/g	8000 ng/g 27%
	Phragmites australis	20%	390 ng/g	72%	1700 ng/g	8500 ng/g 22%
	Phalaris arundinacea	24%	360 ng/g	70%	1600 ng/g	8300 ng/g 25%
	U	21%	360 ng/g	73%	1700 ng/g	8500 ng/g 23%
	Plant uptake	n.d.	n.d.	n.d.	n.d.	n.d.
	Drained water	0.16%	0.52%	0.52%		0.14%

TABLE 12.21 Uptake of Nutrients and Other Substance by Various Plant Species in STWs

References	Plant Species	Biomass (g/m^2)	SLR (kg TS/ m^2/year)	Plant Uptake (% of total input)				
				N	*P*	*Water Content*	*Ash*	*Organic Substances*
Koottatep et al. (2001)	*Typha augustifolia*		250	5-7				
Barbieri et al. (2003)	*Phragmites australis*		10	4 (summer) 2 (winter)				
Panuvatvanich et al. (2009)	*Typha augustifolia*		250	0.58				
Gagnon et al. (2012)	*Phragmites australis*	2669, 4369[a]	30	26 (of TKN)	5.7			
	Typha latifolia	1103, 2781[a]		12	3.4			
	Scirpus fluviatilis	110, 907[a]		1.9	0.2			
Korboulewsky et al. (2012)	*Phragmites australis*			23	17			
	Typha latifolia			15	12			
	Iris pseudacorus			8	6			
Stefanakis and Tsihrintzis (2012c)	*Phragmites australis*	up to 1496	30, 60, 75	<1	<1			

				Plant biomass composition				
Kalisz and Salbut (2007)	*Phragmites australis*	1978-3150 g/m^2		1.63%DM	0.17%DM	39-50% DM	9.3% DM	90.7%DM
	underground parts	476 g/kg		3.15%DM (31.5 g N/kg DM)	0.39%DM (3.9 g P/kg DM)	52.4	13.8% DM	86.2%DM
Kengne et al. (2008)	*Cyperus papyrus*	180 (dry tons/ha/year)	100, 200, 300	2.3, 2.2, 2.2 (TKN, % DM) 289 g N/m^2/year	0.58, 0.63, 0.53 (TP, % DM) 73 g P/m^2/year	54.6, 62.7, 64.9		
	Echinochloa pyramidalis	100 (dry tons/ha/year)		1.8, 1.9, 1.9 (TKN, % DM) 366 g N/m^2/year	0.74, 0.62, 0.59 (TP, % DM) 97 g P/m^2/year	56.1, 63.2, 65.4		
	Natural wetlands			1.8 (TKN, %DM) 1.1 (TKN, %DM)	0.66 (TP, %DM) 0.59 (TP, %DM)	79.6 74.6		
Wang et al. (2009b)	*Phragmites australis*	3273		2%				
	Typha latifolia	1294		2%				
	Iris pseudacorus	1039						

(*Continued*)

TABLE 12.21 Uptake of Nutrients and Other Substance by Various Plant Species in STWs—cont'd

References	Plant Species	Biomass (g/m^2)	SLR (kg TS/ m^2/year)	Plant Uptake (% of total input)				
				N	*P*	*Water Content*	*Ash*	*Organic Substances*
Peruzzi et al. (2009)	*Phragmites australis* (after 12 months of operation)		3.16, 2.32 m^3/m^2	0.217%	0.26%			
Korboulewsky et al. (2012)	*Phragmites australis*	2343, 6212[a]		L: 19.5 [b] R: 14.7 456 kg N/ ha/year	L: 2.1 [b] R: 2.4 147 kg P/ ha/year			
	Typha latifolia	2864, 3478[a]		L: 15.9 [b] R: 10.7 454 kg N/ ha/year	L: 2.1 [b] R: 1.8 62 kg P/ ha/year			
	Iris pseudacorus	1995, 1619[a]		L: 14.2[b] R: 10.8 283 kg N/ ha/year	L: 1.7[b] R: 2.8 45 kg P/ ha/year			
Stefanakis and Tsihrintzis (2012c)	*Phragmites australis*			L: 19.9[b] S: 10.4 R: 170.4	L: 0.418[b] S: 0.281 R: 1.659			

[a]*Above-, belowground dry biomass.*
[b]*L, leaves; S, stems; R, roots (mg/g DM).*

TABLE 12.22 Uptake of Heavy Metals by Various Plant Species in STWs

References	Plant Species	Heavy Metal Uptake by Plants (mg/kg DM)										
		Cr	*Cd*	*Pb*	*Cu*	*Ni*	*Mn*	*Zn*	*Fe*	*Mo*	*Bo*	*Hg*
De Maeseneer (1997)	*P. australis/T. latifolia* above ground parts		0.29	5.99	7.61			146				0.11
	Salix fragilis leaves		6.5		12.1			1432				
	Salix triandra leaves		16		10.9			1691				
Begg et al. (2001)	*P. australis*	10	2	32	155	9	1443	453	3263	1	10	
Peruzzi et al. (2009)	*P. australis* (after 12 months of operation)	1.25, 2.25	n.d.	11, 11	21, 12	57, 46		135, 75				
Stefanakis and Tsihrintzis (2012d)	*P. australis*	52-240	0.18-3.11	11.1-101	3.4-194.6	0.71-35.4	36.1-436.8	87.4-435.4	188.2-2600.1			

As Table 12.21 shows, the actual role of the plants in the removal of nutrients is not significant when compared to the total nitrogen and phosphorus loading. Usually, plant uptake accounts for only a few percent of the total amount. Stefanakis and Tsihrintzis (2012c) reported an uptake of <1% for both nutrients and for SLRs up to 75 kg TS/m^2/year under the Mediterranean climate, while Barbieri et al. (2003) reported values between 2% and 4% for nitrogen uptake for a system in Italy. In STW systems where the water level (20-25 cm) is kept constant within the media layer, the prolonged contact of plant roots with the water allows for higher uptake rates up to 26% for nitrogen and 17% for phosphorus, as found by different studies using STWs with this design modification (Gagnon et al., 2012; Korboulewsky et al., 2012). Plants also uptake HM, but for these compounds there is limited information available concerning their importance in STW systems. However, assimilated amounts seem to be considerable but below the respective limits for land application in case the reed biomass is to be exploited. Generally, among the plant species used, *P. australis* appear to assimilate higher amounts of the various pollutants compared to *Typha* spp. and *Scirpus* spp.

Interesting results are also reported concerning the allocation of the assimilated nutrients and HM in the plant biomass. Kalisz and Sałbut (2007) reported that produced aboveground biomass of *P. australis* was 1978-3150 g/m^2 for each of the 9 years of operation of an STW system at 50 kg DM/m^2/year, while the belowground parts had a dry mass of 476 g/kg. They found that the content of total nitrogen (3.15% DM or 31.5 g N/kg DM) and total phosphorus (0.39% DM or 3.9 g P/kg DM) was almost double in the below- than the aboveground parts.

Korboulewsky et al. (2012) also reported higher biomass production for the belowground parts of *P. australis* (6212 g/m^2), *T. latifolia* (3478 g/m^2), and *I. pseudacorus* (1619 g/m^2), compared to the aboveground parts (2343, 2864, and 1995 g/m^2 for the three plant species, respectively). These authors also found that nitrogen concentration was highest in plant leaves than roots for all three plants: *Phragmites* (19.5 and 14.7 mg N/g DM, respectively), *Typha* (15.9 and 10.7 mg N/g DM, respectively), and *Iris* (14.2 and 10.8 mg N/g DM, respectively). Phosphorus concentration was lower compared to nitrogen in plant biomass. Higher P concentrations were found in leaves for both *Phragmites* and *Typha* (2.1 mg P/g DM) than in roots (2.8 and 2.4 mg P/g DM, respectively), while for *Iris* similar concentrations were found in leaves and roots (1.7 and 1.8 mg P/g DM, respectively). Generally, the amount of N and P stored in the aboveground biomass accounted for 9% and 5%, respectively, which means that the capacity which can be annually removed with harvesting is 456, 454, and 283 kg N/ha/year and 147, 62, and 45 kg P/ha/year for *Phragmites*, *Typha*, and *Iris*, respectively. A similar observation was made by Ozimek et al. (2001), who found that more than 50% of the assimilated nitrogen amount was located in the leaves. Stefanakis and Tsihrintzis (2012c) also reported that the assimilated amount of nitrogen and phosphorus in the leaves (mean values 19.9 and

0.42 mg/g, respectively) was almost double compared to the respective amount in the stems (10.4 and 0.28 mg/g). Moreover, roots contained the highest amount of both nitrogen (191.5 mg/g) and phosphorus (1.7 mg/g). These results indicate that plants mainly utilize nitrogen for their growth needs, while results of this study showed that the order of preference for absorbance and utilization is ammonia nitrogen, nitrate, and nitrite.

Reported results from the literature show that there is also an allocation of assimilated HM in plant biomass. Stefanakis and Tsihrintzis (2012d) found that the majority of the assimilated amount was located in the belowground parts of *P. australis*, followed by leaves and stems. Similar observations were made by Yadav et al. (2010) for *C. indica*, by Cheng et al. (2002) for *Cyperus alternifolius*, and by Mant et al. (2006) and Bonanno and Giudice (2010) for *P. australis*. It should be noted that these studies are dealing with the treatment of wastewater polluted with HM, since there are not available at the present time other studies reporting HM uptake and allocation by plants in STWs. Stefanakis and Tsihrintzis (2012d) also calculated several expressions of the translocation factor (TF). TF_1 (ratio of above- to belowground biomass HM content) values indicated that the accumulated HM amount in the roots of *P. australis* was almost double than the respective amount in the aboveground parts of the plant. TF_2 (ratio of leaf to stem HM content) showed that the assimilated HM amount in the leaves was two to five times higher than the respective amount in stems. Also, TF_3 values (ratio of residual sludge content to root content) implied that although the majority of the assimilated HM was located in plant roots, respective HM content in the residual sludge was even higher.

In the study of Stefanakis and Tsihrintzis (2012d), heavy metal uptake by plants accounted for $<3\%$ of the total heavy metal load with the decreasing trend Mn > Cu > Zn > Cr > Pb > Fe > Ni > Cd. Similar values ($>1\%$ for Cr and Ni in the aboveground biomass) are reported by Yadav et al. (2010), while for the belowground biomass respective figures for Cr and Ni were 1.7-3% and 0.13-0.59%. Mant et al. (2006) also found that uptake of Cr by *P. australis* represented about 2% of the influent load. These values are in agreement with expected uptake rate ($<5\%$) by plants found in the literature (Yeh, 2008). Cheng et al. (2002) reported higher values of 5-15%, but treating polluted wastewater with higher influent concentrations. Also, HM uptake in willow leaves was found to be high for Zn and Cd, indicating that this species is not suitable for STW systems (De Maeseneer, 1997).

Generally, plant presence appears to play a significant role in STWs, with multiple benefits (improved dewatering efficiency, better drained water quality, smaller drained water volume to be treated). Among the various plant species, *P. australis* appears to be the most widely used, based on its specific growing characteristics, high tolerance to the microenvironment of an STW bed and increased transpiration capacity. Other species could also be used, taking into account the local availability and the climatic conditions. The plant presence mainly affects indirectly the system performance (bed stabilization, enhancement of water losses through evapotranspiration, conservation of hydraulic conductivity of the bed, etc.), rather than directly with uptake of pollutants, since they usually assimilate only a small percentage of the influent annual pollutant load. Also, assimilated amounts of pollutants are translocated within the plant parts mainly to the roots and leaves of the plant.

12.10 EFFECTS OF POROUS MEDIA

Investigations concerning the use of different filter media in STWs are very limited. More focus is given in the use of various porous media as substrate in VFCWs for wastewater treatment, as discussed in Part A. Thus, since VFCWs and STWs are practically similar systems, results drawn for VFCWs could also be adopted for STWs. However, some studies have been carried out in STWs as well. Wang et al. (2010b) used an organic substrate (a 50-50% mixture of peat and crushed pine bark) in pilot STWs planted with *P. australis*, *T. latifolia*, and *I. pseudacorus* for the treatment of sludge containing high organic matter content. They report that an organic substrate can be used in STWs, but it may result in an increased concentration of COD in drained water due to organic matter release. However, the presence of plant seems to counterbalance this effect. Furthermore, no clogging problems occurred with this substrate, while the beds planted with *P. australis* showed a higher degradation rate of organic matter.

Stefanakis and Tsihrintzis (2012c) used two different materials as substrate in two similar STWs; igneous and carbonate rocks, obtained from a river bed and a quarry, respectively. The river bed material was rich in Fe, Al, and Ca oxides, while the carbonate mainly contained Ca and small amounts of Fe and Al. They reported that both these units had similar general performance and dewatering efficiency, with only slight differences in the characteristics of the residual sludge and the drained water quality. Both units also showed similar content of HM, with the exception of Ni, while the metal accumulation rate was found to be higher in the bed with the igneous material (Stefanakis and Tsihrintzis, 2012c). Stability and maturity indices were also similar between the two units (Stefanakis et al., 2011b), as also evapotranspiration rates (Stefanakis and Tsihrintzis, 2011). Overall, the absence of statistical significance among the various parameters tested indicated that the material origin-composition does not play a significant role in the system performance.

The same authors also tested different sizes of the substrate media (Stefanakis and Tsihrintzis, 2012c), i.e., they compared two units with the same igneous material but of different size. One unit contained three layers from bottom to top: cobbles ($d_{50}=90.0$ mm), medium ($d_{50}=24.4$ mm), and fine gravel ($d_{50}=6$ mm), each of 10, 15 and 15 cm thickness; the other unit contained two layers: an extended layer of cobbles (25 cm thickness) and a 15-cm medium gravel layer (i.e., no fine gravel layer). Both units showed similar dewatering efficiency, while the unit with the coarse-grained material had drained water with higher COD, nitrogen, and phosphorus concentrations. Moreover, heavy metal accumulation was lower in the coarse-grained bed (Stefanakis and Tsihrintzis, 2012d), while no differences were found in the maturity and stability indices (Stefanakis et al., 2011b). These could be related with the higher volume of drained water and reduced evapotranspiration rate in the bed with the coarse-grained material, compared to the bed with the fine-grained material (Stefanakis and Tsihrintzis, 2011). Generally, a fine-grained material appears to be more appropriate for use in STWs, since the produced drained water is of better quality and of reduced volume for handling. A coarse-grained material means that the porosity of the media is higher and, due to the higher pore volume, more water drains vertically and reaches the outflow, while comparatively increased pollutant concentrations are found in drained water due to lower contact time with the substrate (smaller surface area of the coarse-grained material).

Uggetti et al. (2012a) tested two pilot STW beds: one was planted with *T. latifolia* and contained three media layers (10 cm of stones, 30 cm of fine gravel, and 10 cm of sand, with respective d_{50} of 250, 30, and 5 mm), while the other was similar but the gravel layer was replaced by wood shavings. Results showed that both beds had similar performance in terms of dewatering, mineralization, and drained water quality. The authors suggested that the use of cheap waste materials, like wood shavings, could be beneficial, since it does not affect the system performance but reduces the use of natural resources (use of fresh, clean gravel is avoided), thus, reducing the environmental footprint and the investment cost of the facility.

Generally, although there are no extended studies on STWs, the use of materials with different compositions and origins does not seem to affect the overall performance of the bed. If alternative waste materials are to be used, depending on the local availability and production, additional environmental and economic benefits could be gained.

Chapter 13

Techno-Economic Aspects of Vertical Flow Constructed Wetlands

13.1 COSTING

As for all CW types (and, generally, all treatment systems and methods), economic characterization of VFCWs comprises two distinct areas: the initial investment costs and the operation/maintenance costs. Investment costs represent the main (and higher) expenses of a full-scale CW facility. Since VFCWs have met an increasing interest only during the past 10–15 years, data available for VFCW systems are far more limited than for the other CW types (FWS and HSF).

Although most data regarding cost evaluation come from FWS and HSF CWs (Gunes et al., 2011), there are general remarks that apply for all CW types. Generally, VFCWs tend to have a slightly higher investment cost due to the use of porous media compared to FWS CWs (Kadlec and Wallace, 2009), although VFCWs generally possess lower area demands compared to both FWS and HSF systems (USEPA, 2000a; Luederitz et al., 2001). Additionally, it is easier for FWS and HSF systems to be constructed without the use of pumps for water transfer, especially when the site topography is favorable. On the other hand, due to the relatively large number of beds, VFCW systems may need the use of pumps, although careful design may avoid it. So far, experience dictates that VFCWs are more economically beneficial for relatively lower flow rates (USEPA, 2000a; Kadlec and Wallace, 2009), despite the general lower surface of VFCWs, mainly due to media costs. For example, USEPA (2000a) reports that for flow rates higher than 227 m^3/d, FWS CWs are usually cheaper to build compared to SF CWs. Nevertheless, in many cases, VFCWs appear as the optimum solution for other reasons, e.g., when there is public access to the facility, which makes the FWS CW inappropriate for public health reasons, mosquito problems or due to wildlife issues, or when the available area is limited.

As it is clear to understand, it is practically unfeasible to present global cost estimates for VFCW systems. Since economic status and parameters change

☆"To view the full reference list for the book, click here"

from country to country, there is a respective change and variability in VFCW cost estimation. An estimation could be carried out for the specific components of the VFCW systems, which are similar for all systems worldwide. Generally, investment costs of a full-scale VFCW facility include:

- necessary land
- site investigation and facility design
- earthworks
- liners (e.g., geomembrane and geotextile)
- porous media (e.g., sand, gravel, or other materials)
- vegetation and planting
- wastewater distribution pipes, drainage pipes, aeration pipes
- water control configurations
- configuration of the surrounding area (access roads, facility lighting, fencing, etc.)
- hydraulic equipment (pumps, valves, pressure system, etc.)
- electronic monitoring equipment (optional)
- access roads
- retaining structures
- drainage structures
- warehouse and office buildings.

These costs represent the necessary tasks in the process of constructing a VFCW facility. Of course, the total amount should also include labor cost, potential expenses for legal permits and licensing (depending on the region/country legislation), construction management, contractor fees and profit. The necessary step for the detailed estimate of the total cost of a VFCW facility is the completion and finalization of the system design, including all treatment stages with all respective parameters (dimensions of all items and units, etc.). In practice, once the final design is ready, the final cost calculation includes all the abovementioned parameters.

Among the various expenses, the most expensive items are probably the land purchase (when it is not available), the substrate media, and the liner for the system bottom. As already mentioned in Chapter 1, the main disadvantage of CWs is the higher land area demand compared to conventional treatment systems (Kivaisi, 2001; Brissaud, 2007; Kadlec and Wallace, 2009; Stefanakis and Tsihrintzis, 2012a). If there is readily available land, e.g., if the facility is planned for a municipality or an individual in a rural area, then the cost for land purchase can be either low or totally avoided, since rural land is usually cheaper to purchase than agricultural, e.g., respective prices in the USA are 0.3 and 10 \$/m^2 (Gunes et al., 2011). Furthermore, rural municipalities may owe land which can be used for the facility. Filter media and liner could represent up to 40-50% of the total facility costs (USEPA, 2000a). Hence, it is clear that the tendency should be to use local materials and suppliers as much as possible. Under this aspect, the characteristics of the local economy may significantly affect the final costs of the VFCW facility. As mentioned by Kadlec and Wallace (2009), the basic elements of a VFCW system (and generally for all CW types), i.e., earthworks (excavation and fill), filter media, and plants, are usually supplied by the local market, and their price highly depends on the transportation distance from the supply source (e.g., a local quarry) to the VFCW construction site. Another parameter that varies from country to country is the labor costs.

Current practice and experience shows that, generally, the investment cost for a VFCW facility does not present significant differences from respective costs for alternative treatment methods and technologies. Depending on the specific characteristics of the parameters mentioned above, these VFCW costs usually are lower or comparable, while in some cases they can be slightly higher for the CW system. CWs become cheaper to implement mainly for small populations, i.e., smaller than 1000 pe. For these plants, they do provide an economy of scale compared to common mechanical (high-rate aerobic) treatment methods. For higher populations (e.g., more than 8-10,000 pe), this advantage concerning the investment costs is limited due to higher land requirements compared to other conventional methods.

Although this does not sound very persuasive at a first glance, the main economic advantage that CWs offer is the significantly reduced costs for operation and maintenance compared to common alternative mechanical technologies. This relies on the fact that most common mechanical methods possess high energy demands for their operation compared to CWs where energy may be consumed mainly for the use of pumps for wastewater transferring between the various stages of the facility. This means that, in terms of the life cycle of a VFCW project, the present value of investment and operating costs becomes very attractive compared to common alternative treatment technologies (Kadlec and Wallace, 2009). The main argument, however, towards the implementation of VFCWs is the financial sustainability they provide which is derived from the long-term reliability concerning the technical performance and overall efficiency. It is known that common conventional mechanical methods are subjected to frequent and serious failures during their lifetime. This means that they need repair and service support and, therefore, regular and frequent maintenance, which respectively increases the operation and maintenance costs. As for other CWs, the major maintenance expenses in VFCWs have to do with the management of the pre-treatment unit (e.g., Imhoff or septic tank), the flow distribution/collection system and piping, and the pumps (Cooper, 2005; Rousseau et al., 2005; Gunes et al., 2011). The current experience from HSF and FWS CWs has shown that construction of CW facilities can indeed be cheaper than conventional treatment plants, despite the cost variations from one country to another (Gunes et al., 2011), a remark which obviously also counts for VFCWs.

A UN report (2003) for wastewater treatment technologies mentioned that the investment costs for various treatment methods (CWs, aerated pond, rapid infiltration, slow rate, facultative pond, hyacinth pond) vary between 450 and

3000 $/m^3/d, with CWs being one of the least expensive (500-1000). In terms of operation and maintenance costs, the respective value for CWs is only 0.03-0.09 $/m^3. The significant advantage of CWs over mechanical systems is also reported by UNEP (1997). In this report, the investment costs for CWs in the American continent are considered as more than three times lower, and the operation and maintenance costs more than seven times lower compared to mechanical technologies.

Luederitz et al. (2001) confirmed the above-mentioned remarks. They calculated the total material and energy requirements for full-scale VFCW systems in Germany compared to centralized conventional facilities. Their analysis included all raw materials used, source of energy, potential reuse of the treated effluent, etc. They found that the operation of the CW systems showed 83% lower energy consumption compared to a centralized facility and 72% in case of wastewater discharge to a nearby (20 km distance) centralized conventional treatment plant, which clearly state the reduced operation costs of VFCWs. The same authors also report that the investment costs were more or less comparable for both alternative technologies evaluated, mainly due to the high volume of substrate media needed in the CW beds. It is also worthy to mention that the advantage of VFCWs for their operation is so high that it overcomes within a year of operation the absence of specific advantage in investment costs.

Tsihrintzis and Gikas (2010) presented an extended review concerning the construction, operation, efficiency, and costs of CWs in Greece. They report total costs for four full-scale VFCW facilities serving 600, 800, 1000, and 1200 pe, respectively. Investment costs were 346,234€, 349,000€, 431,373€, and 502,346€ for the four systems, respectively, while annual operation costs were 4410€, 6296€, 6960€, and 6935€ or 7.35€, 7.87€, 6.96€, and 5.78€/pe/year. Same results are also reported by Gkika et al. (2014) for 9 VFCW facilities in Greece for 540-3640 pe (Table 13.1). According to Gkika et al. (2014), annual operation costs in Greece vary between 6.5€ and 13.9€/pe (for 3640-540 pe) for VFCW facilities in Greece. These data clearly indicate the main advantage of VFCWs, i.e., the lower operation costs, which is also characterized by an economy of scale, since the annual operation costs per pe decreases as the design population increases, something also reported elsewhere (Wallace et al., 2006). It is also interesting that the total operation costs were of the order of 0.11€/m^3, which included the operator salary that represented about 86% of these costs (Tsihrintzis and Gikas, 2010).

In Spain, operation costs for VFCWs are higher or equal to 28€/pe/year (Puigagut et al., 2007), which implies that the variability of the various local parameters (e.g., labor costs, energy costs, etc.) in different countries may affect the operation costs of VFCWs. However, these values clearly show that the dominant advantage of VFCWs is the significantly low operation cost compared to conventional treatment technologies. Economic evaluation of CWs and conventional treatment plants in Greece has shown that CW systems possess significantly lower operation and maintenance costs per pe (Tsagarakis

TABLE 13.1 Economic Data for VFCW Systems from Various Countries

Country	Investment Cost (€/pe)	Population Range	Mean Surface Area (m^2/pe)	References
Belgium	FWS: 392	FWS: 1-2000	FWS: 7	Rousseau et al. (2004a)
	HSF: 879-1636	HSF: 152-350	HSF: 5.9-3.7	
	VF: 507	VF: 4-2000	VF: 3.8	
Greece	VF or combined: 577, 436, 431, 419 FWS (Pompia)	600, 800, 1000, 1200	12.2, 10.5, 8, 7.9[a]	Tsihrintzis and Gikas (2010)
Greece	477, 566, 493, 343, 549, 507, 442, 315, 513	600, 800, 1200, 3640, 540, 560, 900, 3000, 1000	2.4, 2.4, 3.3, 11.9, 1.6, 1.5, 2.3, 9.1, 2.1	Gkika et al. (2014)
Spain	HSF: 503	HSF: 15-2000	HSF: 3.6	Puigagut et al. (2007)
	VF: 295	VF: 20-320	VF: 2.9	
	Combined: 220	Combined: 25-2000	Combined: 4.8	
France	VF: 490, 240, 185	100, 400, 1000		Boutin and Liénard (2003)
Czech Republic	VF: 3576€ (mean value)	Up to 8 pe		Vymazal and Kröpfelová (2011)
China	VF: 39.8€/m^2 0.016€/m^3			Chen et al. (2008)

[a]It refers to total area (pretreatment, two stages VF, one stage HSF).

et al., 2003). Same results are also reported for Spain, where, operation costs for WWTPs were calculated at 0.183-0.215€/m^3, including energy consumption, personnel, chemicals, waste management, and maintenance (Molinos-Senante et al., 2010).

The average investment costs for the mentioned four facilities in Greece were 431-577€/pe (Tsihrintzis and Gikas, 2010), values which are similar to the respective values reported for VFCWs in Belgium (507€/pe for 4-2000 pe; Rousseau et al., 2004a), as Table 13.1 shows. Data from Spain present a lower investment cost, mainly due to lower population served (Puigagut et al., 2007). Available data show that there is indeed an economy of scale, since the investment cost per pe decreases or increases at a very lower rate as the design population for the VFCW increases.

Similar results are reported for French systems (Boutin and Liénard, 2003). Investment costs for three VFCWs with respective population of 100, 400, and 1000 pe were found equal to 490, 240, and 185€/pe, which again shows the economy of scale. Compared to wastewater stabilization ponds (420€/pe for 100 pe), the VF system had comparable investment costs. On the other hand, VFCWs possess almost half the investment costs compared to an extended aeration activated sludge plant (ASP) for 400 pe (400€/pe). Operation and maintenance costs for the French VFCW were again lower (20.2€ and 11.2€/pe/year for 400 and 1000 pe, respectively) compared to an ASP (30.1€ and 18.7€/pe/year, respectively).

Construction costs of a small on-site VFCW system for 8 pe in Greece was equal to 8236€ and included a septic tank, a VFCW bed, and a final zeolite filter unit (Tsihrintzis and Gikas, 2010). Respective costs for an on-site HSF CW system (4 pe) were 9431€, showing that the VFCW systems were cheaper and served double population compared to the HSF CW. It is important to have a hint for these small-scale systems since VFCWs are the most popular systems for on-site domestic wastewater treatment up to 20 pe (Rousseau et al., 2004a). Vymazal and Kröpfelová (2011) reported that the average investment costs for a small on-site conventional treatment system of activated sludge in the Czech Republic are about 3461€, while respective costs for electricity and maintenance are 154€ annually. According to these authors, in the Czech Republic, investment costs for a hybrid system (VF+HSF CWs) are comparable (3576€) to those of the conventional method. However, again the main advantage of the system is the significantly lower operation and maintenance costs which are found to be about one third of the respective costs for a conventional system.

Furthermore, the comparison of full-scale FWS CW and VFCW systems in Greece showed that the construction costs of the VFCW system are indeed slightly higher than the FWS (due to the use of substrate materials and pumps), as also mentioned above (Tsihrintzis et al., 2007). However, both CW types were in any case cheaper in both construction and operation compared to equivalent conventional treatment facilities. Same conclusions can be drawn from cost data in Belgium, where VFCWs appear to have higher investment costs per pe but lower area demands compared to FWS CWs (Table 13.1).

Similar remarks are also reported for the Chinese experience. CWs, generally, in China have similar investment costs with conventional treatment methods, but significantly lower operation and maintenance costs (Zhang et al., 2009). A VFCW bed of 600 m^2 was constructed in Beijing, China, for the treatment of polluted river water (Chen et al., 2008). Total investment costs for this facility were 23,954€ or 39.8€/m^2 (land costs were not included). As mentioned above, membrane cover and substrate media accounted for about 51% of total investment costs and engineering fee for 30%. Based on the treatment capacity of 200 m^3/d, total investment costs can also be expressed as 0.016€/m^3. Annual operation and maintenance costs for this facility are

833€, of which 61% is attributed to energy input and 30% to labor costs. Treatment costs in China with conventional methods vary between 0.128-0.145€/m^3, which again shows the significant economic advantages of the VFCW system. It is characteristic that energy consumption in various CW types is estimated at about 10-15% of that of conventional ASPs (Li and Jiang, 1995; Zhang et al., 2009).

The advantages of VFCWs also appear not only in municipal wastewater treatment but also in the treatment of industrial wastewater. A study was carried out in 2013 concerning the implementation of an innovative VFCW system for wastewater treatment of a paper industry in Greece with a daily flow of 600 m^3/d (unpublished data of Dr. Stefanakis). This industry does not have its own treatment facility yet, but it currently discharges its wastewater to a centralized conventional treatment plant, which means that the industry has to pay an annual fee to the operator of this conventional plant. This preliminary study showed that the return of investment will take place within 2 years by saving money through the elimination of the fee paid to the central conventional plant and the reuse of the effluent in the industrial process. Additionally, operational costs (energy consumption and maintenance) are estimated only equal to 0.098€/m^3 of treated wastewater, with respective values for the current solution up to 1.25€/m^3. Moreover, a hybrid system was used for wastewater treatment of a vinery industry in Italy (Masi et al., 2002). This system consisted of an Imhoff tank, two parallel VFCWs, one HSF CW, one FWS CW, and a final pond for a flow of 10 m^3/d. The load applied was equal to 370 pe. Total costs (design and construction) for this facility were 77,300€.

Yang et al. (2007) tried to evaluate both the ecological and economic values of a VFCW system in China, using different methodologies (the contingent valuation method, CVM, and the shadow project approach, SPA). In this approach, the CW ecosystem is considered to have several contributions, like the treated effluent, groundwater recharge, microclimate regulation, gas regulation, recreation activities, scientific research, education, culture, aesthetics, biodiversity, etc. The study was conducted for a VFCW facility (600 m^2), which was used to treat the wastewater from an ornamental fishpond and recharge (and protect) the groundwater. Construction costs of the VFCW were about 33,768€ and the land opportunity cost 2412€/year. The VFCW bed has practically no maintenance costs and most of the maintenance is carried out only for the ornamental pond. Estimated total costs for the VFCW in 20 years are at the height of 82,000€. The first approach (CVM) was used to estimate the nonuse values and nonmarket use values (like cultural heritage, legacy, and biodiversity) of environmental services and goods, while the SPA method includes the total service value (e.g., water costs, electricity, management, etc.). Based on the two approaches, the total economic value of the VFCW for a 20-year period was calculated at 96,480€ (8.04€/m^2/year) by CVM and 2,778,642€ (232€/m^2/year) by SPA. This analysis again showed the overall advantages of the VFCW system such as low investment, minimum operation and maintenance costs, high treatment efficiency, and ecological benefits.

Concerning sludge dewatering in CWs, available data are limited since very few respective studies are published. Uggetti et al. (2011) investigated the investment and operation costs for four different alternatives: (a) STWs, (b) STW+composting, (c) centrifuge+composting, and (d) transport to a centralized treatment plant. Respective calculations were made for 500, 1000, and 2000 pe. Investment costs for STWs were 50,563€, 83,606€, and 159,442€, respectively, which shows that there is an increase, as expected, with increasing treatment capacity. On the contrary, investment costs for the centrifuge showed a smaller decrease (75,000€, 76,000€, and 96,500€). Operation costs were always higher for the centrifuge (7000€/year and 17,000€/year for 500 and 2000 pe, respectively), taking into account the high energy demands for the centrifuge operation. Overall, investment and operation costs were lower for the STWs (0.16-0.18€/m^3) than for the centrifuge (0.28€/m^3) for 500 pe. Generally, the economic analysis presented in this study indicates that STWs are more attractive compared to a common traditional method like centrifuges.

Overall, it is clear that the majority of expenses in the process of implementing a VFCW facility (either for wastewater treatment or for sludge dewatering) arise during the construction phase, mainly due to raw materials used and the liner for the bottom of the CW basin. Thus, the initial investment is the main concern. Compared to conventional treatment technologies like activated sludge, investment costs are more or less comparable; a small advantage may exist for the VFCW, especially for a small population. But again, each case should be investigated and evaluated separately, since the local parameters (economy, remoteness, supplies, labor costs, land value, etc.) significantly affect the overall pricing of the facility. The main and—more or less—standard advantage that VFCWs offer are the significantly reduced operation and maintenance costs compared to conventional technologies. The majority of the expenses during the operation phase derive from energy consumption for the pumps (if any) and for the periodical maintenance and supervision of the facility, which are practically very low (plus final sludge disposal/transportation for STWs).

13.2 ECONOMIC AND ENVIRONMENTAL EVALUATION

It is often considered that CW systems are *a priori* "green systems," meaning that they offer a series of environmental benefits and possess various ecological characteristics which place them in the sustainable category. However, according to Brix (1999), it is not quite feasible or accurate to characterize alternative treatment technologies than conventional ones as green, solely based on their energy consumption and nutrient recycling. There are several factors which should be investigated in order to use the term "green" to characterize a treatment technology, including (Brix, 1999; Shutes, 2001):

- acceptable treatment performance, complying with effluent limits
- process robustness
- minimum pollutant emissions
- minimum or no by-products (e.g., sludge)
- reuse and/or recycling of materials/useful constituents
- minimum energy consumption coupled with use of renewable sources
- minimum or no use of chemicals
- reduced area demands
- reduced environmental nuisance
- environmental benefits

The UK Environmental Agency implemented a project for the estimation of energy consumption and carbon emissions in the wastewater treatment sector (Georges et al., 2009), a parameter which is crucial for the environmental impact estimation. This report estimates carbon emissions produced by different wastewater treatment techniques and methods, aiming at identifying low-carbon strategies, based not only on the requirements of the Water Framework Directive but also on other parameters such as the risk levels of priority pollutants, EDCs, and viruses. Table 13.2 presents the outcome of this report concerning the greenhouse gas emissions (CO_2, CH_4, and N_2O) from different wastewater treatment techniques and methods during the construction and operation stage. Quantification of these gases is essential for the impact of estimation and the global warming potential. It is worthy to mention that the global warming potentials of CH_4 and N_2O are 25 and 298 times higher, respectively, than that of CO_2 (IPCC, 2001, 2007).

As it is clear in Table 13.2, CW systems cause a more or less comparable environmental impact during the construction stage, regarding global warming gas emissions, to most common alternative techniques like ASPs. However, it is also verified that the main advantage of CWs is the fact that they show the lowest greenhouse gas emissions during their operation compared to all related conventional techniques. As said, this fact is related to the minimum consumed energy, mainly for the use of pumps (if necessary).

Life cycle assessment (LCA) is a very popular tool for the quantitative evaluation of several wastewater treatment techniques and methods, including all process stages from the beginning (design), construction, operation, and end (deconstruction) of a treatment facility. In the wastewater treatment field, LCA is applied during the past 15-20 years (Corominas et al., 2013). LCA is considered as a useful tool to identify the most environmentally friendly and sustainable wastewater treatment methods and estimate the environmental impacts of them (Rebitzer et al., 2003), e.g., the greenhouse gas emissions during their entire lifetime.

TABLE 13.2 Greenhouse Gas Emissions (CO_2, CH_4, and N_2O) from Various Wastewater Treatment Techniques and Methods (Georges et al., 2009)

Technique	CO_2 Emissions		Other GGH	By-products
	Investment	*Operation*		
Secondary Treatment				
Activated sludge plants (ASP)	*Medium* Use of concrete/steel tanks 21-10 kg CO_2/1000 m^3 for 2000-100,000 pe	*Medium* 55% of energy consumed for aeration. Depends on size and wastewater composition *Aeration*: 15 kWh/pe/year or 88 kg CO_2/1000 m^3 *Biological breakdown*: 55 g CO_2/pe/d or 224 kg CO_2/1000 m^3	*N_2O*: 0.004 × N load for 53 mg N/L, it is 0.2 kg N_2O/1000 m^3 *CH_4*: 39 g/pe/year or 0.4 kg CH_4/1000 m^3	Sludge
Trickling filters (TF)	*Medium* Use of concrete/steel tanks 21-10 kg CO_2/1000 m^3 for 2000 and 100,000 pe biofilter	*Low* *Biological breakdown*: 55 g CO_2/pe/d or 224 kg CO_2/1000 m^3	*N_2O*: insignificant *CH_4*: insignificant	Biomass of filter media (humus sludge)
Biological nutrient removal (BNR)	*Medium* Use of concrete/steel tanks CO_2 emissions similar to TF	*Medium* Energy consumption for aeration, sludge return, mixing (slightly higher than simple ASP) *Aerobic*: CO_2 emissions similar ASP *Anaerobic*: minimal CO_2 emissions (mainly CH_4)	*Aerobic*: similar to ASP *Anaerobic*: higher CH_4 emissions than in ASP	Sludge
Constructed wetlands	*Medium* 16 kg CO_2/1000 m^3 for 2000 pe (20 years asset life)	*Low* Passive systems, energy for pump use 4-309 mg $CO_2/m^2/h$	*N_2O*: insignificant *CH_4*: 0-93 mg $CH_4/m^2/h$	Reed biomass

Tertiary Treatment				
Granular activated carbon (GAC)	*High* 4 kg CO_2/1000 m^3 for 2000-100,000 pe, but with annual GAC replacement 1000 m^3 clean water requires 24 m^3 GAC. For daily flow 24,500 m^3 (100,000 pe), this means 588 m^3 media/year Based on 950 kg CO_2/m^3, it gives 558 tn CO_2 due to the media only	*Medium* Emissions due to energy in sand filters are 78 (2000 pe) or 66 (100,000 pe) kg CO_2/1000 m^3 Energy for GAC regeneration not included	Negligible	GAC should be periodically replaced and disposed GAC regeneration can cause significant emissions
Sand filters	*Medium* Similar to GAC	*Medium* Energy use: 88 (2000 pe), 102 (10,000 pe), and 89 (100,000 pe) kg CO_2/1000 m^3	Negligible	Periodical backwashing: water and sludge need treatment and disposal
Membrane biological reactor (MBR)	*High* Due to the use of pumps, tanks, membranes	*High* Energy consumption 1.98 kWh/m^3 Submerged MBR: 0.8-4 kWh/m^3 420-2000 kg CO_2/1000 m^3 *Biological breakdown*: also high emissions	N_2O and CH_4 Insignificant	Sludge from the surface: treatment and disposal
Reverse osmosis (RO)	*High* 31 and 2 kg CO_2/1000 m^3 for 2000 and 100,000 pe Requires tertiary treatment	*High* 0.7-0.9 kWh/m^3 (10,000 m^3/d or 50,000 pe) 370-470 kg CO_2/1000 m^3		Concentrated waste stream

(Continued)

TABLE 13.2 Greenhouse Gas Emissions (CO_2, CH_4, and N_2O) from Various Wastewater Treatment Techniques and Methods (Georges et al., 2009)—cont'd

Technique	CO_2 Emissions		Other GGH	By-products
	Investment	*Operation*		
Chemical dosing	*Medium* 27-0.5 kg CO_2/1000 m^3 for 2000 and 100,000 pe	*Low* Energy for pumping only	*N_2O* and *CH_4* Insignificant	None
Advanced oxidation (e.g., UV, H_2O_2)	*Medium*	*High* Depends on size, wastewater composition, strength/duration of UV UV: 112 and 78 kg CO_2/1000 m^3 for 2000 and 100,000 pe *Biological breakdown*: 12 kg CO_2/1000 m^3	*N_2O* and *CH_4* No production	None
Sludge treatment				
Anaerobic digestion	*Medium* Depends on tanks, heaters, and pumps	*Low* For heating: gas combustion produces 25.4 kg CO_2/tn of sludge (without CHP) with CHP: 0.549 kg CO_2/tn	*CH_4* is produced With CHP: minimal emissions Without CHP: 18 kg CH_4/tn of sludge	Biogas: may be used as energy source Digestate

GGH, greenhouse gas emissions.

During the past decade, LCA has been introduced to the natural wastewater treatment systems and, especially, to the field of CWs. This relatively new tool for the CW community appears as an attractive alternative to present the overall reduced environmental impact of CWs over their life cycle in a more technical, sophisticated, and circumstantial way. LCA includes the production and use of all materials (porous media, concrete, fossil fuels) and flows (energy, emissions) used for the construction and the operation of a treatment system till the breaking down of the facility and the final disposal of the treated product (water or solid). LCA is mainly used to compare the overall performance of a CW system and an equivalent conventional wastewater treatment system or other natural or decentralized systems. However, LCA applications for the investigation of the life cycle of CWs and other various options for wastewater treatment are limited, while the majority of CW-based LCA studies focus on HSF or FWS CWs (Dixon et al., 2003; Memon et al., 2007). First studies in CW systems have shown that CWs possess a lower global warming potential in terms of CO_2 emissions, mainly due to reduced energy consumption, and that they are also capable of reducing aquatic toxicity and eutrophication compared to conventional treatment techniques (Dixon et al., 2003; Machado et al., 2007). In conventional treatment plants (i.e., activated sludge systems), environmental impacts of the operation phase are usually higher than those of the construction phase (Dixon et al., 2003; Rebitzer et al., 2003; Ortiz et al., 2007; Renou et al., 2008), while in CW systems, most of life cycle impacts originate from the construction phase (Dixon et al., 2003; Machado et al., 2007).

A summary of the different scenarios tested in various LCA studies including VFCWs is presented in Table 13.3 and the respective results in Table 13.4.

Søvik et al. (2006) report that N_2O emissions from CWs are much smaller compared to conventional treatment plants. VF and HSF CWs were found to emit 2.3 and 1.3-13 g N_2O/pe/year, respectively, based on data measured in CW systems in N. Europe (Estonia, Finland, Norway, Poland). VFCWs showed the highest N_2O emissions among all CW types, while FWS CWs had the highest CH_4 emissions. Teiter and Mander (2005) report 0.33-119 and 0.17-130 mg N_2-N/m^2/h and 35.6-44.7 and 6.4-31 μg N_2O/m^2/h in VFCWs and HSF CWs, respectively, in Estonia. Respective reported values for conventional activated sludge systems are 3.2 (Czepiel et al., 1995), 0-1.9 (Kimochi et al., 1998), or 0.28-141 g N_2O/pe/year (Ahn et al., 2010), depending on the processes involved, while for septic tanks it is 1.8 g N_2O/pe/year (WERF, 2010). Mander et al. (2014) reported that CO_2 emissions are lower in FWS CWs compared to the other types (Table 13.4). On the other hand, lowest CH_4 emissions are measured in VFCWs than in HSF CWs, while N_2O emissions are comparable in all CW types. It is also interesting that hybrid CW systems simultaneously achieve high treatment performance and minimum greenhouse gas emissions. Moreover, implementation of intermittent loading in VFCWs can also limit these emissions (Mander et al., 2014).

Machado et al. (2007) compared two energy-saving systems (a CW facility and a slow rate infiltration plant; SRIP) and a conventional ASP (Table 13.3).

TABLE 13.3 Main Data Used for the Comparison of VFCW Systems and Other Conventional Treatment Techniques in Various LCA and Greenhouse Gas Emission Estimation Studies

References	Alternative Scenarios	PE	Flow (m^3/d)	Area (m^2)
Machado et al. (2007)	(A) Pt+SRIP[a] (B) Pt+Imhoff+VF +HSF (C) Pt+ASP	40 (winter), 120 (summer) 120 500	5 (winter), 15 (summer) 15 60	2000 VF: 317, HSF: 277
Roux et al. (2010)	(A) Pt+ASP (B) Pt+VFCW (C) Direct discharge	1000 806	445 145	2000
Pan et al. (2012)	(A) ASP +mechanical sludge dewatering (landfill) (B) Pt+two-stage VFCW+two STWs	690,000 700	1.01×10^5 (domestic) 68	1000 (VF) +300 (STW)
Fuchs et al. (2007)	(A) Pt+HSF (B) Pt+VF	400 400	60 60	5049 1280
Chen et al. (2011)	(A) VFCW (B) ASP		200 1700	600
Gao et al. (2012)	(A) VFCW		200	600
Lopsik (2013)	(A) Pt+VF+HSF (B) Pt+ASP+Sludge thickening	64 1020		
Uggetti et al. (2011)	(A) STWs (B) Centrifuges	Three scenarios: 500, 1000, 2000	WAS (m^3/year): 275, 550, 1100 SLR: 50 kg TS/m^2/year	167, 331, 662
Uggetti et al. (2012c)	(A) STW (B) STW	12,000 1000	WAS (m^3/year): (A) 6600 and (B) 500 SLR (kg TS/m^2/year): (A) 55-60 and (B) 50	5080 210
Gkika et al. (2014)	Two-stage VF, one HSF, one STW (A) Concrete for anaerobic tank, HDPE liner (B) Earth, HDPE liner (C) Earth, clay cover	1200	180	3329

[a]Pt, pretreatment; SRIP, slow rate infiltration plant; VF, VFCW; HSF, HSF CW; ASP, activated sludge plant; STW, sludge treatment wetlands; WAS, waste activated sludge.

TABLE 13.4 Results of LCA and Greenhouse gas Emission Estimation Studies Comparing VFCW Systems and Other Conventional Treatment Techniques

References	Alternative Scenarios	Results
Machado et al. (2007)	(A) Pt + SRI[1] (B) Pt + Imhoff + VF + HSF (C) Pt + ASP	A and B: lower environmental impact regarding use of construction materials, energy consumption, CO_2 emissions, solid emissions 10% extension of CW,SRI lifetime → 1% decrease in CO_2 emissions 5 – 7% decrease in abiotic depletion Use of HDPE than concrete/steel in ASP → 1% reduction in CO_2 emissions 1% decrease in abiotic depletion Use of HDPE than concrete/steel in Imhofftan *k* → 1% reduction in CO_2 emissions 5% decrease in abiotic depletion
Roux et al. (2010)	(A) Pt + ASP (B) Pt + VFCW (C) Direct discharge	B: lower impact on global warming, ecotoxicity, acidification, abiotic depletion, human toxicity, ozone layer depletion higher impact on eutrophication, due to limited removal of N and P in VFCW
Pan et al. (2011)	(A) Centralized biological treatment + mechanical sludge dewatering (landfill) (B) Pt + two-stage VFCW + two STWs	kg CO_2/kg BOD → A: 7.3, B: 3.18 kg CO_2/d (total) → A: 128.3, B: 38.8 Fractions (CH_4, energy use, N_2O) → A: 71%, 26%, 3%, B: 65.7%, 7%, 27.3% kg CO_2/year (operation) → A: 67.9, B: 20.2 kg CO_2/pe/year (operation) → A: 48.7, B: 7 Energy consumption (kg fossil fuel C/kg C removed) → A: 3.9, B: 0.16

(Continued)

TABLE 13.4 Results of LCA and Greenhouse gas Emission Estimation Studies Comparing VFCW Systems and Other Conventional Treatment Techniques—cont'd

References	Alternative Scenarios	Results
Fuchs et al. (2011)	(A) Pt+HSF (B) Pt+VF	CO_2 (mg/L equivalent) → A: 679.3, B: 312.5 N_2O (mg/L equivalent) → A: 0.5, B: 0.4 CH_4 (mg/L equivalent) → A: 49.9, B: 4.6 B → 60% reduced impact on human health compared to A 50% reduced impact on ecosystem and resources compared to A
Chen et al. (2011)	(A) VFCW (B) ASP	CO_2, CH_4, N_2O emissions (tn, within 20 years) → A: 82.9, 0.42, 0.08, B: 2077, 2.11, 0.20 kg CO_2/m^3 → A: 0.20, B: 0.74 kg CO_2/kg BOD removed → A: 4.75, B: 9.20 kg CO_2/kg COD removed → A: 2.21, B: 4.75 Fractions CO_2, CH_4, N_2O → A: 77.8%, 10.6%, 11.7%, B: 89.0%, 9.8%, 1.2% Direct versus indirect emissions → A: 41.2%, 58.8%, B: 23.9%, 76.1%
Gao et al. (2012)	VFCW	0.45 kg CO_2/m^3, 11.08 kg CO_2/kg BOD removed, 5.16 kg CO_2/kg COD removed Indirect-direct emission: 82.31%, 17.69% of total emissions Total emissions: geotextile 6.30% Substrate media, bricks, concrete and vegetation: 29.25% Hydraulic/electric equipment: 2.29% Electricity: 46.76%
Lopsik (2013)	(A) Pt+VF+HSF (B) Pt+ASP	Higher environmental impact → A: during construction phase B: during operation phase A: negative value in global warming impact, due to the use of LWA, whose production is energy intensive. B: high energy consumption, limited P removal

Mander et al. (2014)	(A)VF (B) HSF (C) FWS	mg $CO_2/m^2/h$ → VF-HSF: 51.9-567 (184.7), FWS: 29.4-176 (92.3)[a] mg $CH_4/m^2/h$ → VF: 0.3-5.4 (2.9), HSF: 0.048-17.5 (6.4), FWS: 0.15-27 (5.9) mg $N_2O/m^2/h$ → VF: 0.003-0.424 (0.14), HSF: 0-0.894 (0.24), FWS: 0.003-0.65 (0.13) CH_4-C/TOC_{in} → VF: 1.17%, HSF: 4.5%, FWS: 16.9% (aver) N_2O-N/TN_{in} → VF: 0.023%, HSF: 0.79%, FWS: 0.13% (aver)
Uggetti et al. (2011)	*Sludge dewatering* (A) STWs (B) Centrifuges for 500, 1000, 1200 pe	Sludge production (m^3/year) → A: 33, 66, 132, B: 41, 82, 165 Energy for pumps (kWh/year) → A: 25, 50, 105, B: 30, 60, 125 Energy for centrifuge (kWh/year) → 140, 280, 560 CH_4 emissions from STWs ($mg/m^2/s$) → <88 Odor emissions ($ou_E/m^2/s$) → 5.7-7.3
Uggetti et al. (2012c)	(A) STW (12,000 pe) (B) STW (1000 pe)	Sludge production (m^3/year) → A: 790, B: 500 Energy for pumps (kWh/year) → A: 630, B: 50 CO_{2eq} emissions ($kg/m^2/d$) → A: 0.09, B: 0.25 CH_4 emissions ($mg/m^2/d$) → A: 1000, B: 1900-3700 N_2O emissions ($mg/m^2/d$) → A: 200, B: 400-750 Carbon footprint (kg CO_{2eq}/1 tn of sludge ww) → A: 1.0, B: 2.1 A (soil basin) kg CO_{2eq} → construction: 0.02, energy consumption: 0.4, CH_4 and N_2O: 0.005 B (concrete basin) kg CO_{2eq} → construction: 1.7, energy consumption: 0.4, CH_4 and N_2O: 0.2
Gkika et al. (2014)	Two-stage VF, one HSF, one STW (A) Concrete for anaerobic tank, HDPE liner (B) Earth, HDPE liner (C) Earth, clay cover	A → higher environmental impact (human health, ecosystem quality, resources) due to the use of concrete and steel for the anaerobic tank C → smaller environmental impact in all categories (climate change, ozone layer, ecotoxicity, land use, minerals, fossil fuels, acidification, carcinogens, radiation)

[a]Min-max (aver).
[1]Pt, pre-treatment; SRIP, slow rate infiltration plant; VF, VFCW; HSF, HSF CW; ASP, activated sludge plant; STW, sludge treatment wetlands; WAS, waste activated sludge.

Materials used in this study were concrete (e.g., for the Imhoff tank), plastic (piping and geomembrane in CW beds), and steel (activated sludge tank). Both CW system and SRIP had very low energy input compared to the ASP and, thus, very low CO_2 emissions, while they operate as carbon sinks too. As expected, the ASP showed the highest environmental impact during its operation and maintenance. However, the deconstruction phase of the CW appears to have a higher impact, indicating that minimization techniques are important for this final stage. An interesting remark is that 10% extension of the operational lifetime of the CW system results in 1% decrease of CO_2 emissions (Table 13.4). Overall, this study showed the positive impact of the CW system in the reduction of global warming impact.

Roux et al. (2010) investigated three different scenarios for domestic wastewater treatment: (a) an ASP with pretreatment (including sewer network, sludge production, transportation, spreading in agricultural land and respective emissions), (b) a two-stage VFCW system with pretreatment (including sewer network, distribution and drainage pipes, liner, plant biomass recovery, etc.), and (c) direct discharge of the wastewater to a surface receiver (Table 13.3). The functional unit chosen for this study was the daily nominal BOD_5 load (48 kg). The VFCW system was found to have the smallest environmental impact on most categories tested (e.g., global warming, ecotoxicity, acidification, abiotic depletion, human toxicity, ozone layer depletion), with the exception of eutrophication due to the limited removal of nutrients. Additionally, the sewer network also had a considerable impact on these categories (Table 13.4).

Pan et al. (2011) compared a VFCW system with a conventional biological plant in China for the estimation of greenhouse gas emissions (Table 13.3). The centralized conventional plant consisted of primary settling, anaerobic-anoxic-aerobic reactors and secondary settling, as also thickening and mechanical dewatering of surplus sludge (including the addition of chemicals and sludge transport to landfill). The VFCW facility included a septic tank, a two-stage VFCW bed, and a sludge treatment wetland. Calculated emission of the conventional plant were 7.26 kg CO_2/kg BOD removed, 48.7 kg CO_2/pe/year, and 67.9 kg CO_2/year, with respective figures for the VFCW plant 3.18 kg CO_2/kg BOD removed, 7 kg CO_2/pe/year, and 20.2 kg CO_2/year (Table 13.4). This study again confirmed that VFCWs possess a significantly lower carbon footprint compared to conventional techniques for their lifetime.

Fuchs et al. (2011) applied an LCA to compare the environmental impact of two different CW types (a VF and a HSF system), regarding nitrogen-related emissions. Both systems were considered to have a pretreatment stage (septic tank) and were designed for population of 400 pe with a flow of 60 m^3/d and a lifetime of 50 years (Table 13.3). Although C and N emissions per m^2 are higher in VFCWs than in HSF CWs, the overall impact of HSF CWs is more than double due to their larger surface area demands for the same equivalent populations and treatment level. It is reported in this study that VFCWs have lower impact than HSF CWs for their entire life cycle due to their improved treatment performance in nitrogen removal and the lower use of materials, land, and equipment during the construction phase (Table 13.4). This finding can be very significant and crucial in the design of CW facilities since it shows that, besides the choice of an appropriate technology, the flow path of the CW is also an important factor toward the reduction of the total environmental impact of the wastewater treatment facility.

Cheng et al. (2011) investigated the overall greenhouse gas emissions from a VFCW bed in China (which was described above; Chen et al., 2008) and compared it with a conventional treatment plant. This VFCW facility treated 200 m^3/d and had a surface of 600 m^2 (Table 13.3). Estimated emissions during the operation phase (20 years) were 82.87 tn CO_2, 0.42 tn CH_4, and 0.08 tn N_2O. CO_2 emissions of the VFCW system were 0.20, 4.75, and 2.21 kg CO_2/m^3 per kilogram BOD removed and per kilogram COD removed, respectively (Table 13.4). Respective values for a conventional ASP were higher (0.74, 9.20, and 4.75), which confirms that VFCWs can be considered as a low-carbon treatment technology. Emission distribution in the VFCW and the conventional plant was 77.8% and 89.0% for CO_2, 10.6% and 9.8% for CH_4, and 11.7% and 1.2% for N_2O, respectively, which again shows that VFCWs emit higher amounts of N_2O, a gas playing a significant role for the global warming phenomenon, as mentioned above. Distribution of indirect (58.8% and 76.1%) and direct (41.2% and 23.9%) greenhouse gas emissions in the VFCW and the conventional plant indicate the significant contribution to these emissions of the operation phase of the conventional plant due to high energy consumption (more than 91% of the indirect emissions). This study also showed that VFCWs are a more favorable treatment technology with many environmental and economic advantages.

Gao et al. (2012) performed an improved assessment of the total greenhouse gas emissions for the same VFCW bed, using a local embodied gas emissions database for the Beijing economy 2002. New values that aroused from this study are 0.45, 11.08, and 5.16 kg CO_2/m^3/kg BOD removed and per kg COD removed, respectively (Table 13.4), which are almost double than those reported by Cheng et al. (2011) due to lower contribution of energy consumption in the indirect emissions and the higher fraction of the local organic substrate used. This alteration also implies that the careful choice of an appropriate database is also significant for the emissions assessment.

Lopsik (2013) compared a hybrid CW system (VF and HSF) with a conventional ASP (Table 13.3). LCA was implemented with two different assessment methods. Parameters included were sewerage system, land use, construction materials, energy consumption, chemical use, plant expected lifetime, and effluent pollutant concentrations. This study showed that the majority of the environmental impacts of the CW facility appear during the construction phase and due to the use of expanded clay (LWA) as filter media. Once again, the minimum energy consumption of the CW during the operational phase resulted in negligible impact during this phase. On the other hand, intense energy consumption and varying effluent quality regulate the impacts of the conventional ASP during the operation phase (Table 13.4). The use of LWA causes a negative impact of the CW to the global warming category, since LWA production demands high energy amounts. Therefore, the use of other energy-saving materials should be promoted in CWs.

An LCA study was performed on a full-scale VFCW facility in Greece (two stages VF, one stage HSF, and a STW) for 1200 pe (Gkika et al., 2014; Table 13.3). Three scenarios were investigated: (a) an anaerobic tank made of concrete and the use of HDPE geomembrane for the CWs bottom, (b) both tank and CWs are covered with HDPE geomembrane, and (c) both tank and CWs are covered with clay material. Results showed that the use of concrete to cover the bottom of the tank causes the greater environmental impact in all related categories for a 20 years lifetime of the facility (Table 13.4).

Interesting results are also obtained from LCA conducted in STWs, although for these systems respective studies are very limited. Uggetti et al. (2011) compared STWs with traditional dewatering technique (centrifuges) for three different populations (500, 1000, and 2000 pe; Table 13.3). This LCA included the transport of the sludge to the dewatering plant (STW or centrifuge) within the treatment line or with transport to a centralized facility (30 km distance), materials needed for construction, and energy consumption. STWs have the minimum environmental impact in all relative categories (abiotic depletion, acidification, eutrophication, global warming). For STWs, the higher impact is caused by the raw materials used for the construction and lower by the energy consumption (Table 13.4). STW contribution to global warming is negligible due to low CH_4 emissions, while with additional treatment (composting), the respective emissions increase. Centrifuges present the highest impact, mainly due to the high energy demands for their operation. Generally, STWs have multiple environmental benefits compared to traditional centrifuge technique. The total environmental impact of STWs is 500 times lower than conventional centrifugation and 2000 times lower than sludge transport to a central treatment facility. However, when sludge transportation is involved in the treatment line, either for a final composting stage or for final disposal, these advantages are limited. Therefore, the choice of the dewatering method should always take into account these parameters, based, of course, on the local conditions and characteristics.

Uggetti et al. (2012c) compared two STW facilities (12,000 and 1000 pe, respectively) to determine the carbon footprint of STWs (Table 13.3). CH_4 emission varied between 1000 and 3700 mg/m^2/d and N_2O between 200 and 750 mg/m^2/d. Overall emissions corresponded to 0.09 and 0.25 CO_{2eq}, while both the methane and the nitrous oxide role in the carbon footprint was practically negligible compared to respective contributions of the use of raw materials and energy consumption. Additionally, the material of the STW basin (concrete or soil-excavated) was found to significantly affect the environmental impact of the system; thus, the proposed solution is either a simple excavation or the use of recycled concrete (Table 13.4).

Generally, the so far experience and respective studies from VFCW facilities show that, besides the economic benefits, this technology is also accompanied with significant environmental advantages. As for all CW systems, the VFCW technology can be characterized as a sustainable alternative for wastewater and sludge treatment, based on the combined benefits of environmental protection, human health protection, use of sustainable resources, technical feasibility, social acceptance, and economic viability (Langergraber, 2013).

As for the cost analysis, the main impact of VFCWs appears during the construction phase due to the raw materials used. During the operation phase, this impact is very-very low. Compared to conventional treatment systems, VFCWs cause a smaller impact during their complete life cycle in all relative categories like construction materials, energy consumption, greenhouse gas emissions and solid emissions, global warming phenomenon, abiotic depletion, toxicity, etc. In this field, further studies are needed to further investigate and quantify the environmental benefits that arise from different construction and operation parameters, which could be helpful in the continuous improvement and optimization of this technology in terms of both treatment performance and ecological footprint.

References

Abdel-Shafya, H.I., El-Khateebab, M.A., 2013. Integration of septic tank and constructed wetland for the treatment of wastewater in Egypt. Desalination and Water Treatment 51 (16–18), 3539–3546.

Abdulla, H., Khafagi, I., El-Kareem, M.A., Dewedar, A., 2007. Bacteriophages in engineered wetland for domestic wastewater treatment. Research Journal of Microbiology 2 (12), 889–899.

Abidi, S., Kallali, H., Jedidi, N., Bouzaiane, O., Hassen, A., 2009. Comparative pilot study of the performances of two constructed wetland wastewater treatment hybrid systems. Desalination 246, 370–377.

Abira, M.A., Van Bruggen, J.J.A., Denny, P., 2005. Potential of a tropical subsurface constructed wetland to remove phenol from pre-treated pulp and papermill wastewater. Water Science and Technology 51 (9), 173–176.

Abou-Elela, S.I., Hellal, M.S., 2012. Municipal wastewater treatment using vertical flow constructed wetlands planted with *Canna*, *Phragmites* and *Cyprus*. Ecological Engineering 47, 209–213.

Aguilar, J.R.P., Cabriales, J.J.P., Vega, M.M., 2008. Identification and characterization of sulfur-oxidizing bacteria in an artificial wetland that treats wastewater from a tannery. International Journal of Phytoremediation 10, 359–370.

Ahmaruzzaman, M., Sharma, D.K., 2005. Adsorption of phenols from wastewater. Journal of Colloid and Interface Science 287, 14–24.

Ahn, J.H., Kim, S., Park, H., Rahm, B., Pagilla, K., Chandran, K., 2010. N2O emissions from activated sludge processes, 2008-2009: results of a national monitoring survey in the United States. Environmental Science and Technology 44, 4505–4511.

Ahuja, L.R., DeCoursey, D.G., Barnes, B.B., Rojas, K.W., 1993. Characteristics of macropore transport studied with the ARS root zone water quality model. Transactions of the American Society of Agricultural and Biological Engineers 36, 369–380.

Ahuja, L.R., Johnsen, K.E., Heathman, G.C., 1995. Macropore transport of a surfaceapplied bromide tracer: model evaluation and refinemen. Soil Science Society of America Journal 59, 1234–1241.

Ahuja, L.R., Ma, Q.L., Rojas, K.W., Boesten, J.J.T.I., Farahani, H.J., 1996. A field test of root zone water quality model - pesticide and bromide behaviour. Pesticide Science 48, 101–108.

Akratos, C.S., Tsihrintzis, V.A., 2007. Effect of temperature, HRT, vegetation and porous media on removal efficiency of pilot-scale horizontal subsurface flow constructed wetlands. Ecological Engineering 29, 173–191.

Akratos, C.S., Tsihrintzis, V.A., Pechlivanidis, I., Sylaios, G.K., Jerrentrup, H., 2006. A free water surface constructed wetland for the treatment of agricultural drainage entering Vassova Lagoon, Kavala, Greece. Fresenius' Environmental Bulletin 15, 1553–1562.

Akratos, C.S., Papaspyros, J.N.E., Tsihrintzis, V.A., 2008. An artificial neural network model and design equations for BOD and COD removal prediction in horizontal subsurface flow constructed wetlands. Chemical Enginnering Journal 143, 96–100.

Akratos, C.S., Papaspyros, J.N.E., Tsihrintzis, V.A., 2009a. Total nitrogen and ammonia removal prediction in horizontal subsurface flow constructed wetlands: use of artificial neural networks and development of a design eEquation. Bioresource Technology 100, 586–596.

Akratos, C.S., Papaspyros, J.N.E., Tsihrintzis, V.A., 2009b. Artificial neural network use in orthophosphate and total phosphorus removal prediction in horizontal subsurface flow constructed wetlands. Biosystems Engineering 102, 190–201.

Aktas, E.S., Imre, S., Ersoy, L., 2001. Characterization and lime treatment of olive mill wastewater. Water Research 35, 2336–2340.

Albuquerque, A., Oliveira, J., Semitela, S., Amaral, L., 2009. Influence of bed media characteristics on ammonia and nitrate removal in shallow horizontal subsurface flow constructed wetlands. Bioresource Technology 100 (24), 6269–6277.

Alexandrino, M., Grohmann, E., Szewzyk, R., Szewzyk, U., 2007. Application of culture-independent methods to assess the bacteria removal efficiency of subsurface flow constructed wetlands. Water Science and Technology 56 (3), 217–222.

Allen, R.G., Pereira, L.S., Howell, T.A., Jensen, M.E., 2011. Evapotranspiration information reporting: I. Factors governing measurement accuracy. Agricultural Water Management 98 (6), 899–920.

Al-Malack, M.H., Abuzaid, N.S., Bukhari, A.A., Essa, M.H., 2002. Characterization, utilization, and disposal of municipal sludge: the state of-the-art. Arabian Journal for Science and Engineering 27 (1B), 3–27.

Álvarez, J.A., Ruíz, I., Soto, M., 2008. Anaerobic digesters as a pretreatment for constructed wetlands. Ecological Engineering 33, 54–67.

Amon, J.P., Agrawal, A., Shelley, M.L., Opperman, B.C., Enright, M.P., Clemmer, N.D., Slusser, T., Lach, J., Sobolewski, T., Gruner, W., Entingh, A.C., 2007. Development of a wetland constructed for the treatment of groundwater contaminated by chlorinated ethenes. Ecological Engineering 30, 51–66.

Andreadakis, A.D., Mamais, D., Gavalaki, E., Kampylafka, S., 2002. Sludge utilisation in agriculture: possibilities and prospects in Greece. Water Science and Technology 46 (10), 231–238.

Angelakis, A.N., Koutsoyiannis, D., Tchobanoglous, G., 2005. Urban wastewater and stormwater technologies in ancient Greece. Water Research 39, 210–220.

Annadurai, G., Babu, S.R., Mahesh, K.P.O., Murugesan, T., 2000. Adsorption and bio-degradation of phenol by chitosan-immobilized *Pseudomonas putida* (NICM 2174). Bioprocess Engineering 22, 493–501.

Appels, L., Baeyens, J., Degrève, J., Dewil, R., 2008. Principles and potential of the anaerobic digestion of waste-activated sludge. Progress in Energy and Combustion Science 34, 755–781.

Arias, C.A., Del Bubba, M., Brix, H., 2001. Phosphorus removal by sands for use as media in subsurface flow constructed reed beds. Water Research 35 (5), 1159–1168.

Arias, C.A., Cabello, A., Brix, H., Johansen, N.-H., 2003a. Removal of indicator bacteria from municipal wastewater in an experimental two-stage vertical flow constructed wetland system. Water Science and Technology 48 (5), 35–41.

Arias, C.A., Brix, H., Johansen, N.-H., 2003b. Phosphorus removal from municipal wastewater in an experimental two-stage vertical flow constructed wetland system equipped with a calcite filter. Water Science and Technology 48 (5), 51–58.

Arias, C.A., Brix, H., Marti, E., 2005. Recycling of treated effluents enhances removal of total nitrogen in vertical flow constructed wetlands. Journal of Environmental Science and Health 40, 1431–1443.

Arienzo, M., Christen, E.W., Quayle, W.C., 2009. Phytotoxicity testing of winery wastewater for constructed wetland treatment. Journal of Hazardous Materials 169, 94–99.

Arraj, A., Bohatier, J., Laveran, H., Traore, O., 2005. Comparison of bacteriophage and enteric virus removal in pilot scale activated sludge plants. Journal of Applied Microbiology 98, 516–524.

Ausland, G., Stevik, T.K., Hanssen, J.F., Køhler, J.C., Jenssen, P.D., 2002. Intermittent filtration of wastewater—removal of fecal coliforms and fecal streptococci. Water Research 36, 3507–3516.

Austin, D., 2006. Influence of cation exchange capacity (CEC) in a tidal flow, flood and drain wastewater treatment wetland. Ecological Engineering 28, 35–43.

Austin, D., Maciolek, D., Davis, B., Wallace, S., 2006. Damköhler number design method to avoid plugging of tidal flow constructed wetlands by heterotrophic biofilms. In: Proceedings of the 10th International Conference on Wetland Systems for Water Pollution Control, 23–29 September, Lisbon, Portugal.

Ávila, C., Salas, J.J., Martín, I., Aragón, C., García, J., 2013. Integrated treatment of combined sewer wastewater and stormwater in a hybrid constructed wetland system in southern Spain and its further reuse. Ecological Engineering 50, 13–20.

Ayaz, S.C., Aktaş, Ö., Findik, N., Akça, L., 2012a. Phosphorus removal and effect of adsorbent type in a constructed wetland system. Desalination and Water Treatment 37, 152–159.

Ayaz, S.C., Aktaş, Ö., Findik, N., Akça, L., Kinaci, C., 2012b. Effect of recirculation on nitrogen removal in a hybrid constructed wetland system. Ecological Engineering 1–5.

Babatunde, A.O., Zhao, Y.Q., 2009. Phosphorus removal in laboratory-scale unvegetated vertical subsurface flow constructed wetland systems using alum sludge as main substrate. Water Science and Technology 60 (2), 483–489.

Babatunde, A.O., Zhao, Y.Q., 2010. Two strategies for improving animal farm wastewater treatment in reed beds. Environmental Technology 31 (12), 1343–1348.

Babatunde, A.O., Zhao, Y.Q., O'Neill, M., O'Sillivan, B., 2007. Constructed wetlands for environmental pollution control: a review of developments, research and practice in Ireland. Environment International 34 (1), 116–126.

Bahlo, K., 2000. Treatment efficiency of a vertical-flow reed bed with recirculation. Journal of Environmental Science and Health A35 (8), 1403–1413.

Bahlo, K., Wach, G., 1995. Naturnahe Abwasserreinigung: Planung und Bau von Pflanzenkläranlagen. Ökobuch Verlag, Freiburg, Germany.

Barbier, E.B., Acreman, M., Knowler, D., 1997. Economic Valuation of Wetlands—A Guide for Policy Makers and Planners. Ramsar Convention Bureau, Gland, Switzerland.

Barbieri, A., Garuti, G., Avolio, F., Bruni, S., 2003. Sludge dewatering using macrophytes in a small wastewater treatment system: a case study of a pilot scale plant in Northern Italy. Journal of Environmental Science and Health. Part A, Toxic/Hazardous Substances & Environmental Engineering 38 (10), 2425–2433.

Barr, M.J., Robinson, H.D., 1999. Constructed wetlands for landfill leachate treatment. Waste Management and Research 17, 498–504.

Barrena, R., Vásquez, F., Sánchez, A., 2006. The use of respiration indices in the composting process: a review. Waste Management & Research 24, 37–47.

Barrena, R., Casals, E., Colón, J., Font, X., Sánchez, A., Puntes, V., 2009. Evaluation of the ecotoxicity of the model nanoparticles. Chemosphere 75, 850–857.

Batty, L.C., Younger, P.L., 2007. The effect of pH on plant litter decomposition and metal cycling in wetland mesocosms supplied with mine drainage. Chemosphere 66, 158–164.

Benami, M., Gross, A., Herzberg, M., Orlohsky, E., Vonshak, A., Gillor, O., 2013. Assessment of pathogenic bacteria in treated graywater and irrigated soils. Science of the Total Environment 458–460, 298–302.

Begg, S.J., Lavigne, L.R., Veneman, L.M.P., 2001. Reed beds: constructed wetlands for municipal wastewater treatment plant sludge dewatering. Water Science and Technology 44 (11), 393–398.

Bertrán, E., Sort, X., Soliva, M., Trillas, I., 2004. Composting winery waste: sludges and grape stalks. Bioresource Technology 95, 203–208.

Białowiec, A., Janczukowicz, W., Randerson, P.F., 2011. Nitrogen removal from wastewater in vertical flow constructed wetlands containing LWA/gravel layers and reed vegetation. Ecological Engineering 37 (6), 897–902.

Bianchi, V., Peruzzi, E., Masciandaro, G., Ceccanti, B., Mora Ravelo, S., Iannelli, R., 2011. Efficiency assessment of a reed bed pilot plant (Phragmites australis) for sludge stabilisation in Tuscany (Italy). Ecological Engineering 37 (5), 779–785.

Blazejewski, R., Murat-Blazejewska, S., 1997. Soil clogging phenomena in constructed wetlands with subsurface flow. Water Science and Technology 35 (5), 183–188.

Bodzek, M., Łobos-Moysa, E., Zamorowska, M., 2006. Removal of organic compounds from municipal landfill leachate in a membrane bioreactor. Desalination 198, 16–23.

Boenigk, J., Arndt, H., 2000. Particle handling during interception feeding by four species of hetertrophic nanoflagellates. Journal of Eukaryotic Microbiology 47, 350–358.

Bolzonella, D., Pavan, P., Battistoni, P., Cecchi, F., 2005. Mesophilic anaerobic digestion of waste activated sludge: influence of the solid retention time in the wastewater treatment process. Process Biochemistry 40, 1453–1460.

Bomo, A.M., Husby, A., Stevik, T.K., Hanssen, J.F., 2003. Removal of fish pathogenic bacteria in biological sand filters. Water Research 37, 2618–2626.

Bonanno, G., Giudice, R. Lo, 2010. Heavy metal bioaccumulation by the organs of *Phragmites australis* (common reed) and their potential use as contamination indicators. Ecological Indicators 10, 639–645.

Bounds, T.R., 1997. Design and performance of septic tanks. In: Bedinger, M.S., Fleming, J.S., Johnson, A.I. (Eds.), Site Characterization and Design of On-site Septic Systems, ASTM STP 1324. In: American Society for Testing and Materials, pp. 217–234.

Boutilier, L., Jamieson, R., Gordon, R., Lake, C., Hart, W., 2009. Adsorption, sedimentation and inactivation of E. coli within wastewater treatment wetlands. Water Research 43, 4370–4380.

Boutin, C., Liénard, A., 2003. Constructed wetlands for wastewater treatment: the French experience. In: 1st International Seminar on the Use of Aquatic Macrophytes for Wastewater Treatment in Constructed Wetlands, 5–10 May, Lisbon, Portugal.

Boutin, C., Liénard, A., Esser, D., 1997. Development of a new generation of reed-bed filters in France: first results. Water Science and Technology 35 (5), 315–322.

Braeckevelt, M., Seeger, E.M., Paschke, H., Kuschk, P., Kaestner, M., 2011a. Adaptation of a Constructed Wetland to simultaneous treatment of monochlorobenzene and perchloroethene. International Journal of Phytoremediation 13, 998–1013.

Braeckevelt, M., Reiche, N., Trapp, S., Wiessner, A., Paschke, H., Kuschk, P., Kaestener, M., 2011b. Chlorobenzene removal efficiencies and removal processes in a pilot-scale constructed wetland treating contaminated groundwater. Ecological Engineering 37, 903–913.

Bragato, C., Brix, H., Malagoli, M., 2006. Accumulation of nutrients and heavy metals in *Phragmites australis* (Cav.) *Trin. ex Steudel* and *Bolboschoenus maritimus* (L.) Palla in a constructed wetland of the Venice lagoon watershed. Environmental Pollution 144 (3), 967–975.

Bragato, C., Achiavon, M., Polese, R., Ertani, A., Pittarello, M., Malagoli, M., 2009. Seasonal variations of Cu, Zn, Ni and Ce concentration in *Phragmites australis* (Cav.) Trin ex steudel in a constructed wetland of North Italy. Desalination 247, 36–45.

Breen, P.F., 1990. A mass balance method for assessing the potential of artificial wetlands for wastewater treatment. Water Research 24 (6), 689–697.

Breen, P.F., 1997. The performance of vertical flow experimental wetland under a range of operational formats and environmental conditions. Water Science and Technology 35 (5), 167–174.

Breen, P.F., Chick, A.J., 1995. Rootzone dynamics in constructed wetlands receiving wastewater: a comparison of vertical and horizontal flow systems. Water Science and Technology 32 (3), 281–290.

Bresler, E., 1973. Simultaneous transport of solutes and water under transient unsaturated flow conditions. Water Resources Research 9, 975–986.

Brissaud, F., 2007. Low technology systems for wastewater perspectives. Water Science and Technology 55 (7), 1–9.

Brisson, J., Chazarenc, F., 2009. Maximizing pollutant removal in constructed wetlands: should we pay more attention to macrophyte species selection? Science of the Total Environment 407, 3923–3930.

Brix, H., 1994a. Use of constructed wetlands in water pollution control: historical development, present status and future perspectives. Water Science and Technology 30 (8), 209–223.

Brix, H., 1994b. Functions of macrophytes in constructed wetlands. Water Science and Technology 29 (4), 71–78.

Brix, H., 1997. Do macrophytes play a role in constructed treatment wetlands? Water Science and Technology 35 (5), 11–17.

Brix, H., 1999. How green are aquaculture, constructed wetlands and conventional wastewater treatment systems? Water Science and Technology 40 (3), 45–50.

Brix, H., 2004. Danish guidelines for small-scale constructed wetland systems for onsite treatment of domestic sewage. In: Proceedings of the 9th International Conference on Wetland Systems for Water Pollution Control, Avignon, France, 26–30 September, 1–8.

Brix, H., Arias, A.C., 2005. The use of vertical flow constructed wetlands for on-site treatment of domestic wastewater: New Danish guidelines. Ecological Engineering 25, 491–500.

Brix, H., Arias, C.A., Del Bubba, M., 2001. Media selection for sustainable phosphorus removal in subsurface flow constructed wetlands. Water Science and Technology 44 (11–12), 47–54.

Brix, H., Arias, C.A., Johansen, N.H., 2003. Experiments in a two-stage constructed wetland system: nitrification capacity and effects of recycling on nitrogen removal. In: Vymazal, J. (Ed.), Wetlands—Nutrients, Metals and Mass Cycling. Backhuys Publishers, Leiden, The Netherlands, pp. 237–258.

Brix, H., Schierup, H.-H., Arias, C.A., 2007. Twenty years experience with constructed wetland systems in Denmark - what did we learn? Water Science and Technology 56 (3), 63–68.

Brooks, A.S., Rozenwald, M.N., Geohring, L.D., Lion, L.W., Steenhuis, T.S., 2000. Phosphorus removal by wollastonite: A constructed wetland substrate. Ecological Engineering 15, 121–132.

Browne, W., Jenssen, P.D., 2005. Exceeding tertiary standards with a pond/reed bed system in Norway. Water Science and Technology 51 (9), 299–306.

Bruch, I., Fritsche, J., Bänninger, D., Alewell, U., Sendelov, M., 2011. Improving the treatment efficiency of constructed wetlands with zeolite-containing filter sands. Bioresource Technology 102, 937–941.

Bubba, M.D., Checchini, L., Pifferi, C., Zanieri, L., Lepri, L., 2004. Olive mill wastewater treatment by a pilot-scale subsurface horizontal flow (SSF-h) constructed wetland. Annali di Chimica 94, 875–887.

Bueno, J.A., Tsihrintzis, V.A., Alvarez, L., 1995. South Florida greenways: a conceptual framework for the ecological reconnectivity of the region. Landscape and Urban Planning 33, 247–266.

Bugg, B.M., 2002. An anion characterization of a constructed wetland used for chlorinated ethane remediation. Thesis, Department of Air Force, Air University, Air Force Institute of Technology, Ohio, USA.

Bulc, T.G., 2006. Long term performance of a constructed wetland for landfill leachate treatment. Ecological Engineering 26, 365–374.

Bulc, T.G., Ojstrsek, A., 2008. The use of constructed wetland for dye-rich textile wastewater treatment. Journal of Hazardous Materials 155 (1–2), 76–82.

Burgoon, P.S., Kirkbride, K.F., Henderson, M., Landon, E., 1997. Reed beds for biosolids drying in the arid North western United States. Water Science and Technology 35 (5), 287–292.

Calheiros, C.S.C., Rangel, A.O.S.S., Castro, P.M.L., 2007. Constructed wetland systems vegetated with different plants applied to the treatment of tannery wastewater. Water Research 41, 1790–1798.

Calheiros, C.S.C., Rangel, A.O.S.S., Castro, P.M.L., 2009a. Treatment of industrial wastewater with two-stage constructed wetlands planted with *Typha latifolia* and *Phragmites australis*. Bioresource Technology 100, 3205–3213.

Calheiros, C.S.C., Duque, A.F., Moura, A., Henriques, I.S., Correia, A., Rangel, A.O.S.S., Castro, P.M.L., 2009b. Changes in the bacterial community structure in two-stage constructed wetlands with different plants for industrial wastewater treatment. Bioresource Technology 100, 3228–3235.

Calheiros, C.S.C., Teixeira, A., Pires, C., Franco, A.R., Duque, A.F., Crispim, L.F.C., Moura, S.C., Castro, P.M.L., 2010. Bacterial community dynamics in horizontal flow constructed wetlands with different plants for high salinity industrial wastewater polishing. Water Research 44, 5032–5038.

Calheiros, C.S.C., Teixeira, A., Pires, C., Franco, A.R., Duque, A.F., Crispim, L.F.C., Moura, S.C., Castro, P.M.L., 2012. Use of constructed wetland systems with *Arundo* and *Sarcocornia* for polishing high salinity tannery wastewater. Journal of Environmental Management 95, 66–71.

Campbell, W.H., 2000. Sludge management—future issues and trends. Water Science and Technology 41 (8), 1–8.

Campbell, D.A., Cole, C.A., Brooks, R.P., 2002. A comparison of created and natural wetlands in Pennsylvania, USA. Wetlands Ecology and Management 10, 41–49.

Canga, E., Dal Santo, S., Pressl, A., Borin, M., Langergraber, G., 2011. Comparison of nitrogen elimination rates of different constructed wetland designs. Water Science and Technology 64 (5), 1122–1129.

Carleton, J.N., Grizzard, T.J., Godrej, A.N., Post, H.E., 2001. Factors affecting the performance of stormwater Treatment Wetlands. Water Research 35 (6), 1552–1562.

Caselles-Osorio, A., García, J., 2007. Impact of different feeding strategies and plant presence on the performance of shallow horizontal subsurface-flow constructed wetlands. Science of the Total Environment 378 (3), 253–262.

Caselles-Osorio, A., Puigagut, J., Segú, E., Vaello, N., Granés, F., García, D., García, J., 2007. Solids accumulation in six full-scale subsurface flow constructed wetlands. Water Research 41, 1388–1398.

Ceccanti, B., Bianchi, V., Iannelli, R., Peruzzi, E., Masciandaro, G., 2011. Phytotreatment of sludges (Phragmites australis) for their reuse in the environment. Water Science and Technology 64 (6), 1233–1238.

Ceustermans, A., Coosemans, J., Ryckeboer, J., 2010. Compost microbial activity related to compost stability. In: Insam, H., Franke-Whittle, I., Goberna, M. (Eds.), Microbes at Work. Springer-Verlag, Heidelberg, pp. 115–134.

Chabaud, S., Andres, Y., Lakel, A., Le Cloirec, P., 2006. Bacteria removal in septic effluent: influence of biofilm and protozoa. Water Research 40, 3109–3114.

Chabot, R., Bouarfa, S., Zimmer, D., Chaumont, C., Duprez, C., 2002. Sugarcane transpiration with shallow water table: sap flow measurements and modelling. Agricultural Water Management 54, 17–36.

Chan, S.Y., Tsang, Y.F., Chua, H., 2008. Domestic wastewater treatment using tidal-flow cinder bed with *Cyperus alternifolius*. Aquatic Ecosystem Health & Management 11 (2), 206–211.

Chang, J.-j., Wu, S.-q., Dai, Y.-r., Liang, W., Wu, Z.-b., 2012. Treatment performance of integrated vertical-flow constructed wetland plots for domestic wastewater. Ecological Engineering 44, 152–159.

Chazarenc, F., Gagnon, V., Comeau, Y., Brisson, J., 2009. Effect of plant and artificial aeration on solids accumulation and biological activities in constructed wetlands. Ecological Engineering 35, 1005–1010.

Chen, G., Yue, P.L., Mujumdar, A.S., 2002. Sludge dewatering and drying. Drying Technology 20 (4–5), 883–916.

Chen, T.Y., Kao, C.M., Yeh, T.Y., Chien, H.Y., Chao, A.C., 2006. Application of a constructed wetland for industrial wastewater treatment: A pilot-scale study. Chemosphere 64, 497–502.

Chen, Z.M., Chen, B., Zhou, J.B., Li, Z., Zhou, Y., Xi, X.R., Lin, C., Chen, G.Q., 2008. A vertical subsurface-flow constructed wetland in Beijing. Communications in Nonlinear Science and Numerical Simulation 13, 1986–1997.

Chen, X., Pauly, U., Rehfus, S., Bester, K., 2009a. Personal care compounds in a reed bed sludge treatment system. Chemosphere 76, 1094–1101.

Chen, X., Pauly, U., Rehfus, S., Bester, K., 2009b. Removal of personal care compounds from sewage sludge in reed bed contained (lysimeter) studies—effects of macrophytes. Science of the Total Environment 407, 5743–5749.

Chen, G.Q., Shao, L., Chen, Z.M., Li, Z., Zhang, B., Chen, H., Wu, Z., 2011. Low-carbon assessment for ecological wastewater treatment by a constructed wetland in Beijing. Ecological Engineering 37, 622–628.

Cheng, S.P., Grosse, W., Karrenbrock, F., Thoennessen, M., 2002. Efficiency of constructed wetlands in decontamination of water polluted by heavy metals. Ecological Engineering 18, 317–325.

Cheng, B., Hu, C.W., Zhao, Y.J., 2011. Effects of plants development and pollutant loading on performance of vertical subsurface flow constructed wetlands. International Journal of Environmental Science and Technology 8 (1), 177–186.

Cheremisinoff, N.P., 2002. Handbook of water and wastewater treatment technologies. Butterworth-Heinemann, Reed Elsevier Group, Woburn, MA, USA.

Choudhary, A.K., Kumar, S., Sharma, C., 2013. Removal of chlorophenolics from pulp and paper mill wastewater through Constructed Wetland. Water Environment Research 85 (1), 54–62.

Coleman, J., Hench, K., Garbutt, K., Sexstone, A., Bissonnette, G., Skousen, J., 2001. Treatment of domestic wastewater by three plant species in constructed wetlands. Water, Air, and Soil Pollution 128, 283–295.

Collins, B., McArthut, J.V., Sharitza, R.R., 2004. Plant effects on microbial assemblages and remediation of acidic coal pile runoff in mesocosm treatment wetlands. Ecological Engineering 23, 107–115.

Cooper, P., 1999. A review of the design and performance of vertical-flow and hybrid reed bed treatment systems. Water Science and Technology 40 (3), 1–9.

Cooper, P., 2001. Nitrification and denitrification in hybrid constructed wetlands systems. In: Vymazal, J. (Ed.), Transformation of nutrients in Natural and Constructed Wetlands. Backhuys Publishers, Leiden, The Netherlands, pp. 256–270.

Cooper, P., 2005. The performance of vertical flow constructed wetland systems with special reference to the significance of oxygen transfer and hydraulic loading rates. Water Science and Technology 51 (9), 81–90.

Cooper, P., 2009. What can we learn from old wetlands? Lessons that have been learned and some that may have been forgotten over the past 20 years. Desalination 247, 12–27.

Cooper, P., Green, B., 1998. United Kingdom. In: Vymazal, J., Brix, H., Cooper, P.F., Green, M.B., Haberl, R. (Eds.), Constructed Wetlands for Wastewater Treatment in Europe. Backhuys Publishers, Leiden, pp. 315–335.

Cooper, P., Cooper, D., Edwards, J., Biddlestone, J., 2003. Treatment of sludges from sewage and agricultural wastewaters in sludge drying reed beds. In: Vymazal, J. (Ed.), Wetlands—Nutrients, Metals and Mass Cycling. Backhuys Publishers, Leiden, The Netherlands, pp. 201–213.

Cooper, P.F., Job, G.D., Green, M.B., Shutes, R.B.E., 1996. Reed Beds and Constructed Wetlands for Wastewater Treatment. Water Research Center Publications, Swindon, UK, 184p.

Cooper, P., Smith, M., Maynard, H., 1997. The design and performance of a nitrifying vertical-flow reed bed treatment system. Water Science and Technology 35 (5), 215–221.

Cooper, P., Griffin, P., Humphries, S., Pound, A., 1999. Design of a hybrid reed bed system to achieve complete nitrification and denitrification of domestic sewage. Water Science and Technology 40 (3), 283–289.

Corominas, L., Foley, J., Guest, J.S., Hospido, A., Larsen, H.F., Morera, S., Shaw, A., 2013. Life cycle assessment applied to wastewater treatment: state of the art. Water Research 47, 5480–5492.

Costanza, R., D'Arge, R., De Groot, R., Farber, S., Grasso, M., Hannon, B., Limburg, K., Naeem, S., O'Neill, R.V., Paruelo, J., Raskin, R.G., Sutton, P., van den Belt, M., 1997. The total value of the world's ecosystem services and natural capital. Nature 387, 253–260.

Cottin, N., Merlin, G., 2010. Fate of chlorinated benzenes in laboratory peat and pozzolana filters. Water, Air, and Soil Pollution 213, 425–435.

Coulibaly, L., Kouakou, J., Savané, I., Gourène, G., 2008. Domestic wastewater treatment with a vertical completely drained pilot scale constructed wetland planted with *Amaranthus hybridus*. African Journal of Biotechnology 7 (15), 2656–2664.

Craun, G.F., 1985. A summary of waterborne illness transmitted through contaminated groundwater. Journal of Environmental Health 48, 122–127.

Crites, R.W., Middlebrooks, J., Reed, S.C., 2006. Natural Wastewater Treatment Systems. CRC Press, Taylor & Francis Group, Boca Raton, FL.

Cronk, J.K., Fennessy, M.S., 2001. Wetland Plants: Biology and Ecology. CRC Press, Boca Raton, Florida, USA.

Cui, L.-H., Liu, W., Zhu, X.-Z., Ma, M., Huang, X.-H., Xia, Y.-Y., 2006. Performance of hybrid constructed wetland systems for treating septic tank effluent. Journal of Environmental Sciences 18 (4), 665–669.

Cui, Y., Sun, T., Zhao, L., Jiang, T., Zhang, L., 2008. Performance of wastewater sludge ecological stabilization. Journal of Environmental Sciences 20, 385–389.

Cui, L.-H., Ouyang, Y., Chen, Y., Zhu, X.Z., Zhu, W.-L., 2009. Removal of total nitrogen by *Cyperus alternifolius* from wastewaters in simulated vertical-flow constructed wetlands. Ecological Engineering 35, 1271–1274.

Cui, L., Ouyang, Y., Lou, Q., Yang, F., Chen, Y., Zhu, W., Luo, S., 2010. Removal of nutrients from wastewater with Canna indica L. under different vertical-flow constructed wetland conditions. Ecological Engineering 36, 1083–1088.

Cui, L.-H., Zhu, X.-Z., Ouyang, Y., Chen, Y., Yang, F.-L., 2011. Total phosphorus removal from domestic wastewater with *Cyperus Alternifolius* in vertical-flow constructed wetlands at the microcosm level. International Journal of Phytoremediation 13, 692–701.

Cui, L., Feng, J., Ouyang, Y., Deng, P., 2012. Removal of nutrients from septic effluent with re-circulated hybrid tidal flow constructed wetland. Ecological Engineering 46, 112–115.

Czepiel, P., Crill, P., Harris, R., 1995. Nitrous oxide emissions from municipal wastewater treatment. Environmental Science and Technology 29 (9), 2352–2356.

Dan, T.H., Quang, L.N., Chiem, N.H., Brix, H., 2011. Treatment of high-strength wastewater in tropical constructed wetlands planted with Sesbania sesban: horizontal subsurface flow versus vertical downflow. Ecological Engineering 37, 711–720.

Davies, T.H., Cottingham, P.D., 1994. The use of constructed wetlands for treating industrial effluent (textile dyes). Water Science and Technology 29 (4), 227–232.

Davis, S.M., Ogden, J.C., 1994. Everglades—The Ecosystem and Its Restoration. St. Lucie Press, DelRay Beach, FL, USA, 826p.

Davison, L., Pont, D., Bolton, K., Headley, T., 2006. Dealing with nitrogen in subtropical Australia: seven case studies in the diffusion of ecotechnological innovation. Ecological Engineering 28, 213–223.

De Biase, C., Reger, D., Schmidt, A., Jechalke, S., Reiche, N., Martínez-Lavanchy, P.M., Rosell, M., Van Afferden, M., Maier, U., Oswald, S.E., Thullner, M., 2011. Treatment of volatile organic contaminants in a vertical flow filter: relevance of different removal processes. Ecological Engineering 37, 1292–1303.

De Castro Furtado, D.F., Philippi, L.S., Suntti, C., Haiml, C., Regi, R., 2012. Use of sludge drying reed beds for dewatering and mineralization of sludge from septic tanks. Revista Brasileira de Geografia Física 06, 1271–1283.

De Feo, G., 2007. Performance of vegetated and non-vegetated vertical flow reed beds in the treatment of diluted leachate. Journal of Environmental Science and Health, Part A 42, 1013–1020.

De Groot, R., Stuip, M., Finlayson, M., Davidson, N. 2006. Valuing Wetlands: Guidance for valuing the benefits derived from wetland ecosystem services. Ramsar Technical Report No. 3, CBD Technical Series No. 27. Ramsar Convention Secretariat, Gland, Switzerland.

De la Rubia, M.A., Perez, M., Romero, L.I., Sales, D., 2006. Effect of solids retention time (SRT) on pilot scale anaerobic thermophilic sludge digestion. Process Biochemistry 41, 79–86.

De la Varga, D., Díaz, M.A., Ruiz, I., Soto, M., 2013. Avoiding clogging in constructed wetlands by using anaerobic digesters as pre-treatment. Ecological Engineering 52, 262–269.

De Maeseneer, J.L., 1997. Constructed wetlands for sludge dewatering. Water Science and Technology 35 (5), 279–285.

De Stefani, G., Tocchetto, D., Salvato, M., Borin, M., 2011. Performance of a floating treatment wetland for in-stream water amelioration in NE Italy. Hydrobiologia 674, 157–167.

Decamp, O., Warren, A., 1998. Bacterivory in ciliates isolated from constructed wetlands (reed beds) used for wastewater treatment. Water Research 32, 1989–1996.

Decamp, O., Warren, A., 2000. Investigation of Escherichia coli removal in various designs of subsurface flow wetlands used for wastewater treatment. Ecological Engineering 14, 293–299.

Decamp, O., Warren, A., Sanchez, R., 1999. The role of ciliated protozoa in subsurface flow wetlands and their potential as bioindicators. Water Science and Technology 40 (3), 91–98.

Del Bubba, M., Arias, C.A., Brix, H., 2003. Phosphorus adsorption maximum of sands for use as media in subsurface-flow constructed reed beds as measured by the Langmuir isotherm. Water Research 37, 3390.

Del Bubba, M., Checchini, L., Pifferi, C., Zanieri, L., Lepri, L., 2004. Olive mill wastewater treatment by a pilot-scale subsurface flow (SSF-h) Constructed Wetland. Annali di Chimica 94, 875–887, Società Chimica Italiana.

Demchak, J., Morrow, J., Skousen, J., 2001. Treatment of acid mine drainage by four vertical flow wetlands in Pennsylvania. Geochemistry: Exploration, Environment, Analysis 1 (1), 71–80.

Deng, H., Ye, Z.H., Wong, M.H., 2004. Accumulation of lead, copper and cadmium by 12 wetland species thriving in metal-contaminated sites in China. Environmental Pollution 132, 29–40.

Dentel, K.S., 2001. Conditioning, thickening, and dewatering: research update/research needs. Water Science and Technology 44 (10), 9–18.

Dhir, B., Sharmila, P., Pardha Saradhi, P., 2009. Potential of aquatic macrophytes for removing contaminants from the environment. Critical Reviews in Environmental Science and Technology 39 (9), 754–781.

Dimitrova, I., Carlsson, H., 2011. Comparison between two techniques for sludge dewatering—trials at Slölunda and Klagshamn WWTPs. Vatten 67, 71–77.

Dittmer, U., Meyer, D., Langergraber, G., 2005. Simulation of a subsurface vertical flow constructed wetland for CSO treatment. Water Science and Technology 51, 225–232.

Dixon, A., Simon, M., Burkitt, T., 2003. Assessing the environmental impact of two options for small-scale wastewater treatment: comparing a reedbed and an aerated biological filter using a life cycle approach. Ecological Engineering 20 (4), 297–308.

Domingos, S., Dallas, S., Germain, M., Ho, G., 2009. Heavy metals in a constructed wetland treating industrial wastewater: distribution in the sediment and rhizome tissue. Water Science and Technology 60, 1425–1432.

Do, J.P., Daileader, K.L., Duran, M., 2007. Anaerobic digestion of biosolids at elevated thermophilic temperatures: inactivation of pathogens for Class A biosolids designation. In: LeBlanc, R.J., Laughton, P.J., Tyagi, R. (Eds.), Moving Forward Wastewater Biosolids Sustainability: Technical, Managerial, and Public Synergy. GMSC, New Brunswick, pp. 409–415.

Dominiak, D., Christensen, M.L., Keiding, K., Nielsen, P.H., 2011. Sludge quality aspects of full-scale reed bed drainage. Water Research 45, 6453–6460.

Dong, Z.D., Sun, T., 2007. A potential new process for improving nitrogen removal in constructed wetlands-Promoting coexistence of partial-nitrification and ANAMMOX. Ecological Engineering 31, 69–78.

Dong, H., Qiang, Z., Li, T., Jin, H., Chen, W., 2012. Effect of artificial aeration on the performance of vertical-flow constructed wetland treating heavily polluted river water. Journal of Environmental Sciences 24 (4), 596–601.

Dotro, G., Larsen, D., Palazolo, P., 2011. Treatment of chromium-bearing wastewaters with constructed wetlands. Water and Environment Journal 25, 241–249.

Drizo, A., Frost, A.C., Grace, J., Smith, A.K., 1999. Physico-chemical screening of phosphate-removing substrates for use in constructed wetland systems. Water Research 33 (17), 3595–3602.

Drizo, A., Forget, C., Chapuis, R.P., Comeau, Y., 2006. Phosphorus removal by electric arc furnace steel slag and serpentinite. Water Research 40, 1547.

Dugan, P., 1993. Wetlands in Danger. Michael Beasley, Reed International Books Limited, London.

Edwards, K.J., Gray, R.K., Cooper, J.D., Biddlestone, J.A., Willoughby, N., 2001. Reed bed dewatering of agricultural sludges and slurries. Water Science and Technology 44 (11), 551–558.

Edwards, K.R., Čižková, H., Zemanová, K., Šantrůčková, H., 2006. Plant growth and microbial processes in a constructed wetland planted with *Phalaris arundinacea*. Ecological Engineering 27, 153.

Eke, P.E., Scholz, M., 2008. Benzene removal with vertical-flow constructed treatment wetlands. Journal of Chemical Technology and Biotechnology 85, 55–63.

Elke, M., Lechner, M., 2012. Comparing the treatment efficiency of different wastewater treatment technologies in Uganda. Sustainable Sanitation Practice 12, 16–21.

Engloner, A.I., 2009. Structure, growth dynamics and biomass of reed (*Phragmites australis*) – A review. Flora 204, 331–346.

Epstein, E., 1997. The Science of Composting. Technomic Pub. Co., Lancaster, PA, USA.

Esser, D., Ricard, B., Fernandes, N., Merlin, G., 2004. Physical-chemical phosphorus removal in vertical flow reed bed treatment systems. In: Proceedings of the 9th IWA International Conference on Wetland Systems for Water Pollution Control, 26–30 September, Avignon, France, Vol. 2, 673–681.

Ettingh, A.C. 2002. Groundwater flow through a Constructed Treatment Wetland. Master Thesis, Air Force Institut of Technology Wright-Patterson, School of Engineering, AFB, OH.

European Commission, DG Environment, 2010a. Working Document, Sludge and Biowaste. Brussels, DG ENV C2/BZ/tb. Available from: <http://www.compostnetwork.info/wordpress/wp-content/uploads/2010/12/101021_ECN_bio-sludge-working-doc_comments.pdf>.

European Commission, DG Environment, 2010b. Environmental, economic and social impacts of the use of sewage sludge on land, Final Report, Part III: Project Interim Reports, Milieu Ltd., Belgium, Brussels.

European Commission, DG Environment, 2000. Working Document on Sludge, 3rd draft, Brussels. Available from http://ec.europa.eu/environment/waste/sludge/pdf/sludge_en.pdf.

European Commission, DG Environment, 2001. Working Document on Biological Treatment of Biowaste. 2nd draft. http://ec.europa.eu/environment/waste/sludge/pdf/sludge_en.pdf.

European Union, 1986. Council Directive on the protection of the environment, and in particular of the soil, when sewage sludge is used in agriculture, 86/278/EEC, Official Journal of the EC No L181/6-12, 4/7/1986.

European Union, 1991. Council Directive concerning urban waste-water treatment. 91/271/EEC, Official Journal of the EC No L135/40, 30/5/1991.

European Environment Agency (EEA), 1997. Sludge Treatment and Disposal: Management Approaches and Experiences. European Environment Agency, Copenhagen, Denmark.

Evans, T.D., 2009. Peak phosphorus – conserving the world's most essential resource. In: Proceedings of the 14th European Biosolids and Organic Resources Conference, 9–11 November, Leeds, UK.

Fan, J., Liang, S., Zhang, B., Zhang, J., 2013. Enhanced organics and nitrogen removal in batch-operated vertical flow constructed wetlands by combination of intermittent aeration and step feeding strategy. Environmental Science and Pollution Research 20, 2448–2455.

Farahani, H.J., Ahuja, L.R., 1996. Evapotranspiration modeling of partial canopy/residue-covered fields. Transactions of the American Society of Agricultural and Biological Engineers 39, 2051–2064.

Farahani, H.J., Bausch, W.C., Aiken, R.M., Ahuja, L.R., 1996. Evapotranspiration in system-wide agricultural models. In: Evapotranspiration and Irrigation Scheduling: Proceedings of the International Conference, American Society of Agricultural and Biological Engineers.

Farahbakshazad, N., Morrison, G.M., 1997. Ammonia removal processes for urine in an upflow macrophyte system. Environmental Science and Technology 31, 3314–3317.

Farahbakshazad, N., Morrison, G.M., 2000. A constructed vertical macrophyte system for the retention of nitrogen in agricultural runoff. Environmental Technology 21, 217–223.

Farahbakshazad, N., Morrison, G.M., 2003. Phosphorus removal in a vertical upflow constructed wetland system. Water Science and Technology 48 (5), 43–50.

Farahbakshazad, N., Morrison, G.M., Filho, E.S., 2000. Nutrient removal in a vertical upflow wetland in Piracicaba, Brazil. AMBIO: A Journal of the Human Environment 29 (2), 74–77.

Farnet, A.M., Prudent, P., Ziarelli, F., Domeizel, M., Gros, R., 2009. Solid-state 13C NMR to assess organic matter transformation in a subsurface wetland under cheese-dairy farm effluents. Bioresource Technology 100, 4899–4902.

Farré, M., Barceló, D., 2001. Toxicity testing of wastewater and sewage sludge by biosensors, bioassays and chemical analysis. Trends in Analytical Chemistry 22 (5), 299–310.

Farré, M., Pasini, O., Alonso, M.C., Castillo, M., Barceló, D., 2003. Toxicity assessment of organic pollution in wastewaters using a bacterial biosensor. Analytica Chimica Acta 426, 155–165.

Farzadkia, M., Norieh, N., 2005. Efficiency of aerobic digester for sludge stabilization in Serkan city sewage treatment plant. In: Proceedings of the 9th International Conference on Environmental Science and Technology, Rhodes Island, Greece, 1-3 September.

Faulwetter, J.L., Gagnon, V., Sundberg, C., Chazarenc, F., Burr, M.D., Brisson, J., Camper, A.K., Stein, O.R., 2009. Microbial processes influencing performance of treatment wetlands: A review. Ecological Engineering 35, 987–1004.

Fediuc, E., Erdei, L., 2002. Physiological and biochemical aspects of cadmium toxicity and protective mechanisms induced in *Phragmites australis* and *Typha latifolia*. Journal of Plant Physiology 159, 265–271.

Fei, M., Li, J., Ting, Z., 2010. Reversing clogging in vertical flow constructed wetlands by backwashing treatment. Advanced Materials Research 129–131, 1064–1068.

Fei, Z., Wei, L., Tao, Y., Shui, P.C., Feng, H., Zhen, B.W., 2011. Removal efficiency and balance of nitrogen in a recirculating aquaculture system integrated with constructed wetlands. Journal of Environmental Science and Health, Part A 46, 789–794.

Feo, G.D., Belgiorno, V., 2005. Treatment of high strength wastewater with vertical flow constructed wetland filters. Water Science and Technology 51, 139–146.

Ferrer, I., Vázquez, F., Font, X., 2012. Long term operation of a thermophilic anaerobic reactor: process stability and efficiency at decreasing sludge retention time. Bioresource Technology 101, 2972–2980.

Finlayson, C.M. Spiers, A.G. (eds), 1999. Global review of wetland resources and priorities for wetland inventory. Supervising Scientist Report 144 / Wetlands International Publication 53, Supervising Scientist, Canberra.

Foladori, P., Ortigara, A.R.C., Ruaben, J., Andreottola, G., 2012. Influence of high organic loads during the summer period on the performance of hybrid constructed wetlands (VSSF - HSSF) treating domestic wastewater in the Alps region. Water Science and Technology 65 (5), 890–897.

Ford, K.L. 2003. Passive treatment systems for acid mine drainage. U.S. Bureau of Land Management Papers. Paper 19. http://digitalcommons.unl.edu/usblmpub/19.

Fratamico, P.M., Whiting, R.S., 1995. Ability of Bdellovibrio bacteriovorus 109J to lyse Gram-negative food-borne pathogenic and spoilage bacteria. Journal of Food Protection 58, 160–164.

Environmental, F.S.A., 2002. Solids Separation Systems for the Pig Industry. Case Study 8, Belt Press, April.

Fuchs, V.J., Gierke, J.S., Mihelcic, J.R., 2007. Hydraulic model comparison of vertical-flow regimes in constructed wetlands. In: Proceedings of the Water Environment Federation, WEFTEC 2007: Session 11 through Session 20. Water Environment Federation, pp. 1039–1055.

Fuchs, V.J., Mihelcic, J.R., Gierke, J.S., 2011. Life cycle assessment of vertical and horizontal flow constructed wetlands for wastewater treatment considering nitrogen and carbon greenhouse gas emissions. Water Research 45, 2073–2081.

Gaboutloeloe, G.K., Chen, S., Barber, M.E., Stöchle, C.O., 2009. Combinations of Horizontal and Vertical Flow Constructed Wetlands to improve nitrogen removal. Water Air and Soil Pollution: Focus 9, 279–286.

Gagnon, V., Chazarenc, F., Kõiv, M., Brisson, J., 2012. Effect of plant species on water quality at the outlet of a sludge treatment wetland. Water Research 46, 5305–5315.

Gagnon, V., Chazarenc, F., Comeau, Y., Brisson, J., 2013. Effect of plant species on sludge dewatering and fate of pollutants in sludge treatment wetlands. Ecological Engineering 61, 593–600, in press.

Gao, J., Pedersen, J.A., 2005. Adsorption of sulfonamide antimicrobial agents to clay minerals. Environmental Science and Technology 39, 9509–9516.

Gao, R.Y., Shao, L., Li, J.S., Guo, S., Han, M.Y., Meng, J., Liu, J.B., Xu, F.X., Lin, C., 2012. Comparison of greenhouse gas emission accounting for a constructed wetland wastewater treatment system. Ecological Informatics 12, 85–92.

García, J., Aguirre, P., Mujeriego, R., Huang, Y., Ortiz, L., Bayona, J.M., 2004. Initial contaminant removal performance factors in horizontal flow reed beds used for treating urban wastewater. Water Research 38, 1669–1678.

García, J., Rousseau, D.P.L., Morató, J., Lesage, E., Matamoros, V., Bayona, J.M., 2010. Contaminant removal processes in subsurface-flow constructed wetlands: a review. Critical Reviews in Environmental Science and Technology 40, 561–661.

García-Pérez, A., Harrison, M., Grant, B., 2011. Recirculating vertical flow constructed wetland for on-site sewage treatment: an approach for a sustainable ecosystem. Journal of Water and Environmental Technology 9 (1), 39–46.

García-Pérez, A., Harrison, M., Grant, B., Chivers, C., 2013. Microbial analysis and chemical composition of maize (*Zea mays*, L.) growing on a recirculating vertical flow constructed wetland treating sewage on-site. Ecological Engineering 114, 351–356.

Gasiunas, V., Strusevicius, Z., Struseviciene, M.S., 2005. Pollutant removal by horizontal subsurface flow constructed wetlands in Lithuania. Journal of Environmental Science and Health, Part A: Toxic/Hazardous Substances and Environmental Engineering 40, 1467–1478.

Georges, K., Thornton, A., Sadler, R., 2009. Transforming Wastewater Treatment to Reduce Carbon Emissions: Resource efficiency programme, Evidence Directorate. Report SC070010/R2. Environmental Agency, Rio House, Waterside Drive, Aztec West, Almondsbury, Bristol, UK, December.

Gerba, C.P., Thurston, J.A., Falabi, J.A., Watt, P.M., Karpiscak, M.M., 1999. Optimization of artificial wetland design for the removal of indicator microorganisms and pathogenic protozoa. Water Science and Technology 40, 363–368.

Ghermandi, A., Van den Bergh, J.C.J.M., Brander, L.M., De Groot, H.L.F., Nunes, P., 2010. Values of natural and human-made wetlands: A meta-analysis. Water Resources Research 46 (W12516), 12 pp.

Ghosh, D., Gopal, B., 2010. Effect of hydraulic retention time on the treatment of secondary effluent in a subsurface flow constructed wetland. Ecological Engineering 36, 1044–1051.

Gikas, G.D., Tsihrintzis, V.A., 2010. On-site treatment of domestic wastewater using a small-scale horizontal subsurface flow constructed wetland. Water Science and Technology 62 (3), 603–614.

Gikas, G.D., Tsihrintzis, V.A., 2012. A small-size vertical flow constructed wetland for on-site treatment of household wastewater. Ecological Engineering 44, 337–343.

Gikas, G.D., Tsihrintzis, V.A., Petalas, C., Skias, S., 2006. Constructed wetlands for wastewater treatment in Korestia, Kastoria district. In: Proceedings of the Conference "Small Scale Wastewater Treatment Facilities" April 8-9, Portaria, Volospp. 201–208 (in Greek).

Gikas, G.D., Tsihrintzis, V.A., Chatzigeorgiou, G., Kapollas, D., 2010. Constructed wetlands for wastewater treatment of Kyprinos settlement. In: Proceedings of the Conference "Small Scale Water and Wastewater Treatment Facilities" May 14-15, Skiathos Inland, pp. 25–31 (in Greek).

Gikas, G.D., Tsihrintzis, V.A., Akratos, C.S., 2011. Performance and modeling of a vertical flow constructed wetland – maturation pond system. Journal of Environmental Science and Health, Part A 46 (7), 692–708.

Gikas, G.D., Tsakmakis, I.D., Tsihrintzis, V.A., 2013. Treatment of olive mill wastewater in pilot-scale natural systems. In: Proceedings 8th International Conference of EWRA "Water Resources Management in an Interdisciplinary and Changing Context", Porto, Portugal, 26-29 June 2013pp. 1207–1216, paper #232.

Giraldi, D., Iannelli, R., 2009a. Measurements of water content distribution in vertical subsurface flow constructed wetlands using a capacitance probe: benefits and limitations. Desalination 243, 182–194.

Giraldi, D., Iannelli, R., 2009b. Short-term water content analysis for the optimization of sludge dewatering in dedicated constructed wetlands (reed bed systems). Desalination 246, 92–99.

Giraldi, D., de'Michieli Vitturi, M., Zaramella, M., Marion, A., Iannelli, R., 2009a. Hydrodynamics of vertical subsurface flow constructed wetlands: tracer tests with rhodamineWT and numerical modelling. Ecological Engineering 35, 265–273.

Giraldi, D., Masciandaro, G., Peruzzi, E., Bianchi, V., Peruzzi, P., Ceccanti, B., Iannelli, R., 2009b. Hydraulic and biochemical analyses on full-scale sludge consolidation reed beds in Tuscany (Italy). Water Science and Technology 60 (5), 1209–1216.

Giraldi, D., Vitturi, M.M., Iannelli, R., 2010. FITOVERT: A dynamic numerical model of subsurface vertical flow constructed wetlands. Environmental Modelling and Software 25, 633–640.

Gkika, D., Gikas, G.D., Tsihrintzis, V.A., 2014. Cost and environmental impact of constructed wetland facilities. In: Proceedings of IWA-Regional Symposium on Water, Wastewater and Environment: Traditions and Culture, 22-24 March 2014, Patras, Greece.

Goldman, G., Starosvetsky, J., Armon, R., 2009. Inhibition of biofilm formation on UF membrane by use of specific bacteriophages. Journal of Membrane Science 342, 145–152.

Gonçalves, R.F., Luduvice, M., von Sperling, M., 2007. Sludge thickening and dewatering. In: Cleverson, V.A., von Sperling, M., Fernandes, F. (Eds.), Sludge Treatment and Disposal. IWA Publishing, London, UK.

Gomez Cerezo, R., Suarez, M.L., Vidal-Abarca, M.R., 2001. The performance of a multi-stage system of constructed wetlands for urban wastewater treatment in a semiarid region of SE Spain. Ecological Engineering 16 (4), 501–517.

Gorsalitz, E.K. 2012. Comparative removal of pharmaceuticals and antimicrobials in conventional and constructed wetland wastewater treatment in cold climate. Thesis, University of Iowa, 2012.

Gottschall, R., Boutin, C., Crolla, A., Kinsley, C., Champagne, P., 2007. The role of plants in the removal of nutrients at a constructed wetland treating agricultural (dairy) wastewater Ontario, Canada. Ecological Engineering 29, 154–163.

Grafias, P., Xekoukoulotakis, N.P., Mantzavinos, D., Diamadopoulos, E., 2010. Pilot treatment of olive pomace leachate by vertical-flow constructed wetland and electrochemical oxidation: An efficient hybrid process. Water Research 44, 2773–2780.

Green, M., Friedler, E., Ruskol, Y., Safrai, I., 1997a. Investigation of alternative method for nitrification in constructed wetlands. Water Science and Technology 35 (5), 63–70.

Green, M.B., Griffin, P., Seabridge, J.K., Dhobie, D., 1997b. Removal of bacteria in subsurface flow wetlands. Water Science and Technology 35, 109–116.

Green, M., Friedler, E., Safrai, I., 1998. Enhancing nitrification in vertical flow constructed wetlands utilizing a passive air pump. Water Research 23 (12), 3513–3520.

Green, M., Shaul, N., Beliavski, M., Sabbah, I., Ghattas, B., Tarre, S., 2006. Minimizing land requirement and evaporation in small wastewater treatment systems. Ecological Engineering 26, 266–271.

Greenway, M., Woolley, A., 1999. Constructed wetlands in Queensland: performance efficiency and nutrient bioaccumulation. Ecological Engineering 12 (1–2), 39–55.

Gross, A., Shmueli, O., Ronen, Z., Raveh, E., 2007. Recycled vertical flow constructed wetland (RVFCW) – a novel method of recycling greywater for irrigation in small communities and households. Chemosphere 66, 916–923.

Gross, A., Sklarz, M.Y., Yakirevich, A., Soares, M.I.M., 2008. Small scale recirculating vertical flow constructed wetland (RVFCW) for the treatment and reuse of wastewater. Water Science and Technology 58 (2), 487–494.

Guidi, G.V., Pini, R., Scotto, A., 1990. Physical characterization of a phytodewatered anaerobic sludge. Biol. Wastes 31, 303–310.

Gunes, K., Tuncsiper, B., Masi, F., Ayaz, S., Leszczynska, D., Hecan, B.F., Ahmad, H., 2011. Construction and maintenance cost analyzing of constructed wetland systems. Water Practice & Technology 6 (3). http://dx.doi.org/10.2166/wpt.2011.043.

Gustavsson, L., Engwall, M., 2012. Treatment of sludge containing nitro-aromatic compounds in reed-bed mesocosms—water, BOD, carbon and nutrient removal. Waste Management 32, 104–109.

Gustavsson, L., Hollert, H., Jönsson, S., van Bavel, B., Engwall, M., 2007. Reed beds receiving industrial sludge containing nitroaromatic compounds. Effects of outgoing water and bed material extracts in the umu-C genotoxicity assay, DR-CALUX assay and on early life stage development in Zebrafish (Danio rerio). Environ. Sci. Pollut. Res 14 (3), 202–211.

Haberl, R., Perfler, R., Mayer, H., 1995. Constructed wetlands in Europe. Water Science and Technology 32 (3), 305–315.

Haberl, R., Perfler, R., Laber, J., Grabher, D., 1998. Austria. In: Vymazal, J., Brix, H., Cooper, P.F., Green, M.B., Haberl, R. (Eds.), Constructed Wetlands for Wastewater Treatment in Europe. Backhuys Publishers, Leiden, pp. 67–76.

Haberl, R., Grego, S., Langergraber, G., Kadlec, R.H., Cicalini, A.-R., Dias, S.M., Novais, J.M., Aubert, S., Gerth, A., Thomas, H., Hebner, A., 2003. Constructed wetlands for the treatment of organic pollutants. Journal of Soils and Sediments 3 (2), 109–124.

Hallberg, K.B., Johnson, D.B., 2005. Microbiology of a wetland ecosystem constructed to remediate mine drainage from a heavy metal mine. Science of the Total Environment 338, 53–66.

Hammer, D.A., Bastian, R.K., 1989. Wetlands ecosystems: natural water purifiers? In: Hammer, D. A. (Ed.), Constructed wetlands for wastewater treatment. Lewis Publisher, USA.

Hansen, J.I., Anderson, F.O., 1981. Effects of Phragmites australis roots and rhizomes on redox potentials, nitrification and bacterial numbers in the sediment. In: 9th Nordic Symposium, Malma, Sweden.

Hardej, M., Ozimek, T., 2002. The effect of sewage sludge flooding on growth and morphometric parameters of Phragmites australis (Cav.) Trin. ex Steudel. Ecological Engineering 18, 343–350.

Harvey, P.J., Campanella, B.F., Castro, P.M.L., Harms, H., Lichtfouse, E., Schäffner, A.R., Smrcek, S., Werch-Reichhart, D., 2002. Phytoremediation of polyaromatic hydrocarbons, anilines and phenols. Environmental Science and Pollution Research 9 (1), 29–47.

Hathaway, J.M., Cook, M.J., Evans, R.O., 2010. Nutrient removal capability of a constructed wetland receiving groundwater contaminated by swine lagoon seepage. Transaction of the American Society of Agricultural and Biological Engineers 53, 1–9.

Hawkes, H.A., 1983. Activated sludge. In: Curds, C.R., Hawkes, H.A. (Eds.), Ecological Aspects of Used Water Treatment1, Academic, London.

He, L.-S., Liu, H.-L., Xi, B.-D., Zhu, Y.-B., 2006a. Enhancing treatment efficiency of swine wastewater by effluent recirculation in vertical-flow constructed wetland. Journal of Environmental Sciences 18 (2), 221–226.

He, L.-S., Liu, H.-L., Xi, B.-D., Zhu, Y.-B., 2006b. Effects of effluent recirculation in vertical-flow constructed wetland on treatment efficiency of livestock wastewater. Water Science and Technology 54 (11–12), 137–146.

He, W., Yongfeng, J., 2009. Bioaccumulation of heavy metals by *Phragmites australis* cultivated in synthesized substrates. Journal of Environmental Sciences 21, 1409–1414.

Headley, T.R., Tanner, C.C., 2012. Constructed Wetlands with floating emergent macrophytes: an innovative stormwater treatment technology. Critical Reviews in Environmental Science and Technology 42, 2261–2310.

Heinss, U., Koottatep, T., 1998. Use of Reed Beds for Faecal Sludge Dewatering—A Synopsis of Reviewed Literature and Interim Results of Pilot Investigations with Septage Treatment in Bangkok, Thailand. Asian Institute of Technology, Bangkok, Thailand, Urban Env. Engineering & Management Program, Swiss Federal Institute for Environmental Science & Technology.

Henrichs, M., Langergraber, G., Uhl, M., 2007. Modelling of organic matter degradation in constructed wetlands for treatment of combined sewer overflow. Science of the Total Environment 380, 196–209.

Henze, M., Gardy Jr., C.P.L., Gujer, W., Marais, G.V.R., Matsuo, T. 1987. Activated Sludge Model No. 1. Scientific and Technical Report No. 1. London, UK, IAWPRC.

Heritage, A., Pistillo, P., Sharma, K.P., Lantzke, I.R., 1995. Treatment of primary-settled urban sewage in pilot-scale vertical flow wetland filters: comparison of four emergent macrophyte species over a 12 month period. Water Science and Technology 32 (3), 295–304.

Herouvim, E., Akratos, C.S., Tekerlekopoulou, A., Vayenas, D.V., 2011. Treatment of olive mill wastewater in pilot-scale vertical flow constructed wetland. Ecological Engineering 37, 931–939.

Hijosa-Valsero, M., Matamoros, V., Martín-Villacorta, J., Bécares, E., Bayona, J.M., 2010a. Assessment of full-scale natural systems for the removal of PPCPs from wastewater in small communities. Water Research 44, 1429–1439.

Hijosa-Valsero, M., Matamoros, V., Sidrach-Cardona, R., Martín-Villacorta, J., Bécares, E., Bayona, J.M., 2010b. Comprehensive assessment of the design configuration of constructed wetlands for the removal of pharmaceuticals and personal care products from urban wastewater. Water Research 44, 3669–3678.

Hogg, D., Barth, J., Favoino, E., Centemero, M., Caimi, V., Amlinger, F., Devliegher, W., Brinton, W., Antler, S., 2002. Comparison of Compost Standards within the EU, North America and Australasia. The Waste and Resources Action Programme, Oxon, UK, pp. 1–97.

Hua, G.F., Zhu, W., Zhao, L.F., Huang, J.Y., 2010. Clogging pattern in vertical-flow constructed wetlands: insight from a laboratory study. Journal of Hazardous Materials 180, 668–674.

Huang, X., Liu, C., Wang, Z., Gao, C., Zhu, G., Liu, L., 2013. The effects of different substrates on ammonium removal in constructed wetlands: a comparison of their physicochemical characteristics and ammonium-oxidizing prokaryotic communities. Clean – Soil, Air, Water 41 (3), 283–290.

Hubbard, R.K., 2010. Floating vegetated mats for improving surface water quality. In: Shad, V. (Ed.), Emerging Environmental Technologies. Springer Science+Business Media.

Hübner, T.M., Tischer, S., Tanneberg, H., Kuschk, P., 2000. Influence of phenol and phenanthrene on the growth of *Phalaris arundinacea* and *Phragmites australis*. International Journal of Phytoremediation 2 (4), 331–342.

Hunt, P.G., Poach, M.E., 2001. State of the art for animal wastewater treatment in constructed wetlands. Water Science and Technology 44, 19–25.

Hunt, P.G., Matheny, T.A., Szogi, A.A., 2003. Denitrification in constructed wetlands used for treatment of swine wastewater. Journal of Environmental Quality 32, 727–735.

Hussain, S.A. 2011. Removal of poultry pharmaceuticals by Constructed Wetlands. Thesis, Department of Bioresource Engineering, Macdonald Campus of McGill University, Montreal, Canada, July.

Iamchaturapatr, J., Yi, S.W., Rhee, J.S., 2007. Nutrient removals by 21 aquatic plants for vertical free surface-flow (VFS) constructed wetland. Ecological Engineering 29 (3), 287–293.

Iannotti, D.A., Pang, T., Toth, B.L., Elwell, D.L., Keener, H.M., Hoitink, H.A.J., 1993. A quantitative respirometric method for monitoring compost stability. Compost Science and Utilization 1, 52–65.

Ibekwe, A.M., Grieve, C.M., Lyon, S.R., 2003. Characterization of microbial communities and composition in constructed dairy wetland wastewater effluent. Applied Environmental Microbiology 69, 5060–5069.

Imfeld, G., Braeckevelt, M., Kuschk, P., Richnow, H.H., 2009. Monitoring and assessing processes of organic chemicals removal in constructed wetlands. Chemosphere 74, 349–362.

IPCC, 2001. CH_4 and N_2O emissions from waste water handling. In: Background Paper for Good Practice Guidance and Uncertainty Management in National Greenhouse Gas Inventories. Intergovernmental Panel on Climate Change Taskforce on National Greenhouse Gas Inventories.

IPCC, 2007. Intergovernmental Panel on Climate Change, Fourth Assessment Report (AR4). Intergovernmental Panel on Climate Change, Geneva, Switzerland, In: Pachauri, R.K., Reisinger, A. (Eds.), p. 104.

Iwasaki, I., 1973. Some notes on sand filtration. Journal of American Water Works Association 29, 1591–1602.

Jamieson, R., Joy, D.M., Lee, H., Kostaschuk, R., Gordon, R., 2005. Transport and deposition of sediment associated *Escherichia coli* in natural streams. Water Research 39, 2665–2675.

Janzen, N., Banzhaf, S., Scheytt, T., Bester, K., 2009. Vertical flow soil filter for the elimination of micro pollutants from storm and waste water. Chemosphere 77 (10), 1358–1365.

Jetten, M.S.M., 2001. New pathways for ammonia conversion in soil and aquatic systems. Plant and Soil 230, 9–19.

Jia, W., Zhang, J., Wu, J., Xie, H., Zhang, B., 2010. Effect of intermittent operation on contaminant removal and plant growth in vertical flow constructed wetlands: a microcosm experiment. Desalination 262, 202–208.

Johnson, D.B., Hallberg, K.B., 2005a. Acid mine drainage remediation options: a review. Science of the Total Environment 338, 3–14.

Johnson, D.B., Hallberg, K.B., 2005b. Biogeochemistry of the compost bioreactor components of a composite acid mine drainage passive remediation system. Science of the Total Environment 338, 81–93.

Johnson, R., Pankow, J., Bender, D., Price, C., Zogorsky, J., 2000. MTBE to what extent will past releases contaminate community water supply wells? Environmental Science and Technology 34, 210A–217A.

Joss, A., Zabczynski, S., Göbel, A., Hoffmann, B., Löffler, D., McArdell, C.S., Ternes, T.A., Thomsen, A., Siegrist, H., 2006. Biological degradation of pharmaceuticals in municipal wastewater treatment: proposing a classification scheme. Water Research 40, 1686–1696.

Jury, W.A., Gardner, W.R., Gardner, W.H., 1991. Soil Physics. John Wiley and Sons Inc., New York, USA.

Justin, M.Z., Zupancic, M., 2009. Combined purification and reuse of landfill leachate by constructed wetland and irrigation of grass and willows. Desalination 24, 157–168.

Justin, M.Z., Vrhovsek, D., Stuhlbacher, A., Bulc, T.G., 2009. Treatment of wastewater in hybrid constructed wetland from the production of vinegar and packaging of detergents. Desalination 246, 100–109.

Kadlec, R.H., Knight, R.L., 1996. Treatment Wetlands. CRC Press/Lewis Publishers, Boca Raton, FL.

Kadlec, R.H., Zmarthie, L.A., 2010. Wetland treatment of leachate from a closed landfill. Ecolological Engineering 36, 946–957.

Kadlec, R.H., Tanner, C.C., Hally, V.M., Gibbs, M.M., 2005. Nitrogen spiralling in subsurface-flow constructed wetlands: Implications for treatment response. Ecological Engineering 25, 365–381.

Kadlec, R.H., Wallace, S.D., 2009. Treatment Wetlands, second ed. CRC Press, Boca Raton, FL.

Kalisz, L., Salbut, J., 2007. Dewatering of anaerobically stabilized sludge on reed beds. Ochrona Srodowiska I Zasobów Naturalnych 30, 105–115.

Kantawanichkul, S., Somprasert, S., 2005. Using a compact combined constructed wetland system to treat agricultural wastewater with high nitrogen. Water Science and Technology 51, 47–53.

Kantawanichkul, S., Pilaila, S., Tanapiyawanich, W., Tikampornpittaya, W., Kamkrua, S., 1999. Wastewater treatment by tropical plants in vertical-flow constructed wetlands. Water Science and Technology 40 (3), 173–178.

Kantawanichkul, S., Pingkul, K., Araki, H., 2008. Nitrogen removal by a combined subsurface vertical down-flow and up-flow constructed wetland System. In: Vymazal, J. (Ed.), Wastewater Treatment, Plant Dynamics and Management in Constructed and Natural Wetlands. Springer Science + Business Media.

Kantawanichkul, S., Kladprasert, S., Brix, H., 2009. Treatment of high-strength wastewater in tropical vertical flow constructed wetlands planted with *Typha angustifolia* and *Cyperus involucratus*. Ecological Engineering 35, 238–247.

Kapellakis, I.E., Paranychianakis, V., Tsagkarakis, K.P., Angelakis, A.N., 2012. Treatment of olive mill wastewater with constructed wetlands. Water (Switzerland) 4 (1), 260–271.

Karathanasis, A.D., Potter, C.L., Coyne, M.S., 2003. Vegetation effects on fecal bacteria, BOD, and suspended solid removal in constructed wetlands treating domestic wastewater. Ecological Engineering 20, 157–169.

Karim, M.R., Manshadi, F.D., Karpiscak, M.M., Gerba, C.P., 2004. The persistence and removal of enteric pathogens in constructed wetlands. Water Research 38, 1831–1837.

Kassenga, G.R., Pardue, J.H., Blair, S., Ferraro, T., 2003. Treatment of chlorinated volatile organic compounds in upflow wetland mesocosms. Ecological Engineering 19, 305–323.

Kayser, K., Kunst, S., 2005. Processes in vertical-flow reed beds—nitrification, oxygen transfer and soil clogging. Water Science and Technology 51 (9), 177–184.

Kayser, K., Kunst, S., Fehr, G., Voermanek, H., 2002. Nitrification in reed beds – capacity and potential control methods. Water Science and Technology 46 (6–7), 363–370.

Keffala, C., Ghrabi, A., 2005. Nitrogen and bacterial removal in constructed wetlands treating domestic wastewater. Desalination 185, 383–389.

Kelesidis, A., Stasinakis, A.S., 2012. Comparative study of the methods used for treatment and final disposal of sewage sludge in European countries. Waste Management 32, 1186–1195.

Kengne, I.M., Akoa, A., Soh, E.K., Tsama, V., Ngoutane, M.M., Dodane, P.H., Koné, D., 2008. Effects of faecal sludge application on growth characteristics and chemical composition of Echinochloa pyramidalis (Lam.) Hitch. and Chase and Cyperus papyrus L. Ecological Engineering 34, 233–242.

Kengne, I.M., Akoa, A., Koné, D., 2009a. Recovery of biosolids from constructed wetlands used for faecal sludge dewatering in tropical regions. Environmental Science and Technology 43, 6816–6821.

Kengne, I.M., Dodane, P.H., Akoa, A., Koné, D., 2009b. Vertical-flow constructed wetlands as sustainable sanitation approach for faecal sludge dewatering in developing countries. Desalination 248, 291–297.

Kengne, I.M., Kengne, E.S., Akoa, A., Bemmo, N., Dodane, P.-H., Koné, D., 2011. Vertical-flow constructed wetlands as an emerging solution for faecal sludge dewatering in developing countries. Journal of Water, Sanitation and Hygiene for Development 01 (1), 13–19.

Kern, J., Idler, C., 1999. Treatment of domestic and agricultural wastewater by reed bed systems. Ecological Engineering 12, 13–25.

Kern, J., Idler, C., Carlow, G., 2000. Removal of fecal coliforms and organic matter from dairy farm wastewater in a constructed wetland under changing climate conditions. Journal of Environmental Science and Health, Part A 35 (8), 1445–1461.

Khan, S., Ahmad, I., Shah, M.T., Rehman, S., Khaliq, A., 2009. Use of constructed wetland for the removal of heavy metals from industrial wastewater. Journal of Environmental Management 90, 3451–3457.

Kim, B.J., Smith, E.D., 1997. Evaluation of sludge dewatering reed beds: a niche for small systems. Water Science and Technology 35 (6), 21–28.

Kim, B.J., Smith, E.D., 1997b. Evaluation of sludge dewatering reed beds: a niche for small systems. Water Science and Technology 35 (6), 21–28.

Kim, J.-W., Choi, H., Pachepsky, Y.A., 2010. Biofilm morphology as related to the porous media clogging. Water Research 44, 1193–1201.

Kimochi, Y., Inamori, Y., Mizuochi, M., Xu, K.Q., Matsumura, M., 1998. Nitrogen removal and N2O emission in a full-scale domestic wastewater treatment plant with intermittent aeration. Journal of Fermentation and Bioengineering 86 (2), 202–206.

Kinsley, C., Crolla, A., Kennedy, K., 2011. Septage Treatment Using Reed Bed Filters. Research Note, Ontario Rural Wastewater Centre, Université de Guelph-Campus d'?lfred, Department of Civil Engineering, University of Ottawa, Ontario, Canada. Online via HYPERLINK, http://www.orwc.uoguelph.ca.

Kivaisi, A.K., 2001. The potential for constructed wetlands for wastewater treatment and reuse in developing countries: a review. Ecological Engineering 16, 545–560.

Knight, R.L., 1997. Wildlife babitat and public use benefits of treatment wetlands. Water Science and Technology 35 (5), 35–43.

Knight, R.L., Kadlec, R.H., Ohlendorf, H.M., 1999. The use of treatment wetlands for petroleum industry effluents. Environmental Science and Technology 33 (7), 973–980.

Knight, R.L., Payne Jr., V.W.E., Borer, R.E., Clarke Jr., R.A., Pries, J.H., 2000. Constructed wetlands for livestock wastewater management. Ecological Engineering 15, 41–55.

Knight, R.L., Clarke Jr., R.A., Bastian, R.K., 2001. Surface flow (SF) treatment wetlands as a habitat for wildlife and humans. Water Science and Technology 44, 27–37.

Knowles, P., Dotro, G., Nivala, J., García, J., 2011. Clogging in subsurface-flow treatment wetlands: Occurrence and contributing factors. Ecological Engineering 37, 99–112.

Kõiv, M., Kriipsalu, M., Vohla, C., Mander, Ü., 2009. Hydrated oil shale ash and mineralized peat as alternative filter materials for landfill leachate treatment in vertical flow constructed wetlands. Fresenius Environmental Bulletin 18 (2), 189–195.

Kolecka, K., Obarska-Pempkowiak, H., 2013. Potential fertilizing properties of sewage sludge treated in the sludge treatment reed beds (STRB). Water Science and Technology 68 (6), 1412–1418.

Komilis, D.P., Tziouvaras, I.S., 2009. A statistical analysis to assess the maturity and stability of six composts. Waste Management 29, 1504–1513.

Komilis, D., Evangelou, A., Voudrias, E., 2011a. Monitoring and optimizing the co-composting of dewatered sludge: a mixture experimental design approach. Journal of Environmental Management 92, 2241–2249.

Komilis, D., Kontou, I., Ntougias, S., 2011b. A modified static respiration assay and its relationship with an enzymatic test to assess compost stability and maturity. Bioresource Technology 102 (10), 5863–5872.

Koottatep, T., Polprasert, C., Oanh, N.T.K., Keinss, U., Montanegro, A., Strauss, M., 2001. Septage dewatering in vertical-flow constructed wetlands located in the tropics. Water Science and Technology 44, 181–188.

Koottatep, T., Surinkul, N., Polprasert, C., Kamal, A.S.M., Koné, D., Montangero, A., Heinss, U., Strauss, M., 2005a. Treatment of septage in CWs in tropical climate: lessons learnt after seven years of operation. Water Science and Technology 51 (9), 119–126.

Koottatep, T., Surinkul, N., Polprasert, C., Kamal, A.S.M., Koné, D., Montangero, A., Heinss, U., Strauss, M., 2005b. Treatment of septage in CWs in tropical climate: lessons learnt after seven years of operation. Water Science and Technology 51 (9), 119–126.

Koottatep, T., Polprasert, C., Hadsol, S., 2006. Integrated faecal sludge treatment and recycling through constructed wetlands and sunflower plant irrigation. Water Science and Technology 54 (11), 155–164.

Kopp, J., Dichtl, N., 2000. Prediction of full-scale dewatering results by determining the water distribution of sewage sludges. Water Science and Technology 42 (9), 141–149.

Kopp, J., Dichtl, N., 2001. Influence of the free water content on the dewaterability of sewage sludges. Water Science and Technology 44 (10), 177–183.

Korboulewsky, N., Wang, R., Baldy, V., 2012. Purification processes involved in sludge treatment by a vertical flow wetland system: focus on the role of the substrate and plants on N and P removal. Bioresource Technology 105, 9–14.

Korkusuz, E.A., Beklioglu, M., Demirer, G.N., 2004. Treatment efficiencies of the vertical flow pilot-scale constructed wetlands for domestic wastewater treatment. Turkish Journal of Engineering and Environmental Sciences 28, 333–344.

Korkusuz, E.A., Beklioglu, M., Demirer, G.N., 2005. Comparison of the treatment performances of blast furnace slag-based and gravel-based vertical flow wetlands operated identically for domestic wastewater treatment in Turkey. Ecological Engineering 24, 187–200.

Korkusuz, E.A., Beklioglu, M., Demirer, G.N., 2007a. Use of blast furnace granulated slag as a substrate in vertical flow reed beds: field application. Bioresource Technology 98, 2089–2101.

Korkusuz, E.A., Meyer, D., Langergraber, G., 2007b. CW2D simulation results of lab-scale vertical flowfilters filled with specialmedia and loaded with municipal wastewater. In: Proceedings of the 2nd International Symposium on Wetland pollutant dynamics and control WETPOL, 16–20 September, Tartu, Estonia.

Kosolapov, D.B., Kuschk, P., Vainshtein, M.B., Vatsourina, A.V., Wiessner, A., Kästner, M., Müller, R.A., 2004. Microbial processes of heavy metal removal from carbon-deficient effluents in constructed wetlands. Engineering in Life Sciences 4 (5), 403–411.

Kotti, I.P. 2012. Modeling vertical flow constructed wetlands using RZWQM. Department of Environmental Engineering. Master Thesis, Democritus University of Thrace, Xanthi, Greece, M.Sc.: 85 (in Greek).

Kotti, I.P., Gikas, G.D., Tsihrintzis, V.A., 2010. Effect of operational and design parameters on removal efficiency of pilot-scale FWS constructed wetlands and comparison with HSF systems. Ecological Engineering 36, 862–875.

Kotti, I.P., Sylaios, G.K., Tsihrintzis, V.A., 2013. Fuzzy logic models for BOD removal prediction in free-water surface constructed wetlands. Ecological Engineering 51, 66–74.

Krugel, S., Nemeth, L., Peddie, C., 1998. Extending thermophilic anaerobic digestion for producing Class A biosolids at the Greater Vancouver Regional District Annacis Island wastewater treatment plant. Water Science and Technology 38, 409–416.

Kucuk, O.S., Sengul, F., Kapdan, I.K., 2003. Removal of ammonia from tannery effluents from a red bed constructed wetland. Water Science and Technology 48, 179–186.

Kumaran, P., Paruchuri, Y.L., 1997. Kinetics of phenol biotransformation. Water Research 31 (1), 11–22.

Kurzbaum, E., Kirzhner, F., Sela, S., Zimmels, Y., Armon, R., 2010a. Efficiency of phenol biodegradation by planktonic *Pseudomonas pseudoalcaligenes* (a constructed wetland isolate) vs. root and gravel biofilm. Water Research 44, 5021–5031.

Kurzbaum, E., Zimmels, Y., Armon, R., 2010b. Isolation of a *Halotolerant Streptomyces sp.* from a constructed wetland that biodegrade phenol and various biopolymers. Actinomycetologica 24 (2), 31–38.

Kuschk, P., Wiebner, A., Kappelmeyer, U., Weibbrodt, E., Kästner, M., Stottmeister, U., 2003. Annual cycle of nitrogen removal by a pilot-scale subsurface horizontal flow in a constructed wetland under moderate climate. Water Research 37, 4236–4242.

Kuschk, P., Wiessner, A., Seeger, E.M., Kästner, M., Kappelmeyer, U., Paredes, D., Shtemenko, N.I., 2012. The status of research in constructed wetlands. In: Vitale, K. (Ed.), Environmental and Food Safety and Security for South-East Europe and Ukraine. In: NATO Science for Peace and Security Series C: Environmental Security, Springer Science+Business Media B.V.

Laber, J., Perfler, R., Haberl, R., 1997. Two strategies for advanced nitrogen elimination in vertical flow constructed wetlands. Water Science and Technology 35 (5), 71–77.

Lafleur, P., 1990. Evaporation from wetlands. Canadian Geographer 34, 79–88.

Langergraber, G. 2001. Development of a simulation tool for subsurface flow constructed wetlands. PhD Thesis, Vienna, Wiener Mitteilungen, Ph.D.: 207.

Langergraber, G., 2003. Simulation of subsurface flow constructed wetlands - results and further research needs. Water Science and Technology 48, 157–166.

Langergraber, G., 2005. The role of plant uptake on the removal of organic matter and nutrients in subsurface flow constructed wetlands: a simulation study. Water Science and Technology 51 (9), 213–223.

Langergraber, G., 2008. Modeling of processes in subsurface flow constructed wetlands: a review. Vadose Zone Journal 7, 830–842.

Langergraber, G., 2013. Are constructed treatment wetlands sustainable sanitation solutions? Water Science and Technology 67 (10), 2133–2140.

Langergraber, G., Šimůnek, J., 2005. Modeling variably saturated water flow and multicomponent reactive transport in constructed wetlands. Vadose Zone Journal 4, 924–938.

Langergraber, G., Šimůnek, J., 2012. Reactive transport modeling of subsurface flow constructed wetlands using the HYDRUS wetland module. Vadose Zone Journal 11 (2). http://dx.doi.org/10.2136/vzj2011.0104.

Langergraber, G., Haberl, R., Laber, J., Pressl, A., 2003. Evaluation of substrate clogging processes in vertical flow constructed wetlands. Water Science and Technology 48 (5), 25–34.

Langergraber, G., Prandtstetten, C., Pressl, A., Rohrhofer, R., Haberl, R., 2007a. Optimization of subsurface vertical flow constructed wetlands for wastewater treatment. Water Science and Technology 55 (7), 71–78.

Langergraber, G., Prandtstetten, C., Pressl, A., Sleytr, K., Leroch, K., Rohrhofer, R., Haberl, R., 2008a. Investigations on nitrogen removal in a two-stage subsurface vertical flow constructed wetland. In: Vymazal, J. (Ed.), Wastewater Treatment, Plant Dynamics and Management in Constructed and Natural Wetlands. Springer Science, Business Media.

Langergraber, G., Leroch, K., Pressl, A., Rohrhofer, R., Haberl, R., 2008b. A two-stage subsurface vertical flow constructed wetland for high-rate nitrogen removal. Water Science and Technology 57 (12), 1881–1887.

Langergraber, G., Giraldo, D., Mena, J., Meyer, D., Peña, M., Toscano, A., Brovelli, A., Korkusuz, E.A., 2009. Recent developments in numerical modelling of subsurface flow constructed wetlands. Science of the Total Environment 407, 3931–3943.

Lantzke, I.R., Heritage, A.D., Pistillo, G., Mitchell, D.S., 1998. Phosphorus removal rates in bucket size planted wetlands with a vertical hydraulic flow. Water Research 32 (4), 1280–1286.

Lantzke, I.R., Mitschell, D.S., Heritage, A.D., Sharma, K.P., 1999. A model of factors controlling orthophosphate removal in planted vertical flow wetlands. Ecological Engineering 12, 93–105.

Larsen, E., Greenway, M., 2004. Quantification of biofilms in a sub-surface flow wetland and their role in nutrient removal. Water Science and Technology 49, 115.

Lauschmann, R.E., Lechner, M., Ertl, T., Langergraber, G., 2013. Experiences with pre-precipitation of phosphorus in a vertical flow constructed wetland in Austria. Water Science and Technology 67 (10), 2337–2341.

Lavrova, S., Koumanova, B., 2010. Influence of recirculation in a lab-scale vertical flow constructed wetland on the treatment efficiency of landfill leachate. Bioresource Technology 101, 1756–1761.

Leader, J.W., Reddy, K.R., Wilkie, A.C., 2005. Optimization of low-cost phosphorus removal from wastewater using co-treatments with constructed wetlands. Water Science and Technology 51 (9), 283–290.

Lee, B.-H., Scholz, M., 2007. What is the role of *Phragmites australis* in experimental constructed wetland filters treating urban runoff? Ecological Engineering 29, 87–95.

Lee, C.-Y., Lee, C.-C., Lee, F.-Y., Tseng, S.-K., Liao, C.-J., 2004. Performance of subsurface flow constructed wetland taking pre-treated swine effluent under heavy loads. Bioresource Technology 92, 173–179.

Lee, B.-H., Scholz, M., Horn, A., 2005. Constructed wetlands: treatment of concentrated storm water runoff (Part A). Environmental Engineering Science 23 (2), 191–202.

Lee, C.-G., Fletcher, T.D., Sun, G., 2009. Nitrogen removal in constructed wetlands. Engineering in Life Sciences 9 (1), 11–22.

Lee, M.S., Drizo, A., Rizzo, D.M., Druschel, G., Hayden, N., Twohig, E., 2010. Evaluating the efficiency and temporal variation of pilot-scale constructed wetlands and steel slag phosphorus removing filters for treating dairy wastewater. Water Research 44, 4077–4086.

Lesage, E., Rousseau, D.P.L., Meers, E., Van de Moortel, A.M.K., Du Laing, G., Tack, F.M.G., De Pauw, N., Verloo, M.G., 2007. Accumulation of metals in the sediment and reed biomass of a combined constructed wetland treating domestic wastewater. Water Air and Soil Polluttion 183, 253–264.

Li, X., Jiang, C., 1995. Constructed wetland systems for water pollution control in north China. Water Science and Technology 32 (3), 349–356.

Li, L.-N., Li, Y.-Z., Da, L.-J., You, W.-H., 2009. Treatment of effluent from the recycled paper mill by using integrated vertical-flow constructed wetland. China Pulp and Paper 28 (9), 43–46.

Li, H., Li, Y., Gong, Z., Li, X., 2013. Performance study of vertical flow constructed wetlands for phosphorus removal with water quenched slag as a substrate. Ecological Engineering 53, 39–45.

Lian, J.J., Xu, S.G., Zhang, Y.M., Han, C.W., 2013. Molybdenum(VI) removal by using constructed wetlands with different filter media and plants. Water Science and Technology 67 (8), 1859–1866.

Liang, W., Wu, Z.-B., Cheng, S.-P., Zhou, Q.-H., Hu, H.-Y., 2003. Roles of substrate microorganisms and urease activities in wastewater purification in a constructed wetland system. Ecological Engineering 21, 191–195.

Liénard, A., Duchéne, P., Gorini, D., 1995. A study of activated sludge dewatering in experimental reed-planted or unplanted sludge drying beds. Water Science and Technology 32 (3), 251–261.

Liira, M., Kõiv, M., Mander, Ü., Mötlep, R., Vohla, C., Kirsimäe, K., 2009. Active filtration of phosphorus on Ca-rich hydrated oil shale ash: does longer retention time improve the process? Environmental Science and Technology 43, 3809–3814.

Liolios, K.A., Moutsopoulos, K.N., Tsihrintzis, V.A., 2012. Modeling of flow and BOD fate in horizontal subsurface flow constructed wetlands. Chemical Engineering Journal 200–202, 681–693.

Liolios, K.A., Moutsopoulos, K.N., Tsihrintzis, V.A., 2014. Comparative modeling of HSF constructed wetland performance with and without evapotranspiration and rainfall. Environmental Processes 1 (2), 171–186. http://dx.doi.org/10.1007/s40710-014-0019-5.

Lin, J.-G., Chang, C.-N., Chang, S.-C., 1997. Enhancement of anaerobic digestion of waste wactivated sludge bz alkaline solubilization. Bioresource Technology 62, 85–90.

Liu, D.H., Lipták, B.G. (Eds.), 1997. Environmental Engineers' Handbook, second ed. CRC Press LLC.

Liu, D., Ge, Y., Chang, J., Peng, C., Gu, B., Chan, G.Y.S., Wu, X., 2008. Constructed wetlands in China: recent developments and future challenges. Frontiers in Ecology and the Environment 7, 261–268.

Liu, L., Liu, C., Zheng, J., Huang, X., Wang, Z., Liu, Y., Zhu, G., 2013. Elimination of veterinary antibiotics and antibiotic resistance genes from swine wastewater in the vertical flow constructed wetlands. Chemosphere 91 (8), 1088–1093.

Lloyd, J.R., Klessab, D.A., Parryc, D.L., Buckc, P., Browna, N.L., 2004. Stimulation of microbial sulphate reduction in a constructed wetland: microbiological and geochemical analysis. Water Research 38 (7), 1822–1830.

Lopsik, K., 2013. Life cycle assessment of small-scale constructed wetland and extended aeration activated sludge wastewater treatment system. International journal of Environmental Science and Technology 10, 1295–1308.

Luederitz, V., Gerlach, F., 2002. Phosphorus removal in different constructed wetlands. Acta Biotechnologica 22 (1–2), 91–99.

Luederitz, V., Eckert, E., Lange-Weber, M., Lange, A., Gersberg, M.R., 2001. Nutrient removal efficiency and resource economics of vertical flow and horizontal flow constructed wetlands. Ecological Engineering 18, 157–171.

Luederitz, V., Gerlach, F., 2002. Phosphorus removal in different constructed wetlands. Acta Biotechnologica 22 (1–2), 91–99.

Ma, Q.L., Ahuja, L.R., Rojas, K.W., Ferreira, V.F., DeCoursey, D.G., 1995. Measured and RZWQM predicted atrazine dissipation and movement in a field soil. Transactions of the American Society of Agricultural and Biological Engineers 38, 471–479.

Ma, Q.L., Ahuja, L.R., Wauchope, R.D., Benjamin, J.G., Burgoa, B., 1996. Comparison of instantaneous equilibrium and equlibrium-kinetic sorption models for simulating simultaneous leaching and runoff of pesticides. Soil Science 161, 646–655.

Ma, L., Scott, H.D., Shaffer, M.J., Ahuja, L.R., 1998a. RZWQM simulations of water and nitrate movement in a manured tall fescue field. Soil Science 163, 259–270.

Ma, L., Shaffer, M.J., Boyd, J.K., Waskom, R., Ahuja, L.R., Rojas, K.W., Xu, C., 1998b. Manure management in an irrigated silage corn field: experiment and modeling. Soil Science Society of America Journal 62, 1006–1017.

Ma, L., Ahuja, L.R., Ascough II, J.C., Shaffer, M.J., Rojas, K.W., Malone, R.W., Cameira, M.R., 2000. Integrating system modeling with field research in agriculture: Applications of Root Zone Water Quality Model (RZWQM). Advances in Agronomy 71, 233–292.

Machado, A.P., Urbano, L., Brito, A.G., Janknecht, P., Salas, J.J., Nogueira, R., 2007. Life cycle assessment of wastewater treatment options for small and decentralized communities. Water Science and Technology 56 (3), 15–22.

Madoni, P., 1994. Estimates of ciliated protozoa biomass in activated sludge and biofilm. Bioresource Technology 48, 245–249.

Maier, U., De Biase, C., Baeder-Bederski, O., Bayer, P., 2009. Calibration of hydraulic parameters for large-scale vertical flow constructed wetlands. Journal of Hydrology 369, 260–273.

Malaviya, P., Singh, A., 2012. Constructed wetlands for management of urban stormwater runoff. Critical Reviews in Environmental Science and Technology 42, 2153–2214.

Mander, Ü., Dotro, G., Ebie, Y., Towprayoon, S., Chiemchaisri, C., Nogueira, S.F., Jamsranjav, B., Kasak, K., Truu, J., Tournebize, J., Mitsch, W.J., 2014. Greenhouse gas emission in constructed wetlands for wastewater treatment: a review. Ecological Engineering 66, 19–35.in Presshttp://dx.doi.org/10.1016/j.ecoleng.2013.12.006.

Mander, Ü., Jenssen, P. (Eds.), 2002. Natural Wetlands for wastewater treatment in cold climates. WIT Press, Southampton, UK.

Mander, Ü., Mauring, T., 1997. Constructed wetlands for wastewater treatment in Estonia. Water Science and Technology 35 (5), 323–330.

Mander, Ü., Dotro, G., Ebie, Y., Towprayoon, S., Chiemchaisri, C., Nogueira, S.F., Jamsranjav, B., Kasak, K., Truu, J., Tournebize, J., Mitsch, W.J., 2014. Greenhouse gas emission in constructed wetlands for wastewater treatment: A review. Ecological Engineering in Press, http://dx.doi.org/10.1016/j.ecoleng.2013.12.006.

Manderia, S., 2012. Assessment of performance evaluation of reed (Phragmites karka) in constructed reed bed (CRB) system for domestic sludge, Ujjain city. International Journal of Scientific and Research Publications 2 (2), 33–36.

Mant, C., Costa, S., Williams, J., Tambourgi, E., 2006. Phytoremediation of chromium by model constructed wetland. Bioresource Technology 97, 1767–1772.

Mantovi, P., Marmiroli, M., Maestri, E., Tagliavini, S., Piccinini, S., Marmiroli, N., 2003. Application of a horizontal subsurface flow constructed wetland on treatment of dairy parlor wastewater. Bioresource Technology 88, 85–94.

Marchand, L., Mench, M., Jacob, D.L., Otte, M.L., 2010. Metal and metalloid removal in constructed wetlands, with emphasis on the importance of plants and standardized measurements: A review. Environmental Pollution 158, 3447–3461.

Martin, M.O., 2002. Predatory prokaryotes: an emerging research opportunity. Journal of molecular microbiology and Biotechnology 4 (5), 467–477.

Masferrer, J.A.R., 2002. Passive treatment of acid mine drainage at the La Extranjera mine (Puertollano, Spain). Mine Water and the Environment 21, 111–113.

Masi, F., 2013. Design considerations for constructed wetlands in dry and hot countries. Sustainable Sanitation Practice 14, 55–61.

Masi, F., Conte, G., Martinuzzi, N., Pucci, B., 2002. Winery high organic content wastewaters treated by constructed wetlands in Mediterranean climate. In: Proceedings of IWA 8th International Conference on Wetland Systems for Water Pollution Control. Arusha, Tanzania, 16-19 September.

Masi, F., Martinuzzi, N., Bresciani, R., Giovannelli, L., Conte, G., 2006. Tolerance to hydraulic and organic load fluctuations in constructed wetlands. In: Proceedings of 10th International

Conference on Wetlands Systems for Water Pollution Control, 23–29 September, Lisboa, Portugal.

Masi, F., Martinuzzi, N., Bresciani, R., Giovannelli, L., Conte, G., 2007. Tolerance to hydraulic and organic load fluctuations in constructed wetlands. Water Science and Technology 56 (3), 39–48.

Massmann, J., Farrier, D.F., 1992. Effects of atmospheric pressure on gas transport in the vadose zone. Water Resources Research 28, 777–791.

Massmann, G., Dünnbier, U., Heberer, T., Taute, T., 2008. Behaviour and redox sensitivity of pharmaceutical residues during bank filtration–Investigation of residues of phenazonetype analgesics. Chemosphere 71, 1476–1485.

Matamoros, V., Bayona, J.M., 2008. Behavior of emerging pollutants in constructed wetlands. Emerging Contaminants from Industrial and Municipal Waste. The Handbook of Environmental Chemistry 5S (2), 199–217.

Matamoros, V., Arias, C., Brix, H., Bayona, J.M., 2007. Removal of pharmaceuticals and personal care products (PPCPs) from urban wastewater in a pilot vertical flow constructed wetland and s sand filter. Environmental Science and Technology 41, 8171–8177.

Matamoros, V., Hijosa, M., Bayona, J.M., 2009. Assessment of the pharmaceutical active compounds removal in wastewater treatment systems at enantiometric level. Ibuprofen and naproxen. Chemosphere 75, 200–205.

Matamoros, V., Nguyen, L.X., Arias, C.A., Nielsen, S., Laugen, M.M., Brix, H., 2012. Musk fragrances, DEHP and heavy metals in a 20 years old sludge treatment reed bed system. Water Research 46, 3889–3896.

Matthess, G., Pekdeger, A., 1985. Survival and transport of pathogenic bacteria and viruses in groundwater. In: Ward, C.H., Giger, W., McCarty, P.L. (Eds.), Groundwater Quality. Wiley, New York, NY, pp. 472–482.

Matthews, E., Fung, I., 1987. Methane emissions from natural wetlands: Global distribution, area, and environmental characteristics of sources. Global Biogeochemical Cycles 1 (1), 61–86.

Mayo, A.W., Bigambo, T., 2005. Nitrogen transformation in horizontal subsurface flow constructed wetlands I: model development. Phys Chem. Earth 30, 658–667.

Mburu, N., Rousseau, D.P.L., van Bruggen, J.J.A., Thumbi, G., Llorens, E., García, J., Lens, P.N.L., 2013. Reactive transport simulation in a tropical horizontal subsurface flow constructed wetland treating domestic wastewater. Science of the Total Environment 449, 309–319.

McGechan, M.B., Moir, S.E., Sym, G., Castle, K., 2005. Estimating inorganic and organic nitrogen transformation rates in a model of a constructed wetland purification system for dilute farm effluents. Biosystems Engineering 91, 61–75.

MEA (Millennium Ecosystem Assessment), 2005. Ecosystems and Human Well-being: Wetlands and Water Synthesis. MEA, World Resources Institute, Washington D.C. (www.millenniumassessment.org).

Meers, E., Rousseau, D.P.L., Blomme, N., Lesage, E., Laing, Du., Tack, F.M.G., Verloo, M.G., 2005. Tertiary treatment of the liquid fraction of pig manure with *Phragmites australis*. Water, Air, and Soil Pollution 160, 15–26.

Meers, E., Tack, F.M.G., Tolpe, I., Michels, E., 2008. Application of a full-scale Constructed Wetland for tertiary treatment of piggery manure: monitoring results. Water, Air, and Soil Pollution 193, 15–24.

Melián, J.A.H., Rodríguez, A.J.M., Araña, J., Díaz, O.G., Henríquez, J.J.B., 2010. Hybrid constructed wetlands for wastewater treatment and reuse in the Canary Islands. Ecological Engineering 36, 891–899.

Melidis, P., Gikas, G.D., Akratos, C.S., Tsihrintzis, V.A., 2010. Dewatering of primary settled urban sludge in a vertical flow wetland. Desalination 250, 395–398.

Memon, F.A., Zheng, Z., Butler, D., Shirley-Smith, C., Lui, S., Makropoulos, C., Avery, L., 2007. Life cycle impact assessment of greywater recycling technologies for new developments. Environmental Monitoring and Assessment 129 (1–3), 27–35.

Mena, J., Villaseñor, J., Gómez, R., De Lucas, A., Langergraber, G., 2007. Simulation of the hydraulics and the removal efficiency of pilot-scale horizontal subsurface flow constructed wetlands with different plants (*Phragmites australis* and *Lythrum salicaria*). In: Proceedings of the 2nd International Symposium on "Wetland pollutant dynamics and control" WETPOL, 16–20 September, Tartu, Estonia.

Merlin, G., Pajean, J.-L., Lissolo, T., 2002. Performances of constructed wetlands for municipal wastewater treatment in rural mountainous area. Hydrobiologia 469, 87–98.

Metcalf & Eddy, 2003. Wastewater Engineering: Treatment and Reuse, fourth ed. McGraw-Hill, Boston, USA.

Metcalf & Eddy, Tchobanoglous, G., 1979. Wastewater Engineering: Treatment, Disposal, Reuse. McGraw-Hill, New York, USA.

Meuleman, F.M.A., van Logtestijn, R., Rijs, B.J.G., Verhoeven, T.A.J., 2003. Water and mass budgets of a vertical-flow constructed wetland used for wastewater treatment. Ecological Engineering 20, 31–44.

Mezzanotte, V., Antonelli, M., Citterio, S., Nurizzo, C., 2007. Wastewater disinfection alternatives: chlorine, ozone, peracetic acid, and UV light. Water Environment Research 79, 2373–2379.

Mitsch, W.J., Wise, K.M., 1998. Water quality, fate of metals, and predictive model validation of a constructed wetland treating acid mine drainage. Water Research 32 (6), 1888–1900.

Mitterer-Reichmann, G., 2012. Treatment wetlands in Austria: Practical experiences in planning, construction and maintenance. Sustainable Sanitation Practice 12, 4–8.

Moir, S.E., Svoboda, I., Sym, G., Clark, J., McGechan, M.B., Castle, K., 2005. An experimental plant for testing methods of treating dilute farm effluents and dirty water. Biosystems Engineering 90, 349–355.

Molinos-Senante, M., Hernández-Sancho, F., Sala-Garrido, R., 2010. Economic feasibility study for wastewater treatment: a cost-benefit analysis. Science of the Total Environment 408, 4396–4402.

Molle, P., Liénard, A., Grasmick, A., Iwema, A., 2004. rench vertical flow constructed wetlands: reed bed behaviour and limits due to hydraulic overloading on first stage filters. In: Proceedings of the 9th Conference on Constructed Wetlands and 6th Conference on Stabilization Ponds, 26 September-1 October, Avignon, France.

Molle, P., Liénard, A., Boutin, C., Merlin, G., Iwema, A., 2005. How to treat raw sewage with constructed wetlands: an overview of the French systems. Water Science and Technology 51 (9), 11–21.

Molle, P., Liénard, A., Grasmick, A., Iwema, A., 2006. Effect of reeds and feeding operations on hydraulic behaviour of vertical flow constructed wetlands under hydraulic overloads. Water Research 40, 606–612.

Molle, P., Prost-Boucle, S., Lienard, A., 2008. Potential for total nitrogen removal by combining vertical flow and horizontal flow constructed wetlands: A full-scale experimental study. Ecological Engineering 34, 23–29.

Molleda, P., Blanco, I., Ansola, G., de Luis, E., 2008. Removal of wastewater pathogen indicators in a constructed wetland in Leon, Spain. Ecological Engineering 33, 252–257.

Momba, M.N.B., 2010. Wastewater protozoan-driven environmental processes for the protection of water sources. In: Bux, F., Momba, M. (Eds.), Biomass. Sciyo, Croatia, September.

Montuelle, B., Goillard, J., Le Hy, J.B., 1992. A combined anaerobic-aerobic process for the co-treatment of effluents from a piggery and a cheese factory. Journal of Agricultural Engineering Research 51, 91–100.

Morari, F., Giardini, L., 2009. Municipal wastewater treatment with vertical flow constructed wetlands for irrigation reuse. Ecological Engineering 35, 643–653.

Moreira, V.R., LeBlanc, B.D., Achberger, E.C., Frederick, D.G., Leonardi, C., 2010. Design and evaluation of a sequential biological treatment system for dairy parlor wastewater in southeastern Louisiana. Applied Engineering in Agriculture 26, 125–136.

Moreno, C., Farahbakshazad, N., Morrison, G.M., 2002. Ammonia removal from oil refinery effluent in vertical upflow macrophyte column systems. Water, Air, and Soil Pollution 135, 237–247.

Morris, M., Herbert, R., 1997. The design and performance of a vertical flow reed bed for the treatment of high ammonia, low suspended solids organic effluents. Water Science and Technology 35 (5), 197–204.

Morvannou, A., Choubert, J.M., Vanclooster, M., Molle, P., 2011. Solid respirometry to characterize nitrification kinetics: A better insight for modelling nitrogen conversion in vertical flow constructed wetlands. Water Research 45, 4995–5004.

Moutsopoulos, K.N., Poultsidis, V., Papaspyros, I.N.E., Tsihrintzis, V.A., 2011. Simulation of hydrodynamics and nitrogen transformation processes in HSF constructed wetlands and porous media using the advection-dispersion-reaction equation with linear sink-source terms. Ecological Engineering 37, 1407–1415.

Munoz, P., Drizo, A., Hession, W.C., 2006. Flow patterns of dairy wastewater constructed wetlands in a cold climate. Water Research 40, 3209–3218.

Mustafa, A., Scholz, M., Harrington, R., Carroll, P., 2009. Long-term performance of a representative integrated constructed wetland treating farmyard runoff. Ecological Engineering 35, 779–790.

Nair, C.I., Jayachandran, K., Shashidhar, S., 2008. Biodegradation of phenol. African Journal of Biotechnology 7 (25), 4951–4958.

Nassar, A.M., Smith MCIWEM, M., Afifi, S., 2006. Sludge dewatering using the reed bed system in the Gaza Strip, Palestine. Water and Environment Journal 20 (1), 27–34.

Nassar, A.M., Smith MCIWEM, M., Afifi, S., 2008. Palestinian experience with sewage sludge utilizing reed beds. Water and Environment Journal 23 (1), 75–82.

Neori, A., Reddy, K.R., Ciskova-Koncalova, H., Agami, M., 2000. Bioactive chemicals and biological-biochemical activities and their functions in rhizospheres of wetland plants. The Botanical Review 66, 350–378.

Newman, J.M., Clausen, J.C., 1997. Seasonal effectiveness of a constructed wetland for processing milkhouse wastewater. Wetlands 17, 375–382.

Newman, J.M., Clausen, J.C., Neafsey, J.A., 2000. Seasonal performance of a wetland constructed to process dairy milkhouse wastewater in Connecticut. Ecological Engineering 14, 181–198.

Nguyen, L.M., 2000. Organic matter composition, microbial biomass and microbial activity in gravel-bed constructed wetlands treating farm dairy wastewaters. Ecological Engineering 16, 199–221.

Nguyen, A.V., Pham, N.T., Nguyen, T.H., Morel, A., Tonderski, K., 2007. Improved septic tank with constructed wetland, a promising decentralized wastewater treatment alternative in Vietnam. In: Proceeding of NOWRA, 16th Annual Technical Education Conference & Exposition, Baltimore, Maryland, USA, March.

Nielsen, S., 2003. Sludge drying reed beds. Water Science and Technology 48 (5), 101–109.

Nielsen, S., 2007. Helsinge sludge reed bed system: reduction of pathogenic microorganisms. Water Science and Technology 56 (3), 175–182.

Nielsen, S., Willoughby, N., 2005. Sludge treatment and drying reed bed systems in Denmark. Water and Environment Journal 19 (4), 296–305.

Nielsen, S., Cooper, D.J., 2011. Dewatering sludge originating in water treatment works in reed bed systems. Water Science and Technology 64 (2), 361–366.

Nielsen, S., 2011. Sludge treatment reed bed facilities—organic load and operation problems. Water Science and Technology 63 (5), 941–947.

Nielsen, S., 2012. Sludge treatment in reed beds systems – development, design, experiences. Sustainable Sanitation Practice 12, 33–39.

Nielsen, S., 2005. Mineralisation of hazardous organic compounds in a sludge reed bed and sludge storage. Water Science and Technology 51 (9), 109–117.

Nielsen, S., Willoughby, N., 2005. Sludge treatment and drying reed bed systems in Denmark. Water and Environment Journal 19 (4), 296–305.

Nilratnisakorn, S., Thiravetyan, P., Nakbanpote, W., 2009. A constructed wetland model for synthetic reactive dye wastewater treatment by narrow-leaved cattails (Typha angustifolia Linn.). Water Science and Technology 60, 1565–1574.

Nivala, J., Rousseau, D.P., 2009. Reversing clogging in subsurface-flow constructed wetlands by hydrogen peroxide treatment: two case studies. Water Science and Technology 59 (10), 2037–2046.

Nivala, J., Hoos, M.B., Cross, C., Wallace, S., Parkin, G., 2007. Treatment of landfill leachate using an aerated, horizontal subsurface-flow constructed wetland. Science of the Total Environment 380, 19–27.

Nivala, J., Knowles, P., Dotro, G., García, J., Wallace, S., 2012a. Clogging in subsurface-flow treatment wetlands: Measurement, modeling and management. Water Research 46, 1625–1640.

Nivala, J., Wallace, S., Headley, T., Kassa, K., Brix, H., 2012b. Oxygen transfer and consumption in subsurface flow treatment wetlands. Ecological Engineering 61 (B), 544–554.

Nivala, J., Headley, T., Wallace, S., Bernhard, K., Brix, H., van Afferden, M., Müller, R.A., 2013. Comparative analysis of constructed wetlands: The design and construction of the ecotechnology research facility in Langenreichenbach, Germany. Ecological Engineering in Press, doi.org/10.1016/j.ecoleng.2012.08.028.

Nokes, S.E., Landa, F.M., Hanson, J.D., 1996. Evaluation of the crop component of the root zone water quality model for corn in Ohio. Transactions of the American Society of Agricultural and Biological Engineers 39, 1177–1184.

Noorvee, A., Põldvere, E., Mander, Ü., 2007. The effect of pre-aeration on the purification processes in the long-term performance of a horizontal subsurface flow constructed wetland. Science of the Total Environment 380, 229–236.

Norton, P.J., Norton, C.J., Tyrrell, W., 1998. The design, construction and cost of an engineered wetland for treatment of acid drainage from sulphide mineral-rich strata. In: Proceedings of the International Mine Water Association (IMWA) Symposium, 7–16 September, Johannersburg, South Afrika.

Núñez, M.E., Martin, M.O., Chan, P.H., Spain, E.M., 2005. Predation, death, and survival in a biofilm: Bdellovibrio investigated by atomic force microscopy. Colloid Surface B 42, 263–271.

Nwoko, C.O., 2010. Trends in phytoremediation of toxic elemental and organic pollutants. African Journal of Biotechnology 9 (37), 6010–6016.

NYSERDA (New York State Energy Research, Development Authority), 2006. Energy Efficient Sludge Treatment with Reed-Bed Technology Demonstration Project—Final Report, NYSERDA 4890, Albany, NY, USA, January.

Obarska-Pempkowiak, H., Kołecka, K., 2006. Experiences with utilization of sewage sludge in reed beds. Srodkowo-Pomorskie Towarzystwo Naukowe Ochrony Srodowiska, Tom 8. Rok 2006, 65-78.

Obarska-Pempkowiak, H., Tuszynska, A., Sobocinski, Z., 2003. Polish experience with sewage sludge dewatering in reed systems. Water Science and Technology 48 (5), 111–117.

O' Hogain, S., 2003. The design, operation and performance of a municipal hybrid reed bed treatment system. Water Science and Technology 48 (5), 119–126.

Olsson, L. 2011. Effect of design and dosing regime on the treatment performance of vertical flow constructed wetlands. Master thesis, Linköpings University, Sweden.

Ong, S.-A., Uchiyama, K., Inadama, D., Yamagiwa, K., 2009. Simultaneous removal of color, organic compounds and nutrients in azo dye-containing wastewater using up-flow constructed wetland. Journal of Hazardous Materials 165, 696–703.

Ong, S.-A., Uchiyama, K., Inadama, D., Ishida, Y., Yamagiwa, K., 2010. Performance evaluation of laboratory scale up-flow constructed wetlands with different designs and emergent plants. Bioresource Technology 101, 7239–7244.

Öövel, M., Tooming, A., Mauring, T., Mander, Ü., 2008. Schoolhouse wastewater purification in a LWA-filled hybrid constructed wetland in Estonia. Ecological Engineering 29, 17–26.

Ortiz, M., Raluy, R.G., Serra, L., Uche, J., 2007. Life cycle assessment of water treatment technologies: wastewater and water-reuse in a small town. Desalination 204, 121–131.

Ouattara, J.-M.P., Coulibaly, L., Tiho, S., Ouattara, A., Gourène, G., 2011. *Panicum maximum* (Jacq.) density effect upon macrofauna structure in sediments of pilot-scale vertical flow constructed wetlands treating domestic wastewater. Ecological Engineering 37, 217–233.

Ouellet-Plamondon, C., Chazarenc, F., Comeau, Y., Brisson, J., 2006. Artificial aeration to increase pollutant removal efficiency of constructed wetlands in cold climate. Ecological Engineering 27, 258–264.

Ouki, S.K., Kavannagh, M., 1999. Treatment of metals-contaminated wastewaters by use of natural zeolites. Water Science and Technology 39 (10–11), 115–122.

Outwater, B.A., 1994. Reuse of Sludge and Minor Wastewater Residuals. CRC Press, Florida.

Ouyang, Y., Luo, S.M., Cui, L.H., Liu, Y.H., Qin, Z., 2010. Estimation of water dynamics in a vertical-flow constructed wetland with a growing plant species. Journal of Soils and Sediments 10, 1219–1228.

Ouyang, Y., Luo, S.M., Cui, L.H., 2011. Estimation of nitrogen dynamics in a vertical-flow constructed wetland. Ecological Engineering 37, 453–459.

Ouyang, Y., Luo, S.M., Cui, L.H., Liu, Y.H., Qin, Z., 2011. Estimation of nitrogen dynamics in a vertical-flow constructed wetland. Ecological Engineering 37, 453–459.

Ozimek, T., Obarska-Pempkowiak, H., Cytawa, S., 2001. The effect of secondary sewage sludge on retention of nitrogen and phosphorus in Phragmites australis growing in constructed wetlands. In: Vymazal, J. (Ed.), Transformations of Nutrients in Natural and Constructed Wetlands. Backhuys Publishers, Leiden, The Netherlands, pp. 177–186.

Pagans, E., Font, X., Sánchez, A., 2006. Emission of volatile organic compounds from composting of different solid wastes: abatement by biofiltration. Journal of Hazardous Materials 131 (1–3), 179–186.

Pagga, U., Brown, D., 1986. The degradation of dyestuffs: Part II. Behaviour of dyestuffs in aerobic biodegradation tests. Chemosphere 15, 479–491.

Paing, J., Voisin, J., 2005. Vertical flow constructed wetlands for municipal wastewater and septage treatment in French rural area. Water Science and Technology 51 (9), 145–155.

Palatsi, J., Gimenez-Lorang, A., Ferrer, I., Florats, X., 2009. Start-up strategies of thermophilic anaerobic digestion of sewage sludge. Water Science and Technology 59 (9), 1777–1784.

Pan, J.R., Huang, C., Cherng, M., Li, K.-C., Lin, C.-F., 2003. Correlation between dewatering index and dewatering performance of three mechanical dewatering devices. Advances in Environmental Research 7, 599–602.

Pan, T., Zhu, X.-D., Ye, Y.-P., 2011. Estimate of life-cycle greenhouse gas emissions from a vertical subsurface flow constructed wetland and conventional wastewater treatment plants: a case study in China. Ecological Engineering 37, 248–254.

Pan, J., Zhang, H., Li, W., Ke, F., 2012. Full-scale experiment on domestic wastewater treatment by combining artificial aeration vertical- and horizontal-flow constructed wetlands system. Water, Air, and Soil Pollution 223, 5673–5683.

Pandley, M.K., Jenssen, P.D., Krogstad, T., Jonassonm, S., 2013. Comparison of vertical and horizontal flow planted and unplanted subsurface flow wetlands treating municipal wastewater. Water Science and Technology 68 (1), 117–123.

Pant, H.K., Reddy, K.R., Lemon, E., 2001. Phosphorus retention capacity of root bed media of subsurface flow constructed wetlands. Ecological Engineering 17, 345–355.

Panuvatvanich, A., Koottatepa, T., Kone, D., 2009. Influence of sand layer depth and percolate impounding regime on nitrogen transformation in vertical-flow constructed wetlands treating faecal sludge. Water Research 43, 2623–2630.

Papaevangelou, V.A., Gikas, G.D., Tsihrintzis, V.A., 2012. Evaluation of evapotranspiration in small on-site HSF constructed wetlands. J. Environ. Sci. Health A 47 (5), 766–785.

Paraskeva, P., Diamadopoulos, E., 2006. Technologies for olive mill wastewater (OMW) treatment: a review. Journal of Chemical Technology and Biotechnology 81, 1475–1485.

Paredes, D., Kuschk, P., Mbwette, T.S.A., Stange, F., Muller, R.A., Koser, H., 2007. New aspects of microbial nitrogen transformations in the context of wastewater treatment — a review. Engineering in Life Sciences 7, 13–25.

Park, W.H., Polprasert, C., 2008. Phosphorus adsorption characteristics of oyster shells and alum sludge and their application for nutrient control in constructed wetland system. Journal of Environmental Science and Health, Part A 43, 511–517.

Paterson, S., Mackay, D., Tam, D., Shiu, W.Y., 1990. Uptake of organic chemicals by plants: a review of processes, correlations and models. Chemosphere 21 (3), 297–331.

Pauli, W., Jax, K., Berger, S., 2001. Protozoa in wastewater treatment: function and importance. In: Beek, B. (Ed.), Biodegradation and Persistence. Springer, Berlin, pp. 202–252.

Pempkowiak, J., Obarska-Pempkowiak, H., 2002. Long-term changes in sewage sludge stored in a reed bed. Science of the Total Environment 297, 59–65.

Perfler, R., Laber, J., Langergraber, G., Haberl, R., 1999. Constructed wetlands for rehabilitation and reuse of surface waters in tropical and subtropical areas – first results from small-scale plots using vertical flow beds. Water Science and Technology 40 (3), 155–162.

Peruzzi, E., Macci, C., Doni, S., Masciandaro, G., Peruzzi, P., Aiello, M., Ceccanti, B., 2009. Phragmites australis for sewage sludge stabilization. Desalination 247, 111–120.

Peruzzi, E., Masciandaro, G., Macci, C., Doni, S., Ceccanti, B., 2011. Pollutant monitoring in sludge treatment wetlands. Water Science and Technology 64 (7), 1558–1565.

Peruzzi, E., Masciandaro, G., Macci, C., Doni, S., Ravelo, S.G.M., Peruzzi, P., Ceccanti, B., 2011. Heavy metal fractionation and organic matter stabilization in sewage sludge treatment wetlands. Ecological Engineering 37, 771–778.

Peruzzi, E., Masciandaro, G., Macci, C., Doni, S., Ceccanti, B., 2011. Pollutant monitoring in sludge treatment wetlands. Water Science and Technology 64 (7), 1558–1565.

Peruzzi, E., Nielsen, S., Macci, C., Doni, S., Iannelli, R., Chiarugi, M., Masciandaro, G., 2013. Organic matter stabilization in reed bed systems: Danish and Italian examples. Water Science and Technology 68 (8), 1888–1894.

Philippi, L.S., Sezerino, P.H., Bento, A.P., Magri, M.E., 2006. Vertical flow constructed wetlands for nitrification of anaerobic pond effluent in Southern Brazil under different loading rates. In: Proceedings of the 10th IWA International Specialized Conference on Wetland Systems for Water Pollution Control, 23–29 September, Lisbon, Portugal.

Platzer, C., 1999. Design recommendations for subsurface flow constructed wetlands for nitrification and denitrification. Water Science and Technology 40 (3), 257–263.

Platzer, C., Mauch, K., 1997. Soil clogging in vertical flow reed beds – mechanisms, parameters, consequences and. . . solutions? Water Science and Technology 35 (5), 175–181.

Poach, M.E., Hunt, P.G., Vanotti, M.B., Stone, K.C., Matheny, T.A., Johnson, M.H., Sadler, E.J., 2003. Improved nitrogen treatment by constructed wetlands receiving partially nitrified liquid swine manure. Ecological Engineering 20, 183–197.

Poach, M.E., Hunt, P.G., Reddy, G.B., Stone, K.C., Johnson, M.H., Grubbs, A., 2004. Swine wastewater treatment by marsh-pond-marsh constructed wetlands under varying nitrogen loads. Ecological Engineering 23, 165–175.

Polprasert, C., Dan, N.P., Thayalakumaran, N., 1996. Application of constructed wetlands to treat some toxic wastewaters under tropical conditions. Water Science and Technology 34 (11), 165–171.

Porst-Boucle, S., Molle, P., 2012. Recirculation on a single stage of vertical flow constructed wetland: treatment limits and operation modes. Ecological Engineering 43, 81–84.

Priestley, C.H.B., Taylor, R.J., 1972. On the assessment of surface heat flux and evaporation using large scale parameters. Monthly Weather Review 100, 81–92.

Prigent, S., Belbeze, G., Paing, J., Andres, Y., Voisin, J., Chazarenc, F., 2013. Biological characterization and treatment performances of a compact vertical flow constructed wetland with the use of expanded schist. Ecological Engineering 52, 12–18.

Prochaska, C.A., Zouboulis, A.I., 2006. Removal of phosphates by pilot-scale vertical-flow constructed wetlands using a mixture of sand and dolomite as substrate. Ecological Engineering 26, 293–303.

Prochaska, C.A., Zouboulis, A.I., 2009. Treatment performance variation at different depths within vertical subsurface-flow experimental wetlands fed with simulated domestic sewage. Desalination 237, 367–377.

Prochaska, C.A., Zouboulis, A.I., Eskridge, K.M., 2007. Performance of pilot-scale vertical-flow constructed wetlands, as affected by season, substrate, hydraulic load and frequency of application of simulated urban sewage. Ecological Engineering 31 (1), 57–66.

Prost-Boucle, S., Molle, P., 2012. Recirculation on a single stage of vertical flow constructed wetland: treatment limits and operation modes. Ecological Engineering 43, 81–84.

Puigagut, J., Villaseñor, J., Salas, J.J., Bécares, E., García, J., 2007. Subsurface-flow constructed wetlands in Spain for the sanitation of small communities: A comparative study. Ecological Engineering 30, 312–319.

Raats, P.A.C., 2001. Developments in soil-water physics since the mid-1960s. Geoderma 100, 355–387.

Ramsar, 2010. The Ramsar Manual, Ramsar and Its Mission, Brief History. http://www.ramsar.org/cda/en/ramsar-march11/main/ramsar/1%5E25044_4000_0. Last accessed September 15, 2012.

Randerson, P., Albuquerque, A., Bialowiec, A., 2012. The influence of evapotranspiration on wastewater constructed wetland treatment efficiency. In: Limnaeus ECO-TECH, Kalmar, Sweden, 26–28 November.

Rani, N., Maheshwari, R.C., Kumar, V., Vijay, V.K., 2011. Purification of pulp and paper mill effluent through Typha and Canna using constructed wetlands technology. Journal of Water Reuse and Desalination 1 (4), 237–242.

Ranieri, E., Verlicchi, P., Young, T.M., 2011. Paracetamol removal in subsurface flow constructed wetlands. Journal of Hydrology 404, 130–135.

Rebitzer, G., Hunkeler, D., Jolliet, O., 2003. LCC—the economic pillar of sustainability: methodology and application to wastewater treatment. Environmental Progress 22 (4), 241–249.

Redder, A., Dürr, M., Daeschlein, G., Baeder-Bederski, O., Koch, C., Müller, R., Exner, M., Borneff-Lipp, M., 2010. Constructed wetlands – are they safe in reducing protozoan parasites? International Journal of Hygiene and Environmental Health 213, 72–77.

Reddy, K.R., D'Angelo, E.M., 1997. Biogeochemical indicators to evaluate pollutant removal efficiency in constructed wetlands. Water Science and Technology 35 (5), 1–10.

Reddy, K.R., Patrick, W.H., Lindau, C.W., 1989. Nitrification-denitrification at the plant root-sediment interface in wetlands. Limnology and Oceanography 34, 1004–1013.

Reddy, K.R., Kadlec, R.H., Flaig, E., Gale, P.M., 1999. Phosphorus retention in streams and wetlands: a review. Critical Reviews in Environmental Science and Technology 29 (1), 83–146.

Redman, J.A., Grant, S.B., Olson, T.M., Hardy, M.E., Estes, M.K., 1997. Filtration of recombinant Norwalk virus particles and bacteriophage MS2 in quartz sand: importance of electrostatic interactions. Environmental Science and Technology 31, 3378–3383.

Reed, S., Middlebrooks, E., Crites, R., 1995. Natural Systems for Waste Management and Treatment. McGraw Hill, New York.

Reemtsma, T., Jekel, M., 1997. Dissolved organics in tannery wastewaters and their alteration by a combined anaerobic and aerobic treatment. Water Research 31, 1035–1046.

Reemtsma, T., Zywicki, B., Stueber, M., Kloepfer, A., Jekel, M., 2002. Removal of sulfur-organic polar micropollutants in a membrane bioreactor treating industrial wastewater. Environmental Science and Technology 36, 1102–1106.

Reese, C.S. 2005. Recirculation Effects on total nitrogen and total phosphorus removal in a full-scale subsurface flow constructed wetland. PhD Thesis, Department of Civil and Environmental Engineering of the College of Engineering, University of Cincinnati, USA.

Regulation (CE) No. 2003/2003 of the European Parliament and of the Council relating to fertilizers. Official Journal L 304, 21/11/2003, pp. 1–194.

Reif, R., Besancon, A., Le Corre, K., Jefferson, B., Lema, J.M., Omil, F., 2011. Comparison of PPCPs removal on a parallel-operated MBR and AS system and evaluation of effluent post-treatment on vertical flow reed beds. Water Science and Technology 63 (10), 2411–2417.

Reinoso, R., Torres, L.A., Bécares, E., 2008. Efficiency of natural systems for removal of bacteria and pathogenic parasites from wastewater. Science of the Total Environment 395, 80–86.

Ren, Y.-X., Zhang, H., Wang, C., Yang, Y.-Z., Qin, Z., Ma, Y., 2011. Effects of the substrate depth on purification performance of a hybrid constructed wetland treating domestic sewage. Journal of Environmental Science and Health, Part A 46, 777–782.

Renou, S., Thomas, J.S., Aoustin, E., Pons, M.N., 2008. Influence of impact assessment methods in wastewater treatment LCA. Journal of Cleaner Production 16, 1098–1105.

Rivera, F., Warren, A., Ramirez, E., Decamp, O., Bonilla, P., Gallegos, E., Calderon, A., Sanchez, J.T., 1995. Removal of pathogens from wastewater by the root zone method (RZM). Water Science and Technology 32, 211–218.

Roe, D., Cardinale, J.A., 2005. Evaluation of Microbial Community Structures and Coliform Persistence in the Alfred Waste Water Treatment Plant Reed Bed Sludge Treatment System: Final Report 2005 Roe. May 23, 2005 through August 15. Available at, https://www.researchgate.net/publication/237328561.

Roen, D.T. 2011. Modeling vertical flow treatment wetland hydraulics to optimize treatment efficiency. M.Sc. Thesis, Department of Systems and Engineering Management, Graduate School

of Engineering and Management, Air Force Institute of Technology, Air University, Air Education and Training Command.
Rossmann, M., de Matos, A.T., Abreu, E.C., de Silva, F.F., Borges, A.C., 2012. Performance of constructed wetlands in the treatment of aerated coffee processing wastewater: removal of nutrients and phenolic compounds. Ecological Engineering 49, 264–269.
Rousseau, D.P.L., Vanrolleghem, P.A., De Pauw, N., 2004a. Constructed wetlands in Flanders: a performance analysis. Ecological Engineering 23, 151–163.
Rousseau, D.P.L., Vanrolleghem, P.A., De Pauw, N., 2004b. Model-based design of horizontal subsurface flow constructed treatment wetlands: a review. Water Research 38, 1484–1493.
Rousseau, D.P.L., Horton, D., Griffin, P., Vanrolleghem, P.A., De Pauw, N., 2005. Impact of operational maintenance on the asset life of storm reed beds. Water Science and Technology 51 (9), 243–250.
Rousseau, D.P.L., Lesage, E., Story, A., Vanrolleghem, P.A., De Pauw, N., 2008. Constructed wetlands for water reclamation. Desalination 218, 181–189.
Roussy, J., Chastellan, P., van Vooren, M., Guibal, E., 2005. Treatment of ink-containing wastewater by coagulation/flocculation using biopolymers. Water 31, 369–376.
Roux, P., Boutin, C., Risch, E., Héduit, A., 2010. Life cycle environmental assessment (LCA) of sanitation systems including sewerage: case of vertical flow constructed wetlands versus activated sludge. In: Proceedings of the 12th IWA International Conference on Wetland Systems for Water Pollution Control, 4-8 October, Venice, Italy, pp. 4–8.
Rubio-Loza, L.A., Noyola, A., 2010. Two-phase (acidogenic-methanogenic) anaerobic thermophilic/mesophilic digestion system for producing Class A biosolids from municipal sludge. Bioresource Technology 101, 576–585.
Ruggeri, L., Gea, T., Artola, A., Sánchez, A., 2008. Influence of different co-substrates biochemical composition on raw sludge co-composting. Biodegradation 19, 403–425.
Rulkens, W.H., 2004. Sustainable sludge management—what are the challenges for the future? Water Science and Technology 49 (10), 11–19.
Rustige, H., Tomac, I., Höner, G., 2003. Investigations on phosphorus retention in subsurface flow constructed wetlands. Water Science and Technology 48 (5), 67–74.
RZWQM Team, 2010. Root Zone Water Quality Model, Version 2.1, Technical Documentation: GPSR Technical Report No. 2. USDA-ARS-GPSR, Fort Collins, CO.
Saeed, T., Sun, G., 2011a. A comparative study on the removal of nutrients and organic matter in wetland reactors employing organic media. Chemical Engineering Journal 171 (2), 439–447.
Saeed, T., Sun, G., 2011b. The removal of nitrogen and organics in vertical flow wetland reactors: Predictive models. Bioresource Technology 102, 1205–1213.
Saeed, T., Sun, G., 2012. A review on nitrogen and organics removal mechanisms in subsurface flow constructed wetlands: Dependency on environmental parameters, operating conditions and supporting media. Journal of Environmental Management 112, 429–448.
Saeed, T., Afrin, R., Al Muyeed, A., Sun, G., 2012. Treatment of tannery wastewater in a pilot-scale hybrid constructed wetland system in Bangladesh. Chemosphere 88, 1065–1073.
Sakadevan, K., Bavor, H.J., 1998. Phosphate adsorption characteristics of soils, slags and zeolite to be used as substrates in constructed wetland systems. Water Research 32 (2), 393–399.
Salati Jr., E., Salati, E., Salati, E., 1999. Wetland projects developed in Brazil. Water Science and Technology 40 (3), 19–25.
Samaras, P., Papadimitriou, C.A., Haritou, I., Zouboulis, A.I., 2008. Investigation of sewage sludge stabilization potential by the addition of fly ash and lime. Journal of Hazardous Materials 154, 1052–1059.

Sands, Z., Gill, L.S., Rust, R., 2000. Effluent treatment reed beds: results after ten years of operation. In: Means, J.F., Hinchee, R.E. (Eds.), Wetlands and Remediation. Battelle Press, Columbus, OH, pp. 273–279.

Sani, A., Scholz, M., Babatunde, A., Wang, Y., 2013. Impact of water quality parameters on the clogging of vertical-flow constructed wetlands treating urban wastewater. Water, Air, and Soil Pollution 224, 1487–1505.

Santona, L., Castaldi, P., Melis, P., 2006. Evaluation of the interactions mechanisms between red muds abd heavy metals. Journal of Hazardous Materials B136, 324–329.

Santos, D., Silva, W., Gomes, A., Simões, R., Pascoa, R., Albuquerque, A., Stefanakis, A.I., 2012. Treatment of cork processing industrial effluent with an innovative system of Constructed Wetland and Ozonation. In: Proceedings of the Symposium on Sustainable Processes and Materials, Research Group on Textile and Paper Materials, 6–7 December, University of Beira Interior, Covilhã, Portugal.

Sawaittayothin, V., Polprasert, C., 2007. Nitrogen mass balance and microbial analysis of constructed wetlands treating municipal landfill leachate. Bioresource Technology 98, 565–570.

Schafer, P.L., Farrell, J.B., Newman, G., Vandenburgh, S., 2002. Advanced anaerobic digestion performance comparisons. In: Proceedings of the Water Environment Federation, WEFTEC 2002, pp. 468–484.

Schelling, M., Conti, S., 1986. Host receptor sites involved in the attachment of Bdellovibrio bacteriovorous and Bdellovibrio stolpii. FEMS Microbiology 2–3, 319–323.

Schijven, J.F., de Bruin, H.A.M., Hassanizadeh, S.M., de Roda, H.A.M., 2003. Bacteriophages and clostridium spores as indicator organisms for removal of pathogens by passage through saturated dune sand. Water Research 37, 2186–2194.

Scholz, M., Xu, J., 2002a. Comparison of constructed reed beds with different filter media and macrophytes treating urban stream water contaminated with lead and copper. Ecological Engineering 18, 385–390.

Scholz, M., Xu, J., 2002b. Performance comparison of experimental constructed wetlands with different filter media and macrophytes treating industrial wastewater contaminated with lead and cooper. Bioresource Technology 83, 71–79.

Schönerklee, M., Koch, F., Perfler, R., Haberl, R., Laber, J., 1997. Tertiary treatment in a vertical flow reed bed system - a full scale pilot plant for 200–600 p.e. Water Science and Technology 35 (5), 223–230.

Schuyt, K., and Brander, L. 2004. The economic values of the world's wetlands. WWF International. Prepared with support from the Swiss Agency for the Environment, Forests and Landscape (SAEFL), Gland/Amsterdam, January.

Schwager, A., Boller, M., 1997. Transport phenomena in intermittent filters. Water Science and Technology 35 (6), 13–20.

Seeger, E.M., Reiche, N., Kuschk, P., Borsdorf, H., Kästner, M., 2011a. Performance evaluation using a three compartment mass balance for the removal of volatile organic compounds in pilot scale constructed wetlands. Environmental Science and Technology 45 (19), 8467–8474.

Seeger, E.M., Kuschk, P., Fazakeas, H., Grathwohl, P., Kästner, M., 2011b. Bioremediation of benzene-, MTBE- and ammonia-contaminated groundwater with pilot-scale constructed wetlands. Environmental Pollution 159, 3769–3776.

Seidel, K. 1953. Pflanzungen zwischen Gewässern und Land. Mitteilungen Max-Planck Gesselschaft, 17–20.

Seidel, K., 1965. Neue Wege zur Grundwasseranreicherung in Krefeld, Vol. II. Hydrobotanische Reinigungsmethode. GWF Wasser/Abwasser 30, 831–833.

Seidel, K., 1966. Reinigung von Gewassern durch hohere Pflanzen. Deutsche Naturwissenschaft 12, 297–298.

Sélas, B., Lakel, A., Andres, Y., Le Cloirec, P., 2002. Wastewater reuse in on-site wastewater treatment: bacteria and virus movement in unsaturated flow through sand filter. Water Science and Technology 47 (1), 59–64.

Sellami, F., Hachicha, S., Chtourou, M., Medhioub, K., Ammar, E., 2008. Maturity assessment of composted olive mill wastes using UV spectra and humification parameters. Bioresource Technology 99 (15), 6900–6907.

Seo, D.C., Hwang, S.H., Kim, H.J., Cho, J.S., Lee, H.J., DeLaune, R.D., Jugsujinda, A., Lee, S.T., Seo, J.Y., Heo, J.S., 2008. Evaluation of 2- and 3-stage combinations of vertical and horizontal flow constructed wetlands for treating greenhouse wastewater. Ecological Engineering 32, 121–132.

Serrano, L., de la Varga, D., Ruiz, I., 2011. Winery wastewater treatment in a hybrid constructed wetland. Ecological Engineering 37 (5), 744–753.

Shackle, V., Freeman, C., Reynolds, B., 2006. Exogenous enzyme supplements to promote treatment efficiency in constructed wetlands. Science of the Total Environment 361, 18–24.

Shamir, E., Thompson, T.L., Karpiscak, M.M., Freitas, R.J., Zauderer, J., 2001. Nitrogen accumulation in a constructed wetland for dairy wastewater treatment. Journal of the American Water Resources Association 37 (2), 315–325.

Shapiro, O.H., Kushmaro, A., 2011. Bacteriophage ecology in environmental biotechnology processes. Current Opinion in Biotechnology 22, 449–455.

Shapiro, O.H., Kushmaro, A., Brenner, A., 2010. Bacteriophage predation regulates microbial abundance and diversity in a full-scale bioreactor treating industrial wastewater. The ISME Journal 4, 327–336.

Sheoran, A.S., Sheoran, V., 2006. Heavy metal removal mechanisms of acid mine drainage in wetlands: A critical review. Minerals Engineering 19, 105–116.

Sheta, A.S., Falatah, A.M., Al-Sewailem, M.S., Khaled, E.M., Sallam, A.S.H., 2003. Sorption characteristics of zinc and iron by natural zeolite and bentonite. Microporous and Mesoporous Materials 61, 127–136.

Shipin, O., Koottatep, T., Khanh, N.T.T., Polprasert, C., 2005. Integrated natural treatment systems for developing communities: low-tech N-removal through the fluctuating microbial pathways. Water Science and Technology 51 (12), 299–306.

Shutes, R.B.E., 2001. Artificial wetlands and water quality improvement. Environment International 26, 441–447.

Shutes, R.B.E., Revitt, D.M., Mungur, A.S., Scholes, L.N.L., 1997. The design of wetland systems for the treatment of urban run off. Water Science and Technology 35 (5), 19–25.

Shutes, B., Revitt, M., Scholes, L., 2010. Constructed wetlands for flood prevention and water reuse. In: Proceedings of the 12th International Conference on Wetland Systems for Water Pollution Control, 4th-8th Oct 2010, Venice, Italy.

Simunek, J., Sejna, M., van Genuchten, M.T., 1999. The HYDRUS-2D software package for simulating the two-dimensional movement of water, heat and multiple solutes in variably-saturated media-Version 2.0, Riverside, Cal., USDA.

Silvan, N., Vasander, H., Karsisto, M., Laine, J., 2003. Microbial immobilisation of added nitrogen and phosphorus in constructed wetland buffer. Applied Soil Ecology 24 (2), 143–149.

Singh, S., Haberl, R., Moog, O., Shrestha, R.R., Shrestha, P., Shrestha, R., 2009. Performance of an anaerobic baffled reactor and hybrid constructed wetland treating high-strength wastewater in Nepal-a model for DEWATS. Ecological Engineering 35, 654–660.

Sklarz, M.Y., Gross, A., Yakirevich, A., Soares, M.I.M., 2009. A recirculating vertical flow constructed wetland for the treatment of domestic wastewater. Desalination 246, 617–624.

Sklarz, M.Y., Gross, A.M., Soares, I.M., Yakirevich, A., 2010. Mathematical model for analysis of recirculating vertical flow constructed wetlands. Water Research 44, 2010–2020.

Sklarz, M.Y., Gillor, O., Gross, A., Yakirevich, A., Soares, M.I.M., 2011. Microbial diversity and community composition in recirculating vertical flow constructed wetlands. Water Science and Technology 64 (11), 2306–2315.

Sklarz, M.Y., Gillor, O., Gross, A., Yakirevich, A., Soares, M.I.M., 2011. Microbial diversity and community composition in recirculating vertical flow constructed wetlands. Water Science and Technology 64 (11), 2306–2315.

Skousen, J., Ziemkiewica, P., 2005. Performance of 116 passive treatment systems for acid mine drainage. In: Proceedings of National Meeting of the American Society of Mining and Reclamation, 19–23 June, Breckenridge, CO.

Sleytr, K., Tietz, A., Langergraber, G., Haberl, R., 2007. Investigation of bacterial removal during the filtration process in constructed wetlands. Science of the Total Environment 380, 173–180.

Smith, E., Gordon, R., Madani, A., Stratton, G., 2005. Cold climate hydrological flow characteristics of constructed wetlands. Canadian Biosystems Engineering 47, 1–7.

Smith Jr., J.E., Surampalli, R.Y., Reimers, R.S., Tyagi, R.D., Lohani, B.N., 2008. Disinfection processes and stability refinements to biosolids treatment technologies. Practice Periodical of Hazardous Toxic and Radioactive Waste Management 12 (1), 10–17.

Snyman, H.G., Forssman, P., Kafaar, A., Smollen, M., 2000. The feasibility of electro-osmotic belt filter dewatering technology at pilot scale. Water Science and Technology 41 (8), 137–144.

Sobolewski, A., 1999. A review of processes responsible for metal removal in wetlands treating contaminated mine drainage. International Journal of Phytoremediation 1 (1), 19–51.

Song, H.L., Nakano, K., Taniguchi, T., Nomura, M., Nishimura, O., 2009. Estrogen removal from treated municipal effluent in small-scale constructed wetland with different depth. Bioresource Technology 100, 2945–2951.

Soto, F., Garcia, M., de Luis, E., Becares, E., 1999. Role of *Scirpus lacustris* in bacterial and nutrient removal from wastewater. Water Science and Technology 40, 241–247.

Søvik, A.K., Augustin, J., Heikkinen, K., Huttunen, J.T., Necki, J.M., Karjalainen, S.M., Kløve, B., Liikanen, A., Mander, Ü., Puustinen, M., Teiter, S., Wachniev, P., 2006. Emission of the greenhouse gases nitrous oxide and methane from constructed wetlands in Europe. Journal of Environmental Quality 35, 2360–2373.

Spinosa, L., 2001. Evolution of sewage sludge regulations in Europe. Water Science and Technology 44 (10), 1–8.

Spinosa, L., 2004. From sludge to resources through biosolids. Water Science and Technology 5 (9), 1–8.

Stefanakis, A.I., 2011. Design parameter optimization of vertical flow constructed wetlands for wastewater treatment and sludge dewatering using pilot-scale units. Department of Environmental Engineering, Democritus University of Thrace, Xanthi, Greece, Doctoral dissertation, 333 p (in Greek).

Stefanakis, A.I., Tsihrintzis, V.A., 2009a. Effect of outlet water level raising and effluent recirculation on removal efficiency of pilot-scale horizontal subsurface Flow Constructed Wetlands. Desalination 248 (1–3), 961–976.

Stefanakis, A.I., Tsihrintzis, V.A., 2009b. Performance of pilot-scale vertical flow constructed wetlands treating simulated municipal wastewater: effect of various design parameters. Desalination 248, 753–770.

Stefanakis, A.I., Tsihrintzis, V.A., 2011. Dewatering mechanisms in pilot-scale sludge drying reed beds: effect of design and operational parameters. Chemical Engineering Journal 172, 430–443.

Stefanakis, A.I., Tsihrintzis, V.A., 2012a. Effects of loading, resting period, temperature, porous media, vegetation and aeration on performance of pilot-scale vertical flow constructed wetlands. Chemical Engineering Journal 181–182, 416–430.

Stefanakis, A.I., Tsihrintzis, V.A., 2012b. Use of zeolite and bauxite in filter media treating the effluent of vertical flow constructed wetlands. Microporous and Mesoporous Materials 155, 106–116.

Stefanakis, A.I., Tsihrintzis, V.A., 2012c. Effect of various design and operation parameters on performance of pilot-scale sludge drying reed beds. Ecological Engineering 38, 65–78.

Stefanakis, A.I., Tsihrintzis, V.A., 2012d. Heavy metal fate in pilot-scale sludge drying reed beds under various design and operation conditions. Journal of Hazardous Materials 213–214, 393–405.

Stefanakis, A.I., Akratos, C.S., Gikas, G.D., Tsihrintzis, V.A., 2009a. Effluent quality improvement of two pilot-scale, horizontal subsurface flow constructed wetlands using natural zeolite (clinoptilolite). Microporous and Mesoporous Materials 124 (1–3), 131–143.

Stefanakis, A.I., Akratos, C.A., Melidis, P., Tsihrintzis, V.A., 2009b. Surplus activated sludge dewatering in pilot-scale sludge drying reed beds. Journal of Hazardous Materials 172 (2–3), 1122–1130.

Stefanakis, A.I., Akratos, C.S., Tsihrintzis, V.A., 2011a. Effect of wastewater step-feeding on removal efficiency of pilot-scale horizontal subsurface flow constructed wetlands. Ecological Engineering 37 (3), 431–443.

Stefanakis, A.I., Komilis, D.P., Tsihrintzis, V.A., 2011b. Stability and maturity of thickened wastewater sludge treated in pilot-scale sludge treatment wetlands. Water Research 45, 6441–6452.

Stefanakis, A.I., Seeger, E., Hübschmann, T., Müller, S., Sinke, A., Thullner, M., 2013. Investigation of phenol and m-cresol biodegradation in horizontal subsurface flow constructed wetlands. In: Proceedings of the 5th International Symposium on Wetland Pollutant Dynamics and Control, 13–17 October, Nantes, France, pp. 13–17.

Stein, O.R., Hook, P.B., 2005. Temperature, plants and oxygen: how does season affect constructed wetland performance? Journal of Environmental Science and Health, Part A 40 (6–7), 1331–1342.

Stenstrom, M.K., Song, S.S., 1991. Effect of oxygen transport limitation on nitrification in the activated sludge process. Research Journal of the Water Pollution Control Federation 63 (3), 208–219.

Stevik, T.K., Aa, K., Ausland, G., Hanssen, J.F., 2004. Retention and removal of pathogenic bacteria in wastewater percolating through porous media: a review. Water Research 38, 1355–1367.

Stewart, F.M., Mulholland, T., Cunningham, A.B., Kania, B.G., Osterlund, M.T., 2008. Floating islands as an alternative to constructed wetlands for treatment of excess nutrients from agricultural and municipal wastes—results of laboratory-scale tests. Land Contamination and Reclamation 16 (1), 25–33.

Strauss, M., Drescher, S., Zurbrügg, C., Montangero, A., Cofie, O., Drechsel, P., 2003. Co-Composting of Faecal Sludge and Municipal Organic Waste. A Literature and State-of-Knowledge Review. Swiss Federal Institute of Environmental Science & Technology (EAWAG), Dept. of Water & Sanitation in Developing Countries (SANDEC), Switzerland, International Water Management Institute (IWMI), Ghana.

Stone, K.C., Poach, M.E., Hunt, P.G., Reddy, G.B., 2004. Marsh-pond-marsh constructed wetland design analysis for swine lagoon wastewater treatment. Ecological Engineering 23, 127–133.

Stott, R., May, E., Matsushita, E., Warren, A., 2001. Protozoan predation as a mechanism for the removal of Cryptosporidium oocysts from wastewaters in constructed wetlands. Water Science and Technology 44, 191–198.

Stott, R., May, E., Mara, D.D., 2003. Parasite removal by natural wastewater treatment systems: performance of waste stabilization ponds and constructed wetlands. Water Science and Technology 48 (2), 97–104.

Stottmeister, U., Wiessner, A., Kuschk, P., Kappelmeyer, U., Kästner, M., Bederski, O., Müller, R.A., Moormann, H., 2003. Effects of plants and microorganisms in constructed wetlands for wastewater treatment. Biotechnology Advances 22, 93–117.

Stottmeister, U., Buddhawong, S., Kuschk, P., Wiessner, A., Mattusch, J., 2006. Constructed wetlands and their performance for treatment of water contaminated with Arsenic and heavy metals. In: Twardowska, I. et al., (Ed.), Soil and Water Pollution Monitoring, Protection and Remediation. Springer, pp. 417–432.

Strauss, M., Heinss, U., Montangero, A., 1999. Excreta news—field research progress in faecal sludge (FS) treatment. SANDEC News 4, 22–27, EAWAG/SANDEC, Switzerland.

Strauss, M., Drescher, S., Zurbrügg, C., Montangero, A., Cofie, O., Drechsel, P. 2003. Co-composting of faecal sludge and municipal organic waste. A literature and state-of-knowledge review. Swiss Federal Institute of Environmental Science & Technology (EAWAG), Dept. of Water & Sanitation in Developing Countries (SANDEC), Switzerland, International Water Management Institute (IWMI), Ghana.

Suárez, S., Carballa, M., Omil, F., Lema, J.M., 2008. How are pharmaceutical and personal care products (PPCPs) removed from urban wastewaters? Reviews in Environmental Science and Bio/Technology 7, 125–138.

Summerfelt, T.S., Adler, R.P., Glenn, M.D., Kretschmann, N.R., 1999. Aquaculture sludge removal and stabilization within created wetlands. Aquacultural Engineering 19, 81–92.

Sun, G., Austin, D., 2007a. A mass balance study on nitrification and deammonification in vertical flow constructed wetlands treating landfill leachate. Water Science and Technology 56 (3), 117–123.

Sun, G., Austin, D., 2007b. Completely autotrophic nitrogen-removal over nitrite in lab-scale constructed wetlands: Evidence from a mass balance study. Chemosphere 68, 1120–1128.

Sun, G., Gray, K.R., Biddlestone, A.J., 1998. Treatment of agricultural and domestic effluents in constructed downflow reed beds employing recirculation. Environmental Technology 19 (5), 529–536.

Sun, G., Gray, K.R., Biddlestone, A.J., 1999a. Treatment of agricultural wastewater in a pilot-scale tidal flow reed bed system. Environmental Technology 20 (2), 233–237.

Sun, G., Gray, K.R., Biddlestone, A.J., Cooper, D.J., 1999b. Treatment of agricultural wastewater in a combined tidal flow-downflow reed bed system. Water Science and Technology 40 (3), 139–146.

Sun, G., Gray, K.R., Biddlestone, A.J., Allen, S.J., Cooper, D.J., 2003. Effect of effluent recirculation on the performance of a reed bed system treating agricultural wastewater. Process Biochemistry 39, 351–357.

Sun, G., Zhao, Q.Y., Allen, S.J., 2005. Enhanced removal of organic matter and ammoniacal-nitrogen in a column experiment of tidal flow constructed wetland system. Journal of Biotechnology 115 (2), 189–197.

Sun, G., Zhao, Q.Y., Allen, S.J., Cooped, D., 2006. Generating "Tide" in pilot-scale constructed wetlands to enhance agricultural wastewater treatment. Engineering in Life Sciences 6 (6), 560–565.

Sun, G., Zhao, Q.Y., Allen, S.J., 2007. An alternative arrangement of gravel media in tidal flow reed beds treating pig farm wastewater. Water, Air, and Soil Pollution 182, 13–19.

Sundaravadivel, M., Vigneswaran, S., 2001. Constructed wetlands for wastewater treatment. Critical Reviews in Environmental Science and Technology 31 (4), 351–409.

Susarla, S., Medina, V.F., McCutcheon, S.C., 2002. Phytoremediation: an ecological solution to organic chemical contamination. Ecological Engineering 18, 647–658.

Tan, K., Jackson, W.A., Anderson, T.A., Pardue, J.H., 2004. Fate of perchlorate-contaminated water in upflow wetlands. Water Research 38, 4173–4185.

Tang, X.Q., Li, J.Z., Li, X.J., Liu, X.G., Huang, S.L., 2008. Optimization of nitrogen and phosphorus removal in vertical subsurface flow constructed wetlands by using polypropylene pellet as part of substrate. Huanjing Kexue/Environmental Science 29 (5), 1284–1288.

Tang, X., Eke, P.E., Scholz, M., Huang, S., 2009. Processes impacting on benzene removal in vertical-flow constructed wetlands. Bioresource Technology 100, 227–234.

Tang, X., Huang, S., Scholz, M., Li, J., 2011. Nutrient removal in vertical subsurface flow constructed wetlands treating eutrophic river water. International Journal of Environmental Analytical Chemistry 91 (7–8), 727–739.

Tanner, C.C., 1996. Plants for constructed wetland treatment systems – A comparison of the growth and nutrient uptake of eight emergent species. Ecological Engineering 7, 59–83.

Tanner, C.C., 2001. Plants as ecosystem engineers in subsurface-flow treatment wetlands. Water Science and Technology 44 (11), 9–17.

Tanner, C.C., Headley, T.R., 2011. Components of floating emergent macrophyte treatment wetlands influencing removal of stormwater pollutants. Ecological Engineering 37, 474–486.

Tanner, C.C., Clayton, J.S., Upsdell, M.P., 1995. Effect of loading rate and planting on treatment of dairy farm wastewaters in constructed wetlands - i. removal of oxygen demand, suspended solids and faecal coliforms. Water Research 29, 17–26.

Tanner, C.C., Sukias, J.P.S., Upsdell, M.P., 1998. Organic matter accumulation during maturation of gravel-bed constructed wetlands treating farm dairy wastewaters. Water Research 32 (10), 3046–3054.

Tanner, C.C., Kadlec, R.H., Gibbs, M.M., Sukias, J.P.S., Nguyen, M.L., 2002. Nitrogen processing gradients in subsurface-flow treatment wetlands -influence of wastewater characteristics. Ecological Engineering 18, 499.

Tanner, C.C., Nguyen, M.L., Sukias, J.P.S., 2005. Nutrient removal by a constructed wetland treating subsurface drainage from grazed dairy pasture. Agricultural Ecosystems Environment 105, 145–162.

Tao, W., Wang, J., 2009. Effects of vegetation, limestone, and aeration on nitritation, anammox and denitrification in wetland treatment systems. Ecological Engineering 35 (5), 836–842.

Tao, M., He, F., Xu, D., Li, M., Wu, Z., 2010. How artificial aeration improved sewage treatment of an integrated vertical-flow constructed wetland. Polish Journal of Environmental Studies 19 (1), 183–191.

Tchobanoglous, G., Kreith, F., 1998. Handobook of Solid Waste Management. McGraw and Hill Professional.

Tee, H.C., Seng, C.E., Noor, A.M., Lim, P.E., 2009. Performance comparison of constructed wetlands with gravel- and rice husk-based media for phenol and nitrogen removal. Science of the Total Environment 407 (11), 3563–3571.

Teiter, S., Mander, Ü., 2005. Emission of N_2O, N_2, CH_4, and CO_2 from constructed wetlands for wastewater treatment and from riparian buffer zones. Ecological Engineering 25, 528–541.

Third, K.A., Olav Sliekers, A., Kuenen, J.G., Jetten, M.S.M., 2001. The CANON system (Completely autotrophic nitrogen removal over nitrate) under ammonium limitation: interaction and completion between three groups of bacteria. Systematic and Applied Microbiology 24, 588–596.

Thomas, R.C., Romanek, C.S., 2002. Passive treatment of low-pH, ferric iron-dominated acid rock drainage in a vertical flow wetland I: acidity neutralization and alkalinity generation. In: National Meeting of the American Society of Mining and Reclamation, 19–23 June, Lexington KY.

Thorsell, J., Levy, R.F., Sigaty, T., 1997. A global overview of wetland and marine protected areas on the world heritage list. In: Natural Heritage Programme, IUCN, Gland, Switzerland. In collaboration with The World Conservation Monitoring Centre, September.

Thurston, J.A., Gerba, C.P., Foster, K.E., Karpiscak, M.M., 2001. Fate of indicator microorganisms, *Giardia* and *Cryptosporodium* in subsurface flow constructed wetlands. Water Research 35 (6), 1547–1551.

Tietz, A., Hornek, R., Langergraber, G., Kreuzinger, N., Haberl, R., 2007a. Diversity of ammonia oxidising bacteria in a vertical flow constructed wetland. Water Science and Technology 56 (3), 241–247.

Tietz, A., Kirschner, A., Langergraber, G., Sleytr, K., Haberl, R., 2007b. Characterisation of microbial biocoenosis in vertical subsurface flow constructed wetlands. Science of the Total Environment 380, 163–172.

Tietz, A., Langergraber, G., Watzinger, A., Haberl, R., Kirschner, K.T.A., 2008. Bacterial carbon utilization in vertical subsurface flow constructed wetlands. Water Research 42 (6–7), 1622–1634.

Torrens, A., Molle, P., Boutin, C., Salgot, M., 2009a. Impact of design and operation variables on the performance of vertical-flow constructed wetlands and intermittent sand filters treating pond effluent. Water Research 43, 1851–1858.

Torrens, A., Molle, P., Boutin, C., Salgot, M., 2009b. Removal of bacterial and viral indicators in vertical flow constructed wetlands and intermittent sand filters. Desalination 247, 170–179.

Toscano, A., Langergraber, G., Consoli, S., Cirelli, G.L., 2009. Modelling pollutant removal in a pilot-scale two-stage subsurface flow constructed wetlands. Ecological Engineering 35, 281–289.

Trias, M., Hu, Z., Mortula, M.M., Gordon, R.J., Gagnon, G.A., 2004. Impact of seasonal variation on treatment of swine wastewater. Environmental Technology 25, 775–781.

Troesch, S., Liénard, A., Molle, P., Merlin, G., Esser, D., 2009a. Treatment of septage in sludge drying reed beds: a case study on pilot-scale beds. Water Science and Technology 60 (3), 643–653.

Troesch, S., Liénard, A., Molle, P., Merlin, G., Esser, D., 2009b. Sludge drying reed beds: a full and pilot-scales study for activated sludge treatment. Water Science and Technology 60 (5), 1145–1154.

Tsagarakis, K.P., Mara, D.D., Angelakis, A.N., 2003. Application of cost criteria for selection of municipal wastewater treatment systems. Water, Air, and Soil Pollution 142, 187–210.

Tsihrintzis, V.A., Gikas, G.D., 2010. Constructed wetlands for wastewater and activated sludge treatment in North Greece: a review. Water Science and Technology 61 (10), 2653–2672.

Tsihrintzis, V.A., Hamid, R., 1997. Modeling and management of urban stormwater runoff quality: a review. Water Resources Management 11 (2), 137–164.

Tsihrintzis, V.A., Vasarhelyi, G.M., Lipa, J., 1995a. Hydrodynamic and constituent transport modeling of coastal wetlands. Journal of Marine Environmental Engineering 1 (4), 295–314.

Tsihrintzis, V.A., Vasarhelyi, G.M., Lipa, J., 1995b. Multiobjective approaches in freshwater wetland restoration and design. Water International 20 (2), 98–105.

Tsihrintzis, V.A., Vasarhelyi, G.M., Lipa, J., 1996. Ballona Wetland: A multi-objective salt marsh restoration plan. Water, Maritime and Energy 118 (2), 131–144.

Tsihrintzis, V.A., John, D.L., Tremblay, P.J., 1998. Hydrodynamic modeling of wetlands for flood detention. Water Resources Management 12 (4), 251–269.

Tsihrintzis, V.A., Akratos, C.S., Gikas, G.D., Karamouzis, D., Angelakis, A.N., 2007. Performance and cost comparison of a FWS and a VFS constructed wetlands system. Environmental Technology 28, 621–628.

Tunçsiper, B., 2009. Nitrogen removal in a combined vertical and horizontal subsurface-flow constructed wetland system. Desalination 247, 466–475.

Tunçsiper, B., Ayaz, S., Akça, L., Gunes, K., 2009. Performance of a pilot-scale, three-stage constructed wetland system for domestic wastewater treatment. Environmental Technology 30 (11), 1187–1194.

Tunçsiper, B., Ayaz, S., Akça, L., 2012. Coliform bacteria removal from septic wastewater in a pilot-scale combined constructed wetland system. Environmental Engineering and Management Journal 11 (10), 1873–1879.

Turonskiy, I.S., Mathai, P.K., 2006. Wastewater Sludge Processing. John Wiley & Sons, Inc., Hoboken, New Jersey, USA.

Tuszynska, A., Obarska-Pempkowiak, H., 2008. Dependence between quality and removal effectiveness of organic matter in hybrid constructed wetlands. Bioresource Technology 99, 6010–6016.

Uggetti, E., Llorens, E., Pedescol, A., Ferrer, I., Castellnou, R., García, J., 2009. Sludge dewatering and stabilization in drying reed beds: characterization of three full-scale systems in Catalonia, Spain. Bioresource Technology 100 (17), 3882–3890.

Uggetti, E., Ferrer, I., Llorens, E., García, J., 2010. Sludge treatment wetlands: a review on the state of the art. Bioresource Technology 101, 2905–2912.

Uggetti, E., Ferrer, I., Molist, J., García, J., 2011. Technical, economic and environmental assessment of sludge treatment wetlands. Water Research 45, 573–582.

Uggetti, E., Ferrer, I., Carretero, J., García, J., 2012a. Performance of sludge treatment wetlands using different plant species and porous media. Journal of Hazardous Materials 217–218, 263–270.

Uggetti, E., Ferrer, I., Nielsen, S., Arias, C., Brix, H., García, J., 2012b. Characteristics of biosolids from sludge treatment wetlands for agricultural reuse. Ecological Engineering 40, 210–216.

Uggetti, E., Ferrer, I., Arias, C., Brix, H., García, J., 2012c. Carbon footprint of sludge treatment reed beds. Ecological Engineering 44, 298–302.

Ulrich, H., Klaus, D., Irmgard, F., Annette, H., Juan, L.-P., Regine, S., 2005. Microbiological investigations for sanitary assessment of wastewater treated in constructed wetlands. Water Research 39, 4849–4858.

UN, 2003. Wastewater Treatment Technologies: A General Review. United Nation, New York, 21 September.

UNEP, 1997. Sourcebook of Alternative Technologies for Freshwater Augmentation in Latin America and the Caribbean. Available at: http://www.unep.or.jp/ietc/publications/techpublications/techpub-8c/.

UNIDO, 2011. Inroduction to treatment of tannery effluents. U.N. I. D. Organisation, Vienne, Austria.

Urbanc-Bercic, O., 1997. Constructed wetlands for the treatment of landfill leachates: the Slovenian experience. Wetlands Ecology and Management 4, 189–197.

USEPA, 1972. Federal Water Pollution Control Act. As amended through P.L. 107–303, November 21, 2002.

USEPA, 1988. Design Manual. Constructed Wetlands and Aquatic Plant Systems for Municipal Wastewater Treatment. U.S. Environmental Protection Agency: Office of Research and Development, Cincinnati, OH, USA.

USEPA, 1993. Standards for the Use and Disposal of Sewage Sludge, 40 CFR Part 503, 1993.

USEPA, 1995. A Handbook of Constructed Wetlands – Volume 1: General Considerations, USEPA Region III with USDA, NRCS.

USEPA, 1997. Wastewater Treatment Manuals – Primary, Secondary and Tertiary Treatment. Environmental Protection Agency, Ardcavan, Wexford, Ireland.

USEPA, 2000a. Wastewater Technology Fact Sheet. Wetlands: Subsurface Flow. EPA 832-F-00-023; U.S. Environmental Protection Agency: Office of Water, Washington, D.C., USA.

USEPA, 2000b. Wastewater Technology Fact Sheet: Free Water Surface Wetlands. EPA 832-F-00-024; U.S. Environmental Protection Agency: Office of Water, Washington, D.C., USA.

USEPA, 2000c. Decentralized Systems Technology Fact Sheet, Septic System Tank. EPA 832-F-00-040; U.S. Environmental Protection Agency: Office of Water, Washington, D.C., USA.

USEPA, 2000d. Biosolids Technology Fact Sheet: Centrifuge Thickening and Dewatering: EPA 832-F-00-053. U.S. Environmental Protection Agency: Office of Water, Washington, D.C., USA.

USEPA, 2002. Onsite Wastewater Treatment Systems Manual. EPA/625/R-00/008, U.S. Environmental Protection Agency: Office of Water, Washington, D.C., USA.

USEPA, 2003. Environmental Regulations and Technology: Control of Pathogens and Vector Attraction in Sewage Sludge: EPA/625/R-92/013, revised July, 2003, Cincinnati, OH.

USEPA, 2004. Constructed Treatment Wetlands. EPA 843-F-03-013. U.S. Environmental Protection Agency: Office of Water, Washington, D.C., USA.

Vacca, G., Wand, H., Nikolausz, M., Kuschk, P., Kästner, M., 2005. Effects of plants and filter materials on bacteria removal in pilot-scale constructed wetlands. Water Research 39, 1361–1373.

Vaccari, D.A., Storm, P.F., Alleman, J.E., 2006. Environmental Biology for Engineers and Scientists. John Wiey & Sons, Inc., Hoboken, New Jersey.

van Afferden, M., Rahman, K.Z., Mosig, P., De Biase, C., Thullner, M., Oswald, S.E., Müller, R.A., 2011. Remediation of groundwater contaminated with MTBE and benzene: the potential of vertical-flow soil filter systems. Water Research 45, 5063–5074.

Vandaele, S., Thoeye, C., Van Eygen, B., De Gueldre, G., 2000. Small wastewater treatment plants in Flanders (Belgium): standard approach and experiences with constructed reed beds. Water Science and Technology 41 (1), 57–63.

Vandenburgh, S.R., Ellis, T.G., 2002. Effect of varying solids concentration and organic loading on the performance of temperature phased anaerobic digestion process. Water Environment Research 74 (2), 142–148.

van de Graaf, A.A., Mulder, A., Debruijn, P., Jetten, M.S.M., Robertson, L.A., Kuenen, J.G., 1995. Anaerobic oxidation of ammonium is a biologically mediated process. Applied Environmental Microbiology 61, 1246–1251.

van de Moortel, A.M.K., Meers, E., De Pauw, N., Tack, F.M.G., 2010. Effects of vegetation, season and temperature on the removal of pollutants in Experimental floating treatment wetlands. Water, Air, and Soil Pollution 212 (1–4), 281–297.

Van Genuchten, M.T., 1982. A comparison of numerical solutions of the one-dimensional unsaturated-saturated flow and mass transport equations. Advanced Water Resources 5, 47–55.

Van Shoe, P.M., Young, L.Y., 2000. Biodegradation of phenol: mechanisms and applications. Bioremediation Journal 4 (1), 1–18.

Vandenburgh, S.R., Timothy, G.E., 2002. Effect of varying solids concentration and organic loading on the performance of temperature phased anaerobic digestion process. Water Environment Research 74 (2), 142–148.

Varado, N., Braud, I., Ross, P.J., 2006. Development and assessment of an efficient vadose zone module solving the 1D Richards' equation and including root extraction by plants. Journal of Hydrology 323, 257–258.

Vera, L., Martel, G., Márquez, M., 2013. Two years monitoring of the natural system for wastewater reclamation in Santa Lucía, Gran Canaria Island. Ecological Engineering 21–30, .
Verhoeven, J.T.A., Meuleman, A.F.M., 1995. Wetlands for wastewater treatment: Opportunities and limitations. Ecological Engineering 12, 5–12.
Verhoeven, J.T.A., Meuleman, A.F.M., 1999. Wetlands for wastewater treatment: Opportunities and limitations. Ecological Engineering 12, 5–12.
Verstraete, W., Vaerenbergh, E.V., 1986. Aerobic activated sludge. In: Rehm, H.J., Reed, G. (Eds.), Biotechnology. VCH, Germany.
Vesilind, A.P., Hsu, C.-C., 1997. Limits of sludge dewaterability. Water Science and Technology 36 (11), 87–91.
Vincent, J., Molle, P., Wiesniewski, C., Liénard, A., 2011. Sludge drying reed beds for septage treatment: towards design and operation recommendations. Bioresource Technology 102 (17), 8327–8330.
Visesmanee, V., Polprasert, C., Parkpian, P., 2008. Long-term performance of subsurface-flow constructed wetlands treating Cd wastewater. Journal of Environmental Science and Health, Part A 43, 765–771.
Vohla, C., Alas, R., Nurk, K., Baatz, S., Mander, Ü., 2007. Dynamics of phosphorus, nitrogen and carbon removal in a horizontal subsurface flow constructed wetland. Science of the Total Environment 380 (1–5), 66–74.
Vohla, C., Kõiv, M., Bavor, H.J., Chazarenc, F., Mander, Ü., 2011. Filter materials for phosphorus removal from wastewater in treatment wetlands – A review. Ecological Engineering 37, 70–89.
von Felde, K., Kunst, S., 1997. N- and COD-removal in vertical-flow systems. Water Science and Technology 35 (5), 79–85.
Vymazal, J., 2001. Transformations of Nutrients in Natural and Constructed Wetlands. Backhuys Publishers, Leiden, The Netherlands.
Vymazal, J., 2002. The use of sub-surface constructed wetlands for wastewater treatment in the Czech Republic: 10 years experience. Ecological Engineering 18, 633–646.
Vymazal, J., 2005a. Horizontal sub-surface flow and hybrid constructed wetlands systems for wastewater treatment. Ecological Engineering 25, 478–490.
Vymazal, J., 2005b. Removal of enteric bacteria in constructed treatment wetlands with emergent macrophytes: a review. Journal of Environmental Science and Health, Part A 40, 1355–1367.
Vymazal, J., 2007. Removal of nutrients in various types of constructed wetlands. Science of the Total Environment 380 (1–3), 48–65.
Vymazal, J., 2009. The use constructed wetlands with horizontal sub-surface flow for various types of wastewater. Ecological Engineering 35, 1–17.
Vymazal, J., 2011. Constructed Wetlands for wastewater treatment: five decades of experience. Environmental Science and Technology 45, 61–69.
Vymazal, J., 2013. Vegetation development in subsurface flow constructed wetlands in the Czech Republic. Ecological Engineering 61 (B), 575–581.
Vymazal, J., Krása, P., 2003. Distribution of Mn, Al, Cu and Zn in a constructed wetland receiving municipal sewage. Water Science and Technology 48 (5), 299–305.
Vymazal, J., Kröpfelová, L., 2011. A three-stage experimental constructed wetland for treatment of domestic sewage: First 2 years of operation. Ecological Engineering 37, 90–98.
Vymazal, J., Brix, H., Cooper, P.F., Green, M.B., Haberl, R., 1998. Constructed Wetlands for wastewater treatment in Europe. Backhuys Publishers, Leiden, The Netherlands.
Vymazal, J., Greenway, M., Tonderski, K., Brix, H., Mander, Ü., 2006. Constructed Wetlands for Wastewater Treatment. In: Verhoeven, J.T.A., Beltman, B., Bobbink, R., Whigham, D.F.

(Eds.), Wetlands and Natural Resource Management. Ecological Studies190, Springer-Verlag, Berlin, Germany, pp. 69–94.

Wakeman, R.J., 2007. Separation technologies for sludge dewatering. J. Hazard. Mater 144 (3), 614–619.

Waldron, J.M. 2007. Characterization of chlorinated ethane degradation in a Vertical Flow Constructed Wetland. M.Sc. Thesis, Department of Systems and Engineering Management, Graduate School of Engineering and Management, Air Force Institute of Technology, Air University, Ohio, USA.

Walker, D.J., Hurl, S., 2002. The reduction of heavy metals in a storm water wetland. Ecological Engineering 18 (4), 407–414.

Wallace, S.D. 2001. Patent: system for removing pollutants from water. United States: US 6,200,469 B1.

Wallace, S., Kadlec, R., 2005. BTEX degradation in a cold-climate wetland system. Water Science and Technology 51, 165–171.

Wallace, S.D., Knight, R.L., 2006. Small Scale Constructed Wetland Treatment Systems. Feasibility, Design Criteria, and O&M Requirements. Water Environ. Res. Foundation, Alexandria, VA, USA.

Wallace, S., Liner, M., 2011. Design and performance of the wetland treatment system at Buffalo Niagara International Airport. International Water Association's Specialist Group on Use of Macrophytes in Water Pollution Control 38, 36–42.

Wallace, S., Parkin, G., Cross, C., 2000. Cold climate Wetlands: design and performance. In: Proceedings of the 7th International Conference on Wetland Systems for Water Pollution Control, 11–16 November, Lake Buena Vista, Florida.

Wallace, S., Higgins, J., Crolla, A., Kinsley, C., Bachand, A., Verkuijl, S., 2006. High-rate ammonia removal in aerated Engineered Wetlands. In: Proceedings of the 10th International Conference on Wetland Systems for Water Pollution Control, 23–29 September, Lisbon, Portugal.

Walsh, M.L., 1995. Sludge handling and disposal—an American perspective. Water Quality International 2, 20–23.

Wand, H., Vacca, G., Kuschk, P., Krüger, M., Kästner, M., 2007. Removal of bacteria by filtration in planted and non-planted sand columns. Water Research 41, 159–167.

Wang, Y., Tian, Y., Han, B., Zhao, H.-B., Bi, J.-N., Cai, B.-L., 2007. Biodegradation of Biodegradation of phenol by free and immobilized *Acinetobacter* sp. strain PD12. Journal of Environmental Sciences 19, 222–225.

Wang, X., Han, B.-p., Shi, Y.-z., Pang, Z.-q., 2009a. Advanced wastewater treatment by integrated vertical flow constructed wetland with Vetiveria zizanioides in north China. Procedia Earth and Planetary Science 1, 1258–1262.

Wang, R., Korboulewsky, N., Prudent, P., Baldy, V., Bonin, G., 2009b. Can vertical-flow wetland systems treat high concentrated sludge from a food industry? A mesocosm experiment testing three plant species. Ecological Engineering 35, 230–237.

Wang, Q., Yang, L., Wu, Z., 2010a. Treatment efficiency of integrated vertical-flow constructed wetland for saline wastewater. Wuhan University Journal of Natural Sciences 15 (6), 544–548.

Wang, R., Korboulewsky, N., Prudent, P., Domeizel, M., Rolando, C., Bonin, G., 2010b. Feasibility of using an organic substrate in a wetland system treating sewage sludge: impact of plant species. Bioresource Technology 101, 51–57.

Wang, Z., Liu, C.-X., Li, P.-Y., Dong, J., Liu, L., Zhu, G.-F., 2012a. Study on phosphorus removal capability of constructed wetlands filled with broken bricks. Huanjing Kexue/Environmental Science 33 (12), 4373–4379.

Wang, R., Baldy, V., Périssol, C., Korboulewsky, N., 2012b. Influence of plants on microbial activity in a vertical-downflow wetland system treating waste activated sludge with high organic matter concentrations. Journal of Environmental Management 95, S158–S164.

Wang, Z., Dong, J., Liu, L., Zhu, G., Liu, C., 2013. Screening of phosphate-removing substrates for use in constructed wetlands treating swine wastewater. Ecological Engineering 54, 57–65.

Wanko, A., Forquet, N., Mose, R., Sadowski, A.G., 2006. Oxygen transfer rates estimation for sand bed design: Model calibration and validation. In: Proceedings of the 10th IWA Specialized Group Conference on Wetland Systems for Water Pollution Control, 23–29 September, Lisbon, Portugal.

Weedon, C.M., 2003. Compact vertical flow constructed wetland systems – first two years' performance. Water Science and Technology 48 (5), 15–23.

Weedon, C.M., 2010. A decade of compact vertical flow constructed wetlands. Water Science and Technology 62 (12), 2790–2800.

Welz, P.J., Ramond, J.B., Cowan, D.A., Burton, S.G., 2012. Phenolic removal processes in biological sand filters, sand columns and microcosms. Bioresource Technology 119, 262–269.

Wen, Y., Xu, C., Liu, G., Chen, Y., Zhou, Q., 2012. Enhanced nitrogen removal reliability and efficiency in integrated constructed wetland microcosms using zeolite. Frontiers of Environmental Science & Engineering 6 (1), 140–147.

WERF, 2010. Evaluation of Greenhouse Gas Emissions from Septic Systems: Final Report from Project DEC1R09. Water Environment Research Foundation, Alexandria, VA.

Westholm, L.J., 2006. Substrates for phosphorus removal—potential benefits for on-site wastewater treatment?. Water Research 40, 23–36.

WHO, 1989. Health guidelines for the use of wastewater in agriculture and aquaculture. WHO Tech. Report Series No. 778.

Wilhelm, M., Adams, V., Curtis, J., 2002. Carbon adsorption and air-stripping removal of MTBE from river water. Journal of Environmental Engineering 128 (9), 813–823.

Williams, H.N., Piñeiro, S., 2007. Ecology of the Predatory Bdellovibrio and Like Organisms. In: Jurkevitch, E. (Ed.), Microbiology Monographs 4: Predatory Prokaryotes. Springer-Verlag, Berlin Heidelberg.

Winter, K.-J., Goetz, D., 2003. The impact of sewage composition on the soil clogging phenomena of vertical flow constructed wetlands. Water Science and Technology 48 (5), 9–14.

Withey, S., Cartmell, E., Avery, L.M., Stephenson, T., 2005. Bacteriophages—potential for application in wastewater treatment processes. Science of the Total Environment 339, 1–18.

Wolf, A.A., Drake, B.G., Erickson, J.E., Megonigal, J.P., 2007. An oxygen-mediated positive feedback between elevated carbon dioxide and soil organic matter decomposition in a simulated anaerobic wetland. Global Change Biology 13, 2036–2044.

Wood, R.B., McAtamney, C.F., 1996. Constructed wetlands for waste water treatment: the use of laterite in the bed medium in phosphorus and heavy metal removal. Hydrobiologia 340, 323–331.

Wood, S.L., Wheeler, E.F., Berghage, R.D., 2000. Removal of dimethyl disulfide and p-cresol from swine facility wastewater using constructed subsurface-flow wetlands. Transactions of the American Society of Agricultural and Biological Engineers 43, 973–979.

Wright, E.P. and Graves, R.E. 1998. Guideline for milking center wastewater. New York, Cornell University, Ithaca. NRAES- 115 (DPC-15).

Wu, L., Ma, L.Q., Martinez, G.A., 2000. Comparison of methods for evaluating stability and maturity of biosolids compost. Journal of Environmental Quality 29, 424–429.

Wu, Z.-B., Xu, G.-L., Zhou, P.-J., Zhang, B.-Z., Cheng, S.-P., Fu, G.-P., He, F., 2004. Removal effects of nitrogen in integrated vertical flow constructed wetland sewage treating system. Journal of Agro-Environment Science 23 (4), 757–760.

Wu, S., Zhang, D., Austin, D., Dong, R., Pang, C., 2011a. Evaluation of a lab-scale tidal flow constructed wetland performance: oxygen transfer capacity, organic matter and ammonium removal. Ecological Engineering 37, 1789–1795.

Wu, J., He, F., Xu, D., Wang, R., Zhang, X., Xiao, E., 2011b. Phosphorus removal by laboratory-scale unvegetated vertical-flow constructed wetland systems using anthracite, steel slag and related blends as substrate. Water Science and Technology 63 (11), 2719–2724.

Wu, Y., Li, T., Yang, L., 2012. Mechanisms of removing pollutants from aqueous solutions by microorganisms and their aggregates: a review. Bioresource Technology 107, 10–18.

Wynn, T.M., Liehr, S.K., 2001. Development of a constructed subsurface-flow wetland simulation model. Ecological Engineering 16, 519–536.

Xiao, E.-R., Wei, L., He, F., Cheng, S.-P., Wu, Z.-B., 2010. Performance of the combined SMBR–IVCW system for wastewater treatment. Desalination 250, 781–786.

Xie, X.-L., He, F., Xu, D., Dong, J.-K., Cheng, S.-P., Wu, Z.-B., 2012. Application of large-scale integrated vertical-flow constructed wetland in Beijing Olympic forest park: design, operation and performance. Water and Environment Journal 26, 100–107.

Xin, Z.-J., Li, X.-Z., Nielsen, S.N., Yan, Z.-Z., Zhou, Y.-Q., Jia, Y.-Y., Guo, W.-Y., Sun, Y.-G., 2012. Effect of stubble heights and treatment duration time on the performance of water dropwort floating treatment wetlands (FTWS). Ecological Chemistry and Engineering S 19 (3), 315–330.

Xinshan, S., Qin, L., Denghua, Y., 2010. Nutrient removal by hybrid subsurface flow constructed wetlands for high concentration ammonia nitrogen wastewater. Procedia Environmental Sciences 2, 1461–1468.

Xu, D., Xu, J., Wu, J., Muhammad, A., 2006. Studies on the phosphorus sorption capacity of substrates used in constructed wetland systems. Chemosphere 63, 344–352.

Xu, J.C., Chen, G., Huang, X.F., Li, G.M., Liu, J., Yang, N., Gao, S.N., 2009. Iron and manganese removal by using manganese ore constructed wetlands in the reclamation of steel wastewater. Journal of Hazardous Materials 169, 309–317.

Xu, D., Li, Y., Howard, A., Guan, Y., 2013. Effect of earthworm *Eisenia fetida* and wetland plants on nitrification and denitrification potentials in vertical flow constructed wetland. Chemosphere 92, 201–206.

Xuan, Z. 2009. Nutrient and pathogen removal in a subsurface upflow wetland system using green sorption media. M.Sc. Thesis, Department of Civil, Environmental, and Construction Engineering, College of Engineering and Computer Science, University of Central Florida, Orlando, Florida.

Yadav, A.K., Kumar, N., Sreekrishnan, T.R., Satya, S., Bishnoi, N.R., 2010. Removal of chromium and nickel from aqueous solution in constructed wetland: mass balance, adsorption-desorption and FTIR study. Chemical Engineering Journal 160, 122–128.

Yair, S., Yaacov, D., Susan, K., Jurkevitch, E., 2003. Small eats big: ecology and diversity of Bdellovibrio and like organisms, and their dynamics in predator–prey interactions. Agronomie 23, 1–7.

Yalcuk, A., 2011. Removal of phenol from olive mill wastewater in constructed wetlands using different bedding media. Ekoloji 20 (80), 1–5.

Yalcuk, A., Ugurlu, A., 2009. Comparison of horizontal and vertical constructed wetland systems for landfill leachate treatment. Bioresource Technology 100, 2521–2526.

Yalcuk, A., Pakdil, N.B., Turan, S.Y., 2010. Performance evaluation on the treatment of olive mill waste water in vertical subsurface flow constructed wetlands. Desalination 262, 209–214.

Yan, J., Jianping, W., Jing, B., Daoquan, W., Zongding, H., 2006. Phenol biodegradation by the yeast *Candida tropicalis* in the presence of *m*-cresol. Biochemistry Engineering 29, 227–234.

Yan, C., Zhang, H., Li, B., Wang, D., Zhao, Y., Zheng, Z., 2012. Effects of influent C/N ratios on CO_2 and CH_4 emissions from vertical subsurface flow constructed wetlands treating synthetic municipal wastewater. Journal of Hazardous Materials 203–204, 188–194.

Yang, L., Hu, C.C., 2005. Treatments of oil-refinery and steel-mill wastewaters by mesocosm constructed wetland systems. Water Science and Technology 51 (9), 157–164.

Yang, Q., Chen, Z.-H., Zhao, J.-G., Gu, B.-H., 2007. Contaminant removal of domestic wastewater by constructed wetlands: effects of plant species. Journal of Integrative Plant Biology 49 (4), 437–446.

Yang, W., Chang, J., Xu, B., Peng, C., Ge, Y., 2008. Ecosystem service value assessment for constructed wetlands: a case study in Hangzhou. China Ecological Economics 68, 116–125.

Yang, G.C.C., Chen, M.-C., Yeh, C.-F., 2011. Dewatering of a biological industrial sludge by electrokinetics-assisted filter press. Separation and Purification Technology 79 (2), 177–182.

Ye, F., Li, Y., 2009. Enhancement of nitrogen removal in towery hybrid constructed wetland to treat domestic wastewater for small rural communities. Ecological Engineering 35, 1043–1050.

Ye, J., Wang, L., Li, D., Han, W., Ye, C., 2012. Vertical oxygen distribution and oxygen source analysis for vertical-flow constructed wetlands treating domestic wastewater. Ecological Engineering 41, 8–12.

Yeh, T.Y., 2008. Removal of metals in constructed wetlands: a review. Practice Periodical of Hazardous, Toxic, and Radioactive Waste Management 12 (2), 96–101.

Yeh, T.Y., Chou, C.C., Pan, C.T., 2009. Heavy metal removal within pilot-scale constructed wetlands receiving river water contaminated by confined swine operations. Desalination 249, 368–373.

Yousefi, Z., Davies, C.M., Bavor, H.J., 2004. Bacteriophage technique for assessing viral removal in constructed wetland and detention pond systems. Iranian Journal of Environmental Health Science & Engineering 1 (2), 8–15.

Yuan, L.-J., Han, R.-R., Han, W., 2008. Characteristics of COD and nitrogen removal of integrated vertical-flow constructed wetland with intermittent feeding. Journal of Xi'an University of Architecture and Technology 40 (4), 521–526.

Zakova, Z., Zak, P., 2003. Small on-site constructed wetlands for wastewater treatment in sensitive areas with scattered settlements. In: Vymazal, J. (Ed.), Wetlands-nutrients, Metals and Mass Cycling. Backhuys Publishers, Leiden, The Netherlands, pp. 169–177.

Zapater, M., Gross, A., Soares, M.I.M., 2011. Capacity of an on-site recirculating vertical flow constructed wetlands to withstand disturbances and highly variable influent quality. Ecological Engineering 37, 1572–1577.

Zazo, J.A., Paul, J.S., Jaffe, P.R., 2008. Influence of plants on the reduction of hexavalent chromium in wetland sediments. Environmental Pollution 156, 29–35.

Zhang, K., Farahbakhsh, K., 2007. Removal of native coliphages and coliform bacteria from municipal wastewater by various wastewater treatment processes: implications to water reuse. Water Research 41, 2816–2824.

Zhang, X.-L., Zhang, S., He, F., Cheng, S.-P., Liang, W., Wu, Z.-B., 2007. Differentiate performance of eight filter media in vertical flow constructed wetland: Removal of organic matter, nitrogen and phosphorus. Fresenius Environmental Bulletin 16 (11B), 1468–1473.

Zhang, D., Gersberg, R.M., Keat, T.S., 2009. Constructed wetlands in China. Ecological Engineering 35, 1367–1378.

Zhang, L.-Y., Zhang, L., Liu, Y.-D., Shen, Y.-W., Liu, H., Xiong, Y., 2010. Effect of limited artificial aeration on constructed wetland treatment of domestic wastewater. Desalination 250, 915–920.

Zhang, Y., Zhang, Y., Huang, M., Xu, H., He, Y., 2012a. Pollutants removal and analysis of microbial community of malodorous river in BFG-IVCWs. Advanced Materials Research 518–523, 3045–3052.

Zhang, D.Q., Gersberg, R.M., Zhu, J., Hua, T., Jinadasa, K.B.S.N., Tan, S.K., 2012b. Batch versus continuous feeding strategies for pharmaceutical removal by subsurface flow constructed wetland. Environmental Pollution 167, 124–131.

Zhang, C., Deng, J., Zhang, W., 2012. Dewatering and mineralization of sludge from secondary sedimentation tank in a constructed sludge drying reed bed. Applied Mechanics and Materials 209–211, 1111–1115.

Zhao, Q.Y., Sun, G., Lafferty, C., Allen, S.J., 2004a. Optimising the performance of a lab-scale tidal flow reed bed system treating agricultural wastewater. Water Science and Technology 50 (8), 65–72.

Zhao, Q.Y., Sun, G., Allen, J.S., 2004b. Purification capacity of a highly loaded laboratory scale tidal flow reed bed system with effluent recirculation. Science of the Total Environment 330, 1–8.

Zhao, W.Y., Zhen, B.W., Zhou, Q.H., Cheng, S.P., Fu, G.P., He, F., 2004c. Removal of dibutyl phthalate by a staged, vertical-flow constructed wetland. Wetlands 24 (1), 202–206.

Zhao, Y.Q., Sun, G., Allen, S.J., 2005. Anti-sized reed bed system for animal wastewater treatment: a comparative study. Water Research 38 (12), 2907–2917.

Zhao, Y.Q., Babatunde, A.O., Razali, M., Harty, F., 2008. Use of dewatered alum sludge as a substrate in reed bed treatment systems for wastewater treatment. Journal of Environmental Science and Health, Part A 43, 105–110.

Zhao, L., Zhu, W., Tong, W., 2009. Clogging processes caused by biofilm growth and organic particle accumulation in lab-scale vertical flow constructed wetlands. Journal of Environmental Sciences 21, 750–757.

Zhao, Y.J., Liu, B., Zhang, W.G., Ouyang, Y., An, S.Q., 2010. Performance of pilot-scale vertical-flow constructed wetlands in responding to variation in influent C/N ratios of simulated urban sewage. Bioresource Technology 101, 1693–1700.

Zhao, Y.Q., Babatunde, A.O., Hu, Y.S., Kumar, J.L.G., Zhao, X.H., 2011a. Pilot field-scale demonstration of a novel alum sludge-based constructed wetland system for enhanced wastewater treatment. Process Biochemistry 46 (1), 278–283.

Zhao, Y.J., Hui, Z., Chao, X., Nie, E., Li, H.J., He, J., Zheng, Z., 2011b. Efficiency of two-stage combinations of subsurface vertical downflow and upflow constructed wetland systems for treating variation in influent C/N ratios of domestic wastewater. Ecological Engineering 37, 1546–1554.

Zhao, F., Xi, S., Yanh, X., Yang, W., Li, J., Gu, B., He, Z., 2012. Purifying eutrophic river waters with integrated floating island systems. Ecological Engineering 40, 53–60.

Zhen, W., Dong, J., Liu, L., Zhu, G., Liu, C., 2013. Study of oyster shell as a potential substrate for constructed wetlands. Water Science and Technology 67 (10), 2265–2272.

Zhou, S., Hou, H., Hosomi, M., 2008. Nitrogen removal, N_2O emission, and NH_3 volatilization under different water levels in a Vertical Flow Treatment System. Water, Air, and Soil Pollution 191, 171–182.

Zhou, Q., He, F., Zhang, L., Wang, Y., Wu, Z., 2009. Characteristics of the microbial communities in the integrated vertical-flow constructed wetlands. Journal of Environmental Sciences 21, 1261–1267.

Zhou, J., Guo, X., Han, Y., Liu, J., Liang, H., 2012. Study of a novel vertical flow constructed wetland system with drop aeration for rural wastewater treatment. Water, Air, and Soil Pollution 223, 889–900.

Zhou, Y., Tigane, T., Li, X., Truu, M., Truu, J., Mander, Ü., 2013. Hexachlorobenzene dechlorination in constructed wetland mesocosms. Water Research 47 (1), 102–110.

Zhu, T., Jenssen, P.D., Maehlum, T., Krogstad, T., 1997. Phosphorus sorption and chemical characteristics of lightweight aggregates (LWA)—potential filter media in treatment wetlands. Water Science and Technology 35 (5), 103–108.

Zhu, S.-X., Ge, H.-L., Ge, Y., Cao, H.-Q., Liu, D., Chang, J., Zhang, C.-B., Gu, B.-J., Chang, S.-X., 2010. Effects of plant diversity on biomass production and substrate nitrogen in a subsurface vertical flow constructed wetland. Ecological Engineering 36, 1307–1313.

Zhu, L., Takala, J., Hiltunen, E., Li, Z., Kristianto, Y., 2013. Comparison of vertical-flow constructed wetlands with and without supplementary aeration treating decentralized domestic wastewater. Environmental Technology 34 (1), 53–60.

Zurita, F., De Anda, J., Belmont, M.A., 2009. Treatment of domestic wastewater and production of commercial flowers in vertical and horizontal subsurface-flow constructed wetlands. Ecological Engineering 35, 861–869.

Zurita, F., Belmont, M.A., De Anda, J., White, J.R., 2011. Seeking a way to promote the use of constructed wetlands for domestic wastewater treatment in developing countries. Water Science and Technology 63 (4), 654–659.

Zwara, W., Obarska-Pempkowiak, H., 2000. Polish experience with sewage sludge utilization in reed beds. Water Science and Technology 41 (1), 65–68.

Nomenclature

AHTN	tonalide
Al	aluminium
AL/AR Py-GC	pyrolytic index of energy storage (aliphatic/aromatic) pyrolytic index of energy storage (aliphatic/aromatic compounds)
ALR	removed areal load ($mg/m^2/d$)
ANAMMOX	anaerobic ammonium oxidation
ANNs	artificial neural networks
As	arsenic
AS	activated sludge
ASM1	activated sludge model 1
ASP	activated sludge plant
B/E3 Py-GC	pyrolytic index of humification (benzene to toluene), which increases with the maturity of the organic matter
BLO	Bdellovibrio-like organisms
Bo	boron
BOD	biochemical oxygen demand
BTEX	benzene, toluene, ethylbenzene, xylene
C	concentration of single soluble component in water
C/N	carbon/nitrogen
Ca	calcium
CANON	completely autotrophic nitrogen-removal over nitrite
Cd	cadmium
CEC	cation exchange capacity
CFU	colony forming units
C_{half}	Monod half saturation constant
Co	Cobalt
COD	chemical oxygen demand
Cr	chromium
Cr(III)	trivalent chromium
CR(VI)	hexavalent chromium
CSTR	continuous stirred-tank reactor
Cu	copper
CW2D	constructed wetland 2-dimensions
CWs	constructed wetlands
D	diffusion coefficient for oxygen in the bed
d	dispersion coefficient
Da	Damköhler number
DAF	dissolved air flotation

DEHP	di(2-ethylhexyl) phthalate
DHA	dehydrogenase activity
Dhase	dehydrogenase
dm	dry mass
DO	dissolved oxygen
D_o	molecular diffusion coefficient
DRI	dynamic respiration index
DRI_{max}	oxygen uptake rate (mg O_2/kg DM/h)
DS	dry solids
EC	electrical conductivity
ECDs	endocrine disruptors
EFF	effluent
ET	evapotranspiration
FC	faecal coliform
Fe	iron
FS	faecal streptococcus
FTWs	floating treatment wetlands
FWS	free water surface
GI	germination index
h	depth
H	pressure head
HDPE	high-density polyethylene
Hg	mercury
HHCB	galaxolide
HLR	hydraulic loading rate
HM	heavy metals
HPC	heterotrophic plate count
HRT	hydraulic residence time
HSF	horizontal subsurface flow
I/O	Input/Output
IN	influent
IP	inorganic phosphorus
IVFCWs	integrated VFCWs
K	hydraulic conductivity
k	rate of first-order rate constant
K_{ANS}	saturation coefficient (mg/L)
KL_a	oxygen transport coefficient
K_{max}	Monod maximum pollutant removal rates
k_u	maximum specific substrate utilization rate (g substrate/g VS d)
LAS	linear alkylbenzene sulphonates
LCA	life cycle assessment
LECA	light expanded clay aggregates
LESA	lightweight expanded shale aggregate
LN	leaching rate of the *N* species (mg/h)
LWA	lightweight aggregates
MBR	membrane bioreactor
MC	specific moisture capacity
Mg	magnesium

M_{LA}	specific mass loading rate (g substrate/m^2/d medium)
MLSS	mixed liquor suspended solids
Mn	manganese
Mo	molybdenum
MPN	most probable number
MSW	municipal solid waste
MTBE	methyl tertiary-butyl ether
N	nitrogen
NH_4^+**–N**	ammonium nitrogen
Ni	nickel
NPE	nonylphenolethoxylates
OH	hydroxyl group
O/N Py-GC	pyrolytic index of mineralization (pyrrole to furfural) derived by pyrolysis-gas chromatography
OD	oxygen demand
OLR	organic loading rate
OM	organic matter
OMW	olive mill wastewater
OP	*ortho*-phosphate
OTC	oxygen transfer capacity
P	phosphorus
PAH	polycyclic aromatic hydrocarbon
Pb	lead
PCBs	polychlorinated biphenyls
PCE	perchloroethene
PCPs	personal care products
pe	person equivalent
PFR	plug flow reactor
PRI	primary sludge
Pt	pretreatment
q	specific flow rate
Q	water discharge
R	rate of N species uptake (mg/h)
r	reaction rate
RD	number of resting days (d)
S	absorbed concentration
SAPS	successive alkalinity producing systems
SAS	surplus-activated sludge
SCD	secondary sludge
Se	Selenium
SF	subsurface flow
Si	silica
SLR	sludge loading rate
SMLR	stepwise multiple linear regression
SRI	static respiration index
SRIP	slow rate infiltration plant
SS	suspended solids
STWs	sludge treatment wetlands

SVI	sludge volume index
t	time
TC	total coliform
TCE	trichlorethene
$\mathbf{TF}_{1,2,\ldots}$	translocation factor
TFi	trickling filter
TFWs	tidal flow wetlands
TKN	total Kjeldahl nitrogen
TN	total nitrogen
TON	total organic nitrogen
TP	total phosphorus
TPF	triphenylformazan
TSS	total suspended solids
TTC	triphenyltetrazolium chloride
U	unplanted
UASB	upflow anaerobic sludge blanket
UV	ultraviolet
V	water volume
VF	vertical flow
VLP	virus-like particles
VOCs	volatile organic compounds
VS	volatile solids
VSS	volatile suspended solids
WAS	waste-activated sludge
WSC	water soluble carbon
WWTP	wastewater treatment plants
X_a	maximum specific concentration of active cells (g VS/m^2 medium); VS: are the volatile solids, (g)
XH	heterotrophic bacteria
Y_A	cellular yield coefficient
Zn	zinc
δ	coefficient that controls the intake of a N species
ε	porosity
θ	volumetric water content
λ	dispersivity
μ_{max}	maximum autotrophic growth rate (1/h)
ρ	bulk density
τ	tortuosity factor
v	average pore-water velocity
ω_i	first-order exchange rate

Index

Note: Page numbers followed by *f* indicate figures and *t* indicate tables.

D

T

U

V